38th International Symposium on Remote Sensing of Environment (ISRSE 38) and 21st William T. Pecora Memorial Remote Sensing Symposium (Pecora 21)

Earth Observation – Continuous Monitoring of Our Changing Planet: From Sensors to Decisions

International Archives of the Photogrammetry, Remote Sensing and Spatial Information Sciences
XLII-3/W9

Baltimore, Maryland, USA
6-11 October 2019

ISBN: 978-1-7138-0217-4

Printed from e-media with permission by:

Curran Associates, Inc.
57 Morehouse Lane
Red Hook, NY 12571

Some format issues inherent in the e-media version may also appear in this print version.

Printed with permission by Curran Associates, Inc. (2020)

International Center for Remote Sensing of the Environment, Inc. (ICRSE)
1955 E. Sixth Street, Suite 115
Tucson, AZ 85719
USA

Phone: (520) 621-8567
Fax: (520) 621-7834

isrse@symposia.org

Additional copies of this publication are available from:

Curran Associates, Inc.
57 Morehouse Lane
Red Hook, NY 12571 USA
Phone: 845-758-0400
Fax: 845-758-2633
Email: curran@proceedings.com
Web: www.proceedings.com

TABLE OF CONTENTS

FROM PIXEL TO YIELD: FORECASTING POTATO PRODUCTIVITY IN LEBANON AND IDAHO

H. Abou Ali [1], D. Delparte [1,*], L. M. Griffel [2]

[1] Idaho State University, Department of Geosciences, 83209 Pocatello, Idaho, USA - (abouhana, delparte)@isu.edu
[2] Biofuels and Renewable Energy Technology Dept, Idaho National Lab, 83415 Idaho Falls, Idaho, USA – mike.griffel@inl.gov

KEY WORDS: yield prediction, vegetation indices, PlanetScope, remote sensing, precision agriculture, crop monitoring

ABSTRACT:

Idaho and Lebanon rely on potatoes as an economically important crop. NDVI (Normalized Difference Vegetation Index), GNDVI (Green Normalized Difference Vegetation Index), SAVI (Soil Adjusted Vegetation Index), and MSAVI2 (Modified Soil Adjusted Vegetation Index 2) indices were calculated from PlanetScope satellite imagery for the 2017 growing season cloud free days. Variations in vegetation health were tracked over time and correlated to yield data provided by growers in Idaho. Based on ordinary least squares regression an Idaho yield forecast model was developed. Vegetation response during the growth stage at which potato tubers were filling out was significant in predicting yield for both the Norkotah and Russet potato variety. This corresponded to a week with high recorded temperatures that impacted the health status of the crops. The yield forecasting model was validated with a cross validation approach and then applied to potato fields in Lebanon. The Idaho model successfully displayed yield variation in crops for Lebanon. Spectral indices along with field topography allow the prediction of yield based on the crop type and variety.

1. INTRODUCTION

Agriculture is an important sector in the global economy and is a crucial component for fighting hunger and food insecurity. According to the United Nations Food and Agriculture Organization (2017), the expectation is that the world population will reach 10 billion by 2050 and there is a need to produce around 50% more food than in 2012. Idaho and Lebanon rely on potatoes as an economically important crop. Potatoes are responsible for employing 46% of the Idahoan agricultural processing workforce (Lewin et al. 2011) while in Lebanon potatoes account for 56% of vegetable production in the country, mainly in the Bekaa Valley and North Lebanon states (Hatoum 2005).

Precision agriculture uses information technology to better manage crop production by taking into consideration the variations within the field to increase profitability and sustainability (Zhang 2016). By utilizing remote sensing data from satellites systems, farmers can leverage cost-effective technologies to mitigate crop threats with targeted approaches for grower decision making such as variable rate fertilizer application, timely irrigation, early disease detection, and pest control. Introducing the concept of precision agriculture plays a major role in empowering local farmers and stakeholders by educating them about new technologies that have the potential to improve their crop productivity and enhance their economic sustainability.

New satellite systems have emerged in the past five years that have high spatial, temporal (PlanetScope), and spectral resolution (Sentinel-2) that have the potential to revolutionize the field of precision agriculture by aiding decision making. These advances in satellite imagery provide a valuable resource for crop monitoring and specifically yield modeling and prediction.

Crop yield forecasting is crucial for farmers as it helps with management decisions regarding harvesting, storage, pricing and marketing. The availability of satellite imagery with higher spatial and temporal resolution provides detailed information about crop yield over the duration of the growing season. A previous research study assessed the performance of different sensors with varying resolutions (5 m for RapidEye, 3 m for PlanetScope, 10 m for Sentinel-2 and 30 m for Landsat) for yield monitoring and concluded that higher resolution imagery gave more accurate results (Burke and Lobell 2017). Regression models are commonly used in yield prediction. Linear regression uses one variable at a time to explain and predict the yield using separate equations for each (Aditya Shastry 2017) assuming that there is a linear relation between the yield and the explanatory variable, model residuals are almost normally distributed such that there is no clustering in the data (Sellam and Poovammal 2016). In their paper, Rembold et al. (Rembold et al. 2013), used linear regression with NDVI (Normalized Difference Vegetation Index) from SPOT (Satellite Pour l'Observation de la Terre) imagery to explain yield values and obtained an R^2 of 0.930 and 0.799 for the Morocco and Egypt study areas, respectively. Al-Gaadi et al. (2016) also used linear regression but used Landsat-8 and Sentinel-2 imagery for computing both NDVI and SAVI (Soil Adjusted Vegetation Index). They generated a linear regression equation per index per sensor per field and concluded that higher resolution imagery yielded better regression models where the Landsat-8 imagery resulted in an R^2 range between 0.39 and 0.65 compared to 0.47 and 0.65 for the Sentinel-2 dataset. While previous studies on modelling yield prediction used Landsat (Song et al. 2016) and Sentinel-2 (Al-Gaadi et al. 2016), this paper offers a new approach utilizing high spatial and temporal resolution PlanetScope (3 m, daily) analytical scenes in comparison with Sentinel-2A (20 m, 10 days) to monitor potato crop health over the 2017 growing season and predict crop yield. The most common vegetation indices used for forecasting crop yield status are NDVI, Green Normalized Difference Vegetation Index (GNDVI), SAVI and Modified Soil Adjusted Vegetation Index 2 (MSAVI2). These indices were all tested and compared to potato yield data from ten fields (402.98 ha) in an ordinary

* Corresponding author

least squares (OLS) regression approach to choose the best performing variable for yield prediction. The critical week for each of the different potato varieties corresponded to the potato growth stage where tubers are in the process of filling out beneath the soil. Using the averaged SAVI and slope values for the critical weeks provided the best fit model for predicting yield in the Idaho fields. Upon considering the offset in the growing seasons between Idaho and Lebanon as well as the potato varieties, the adjusted Idaho model was applied to forecast potato yield in Lebanon.

2. METHODS

2.1 Study Areas

For the purpose of this research, we selected two study areas (Figure 1 and Figure 2) that shared topographic, geographic and environmental features including related crop threats. Both study areas have cold wet winters and hot dry summers and are within river plains and part of the North Temperate Zone. The Idaho study area is located at a latitude of 43.9º and altitude of 1502 m above mean sea level with a wet season from October through June with a high average temperature during the 2017 season of 24ºC. The Lebanon study area is at 10º lower latitude than the Idaho study area, located at 33.9º latitude with an elevation of 872 m above sea level and a high average temperature of 34ºC during the 2017 growing season. Moreover, potatoes are the dominant crop in both areas.

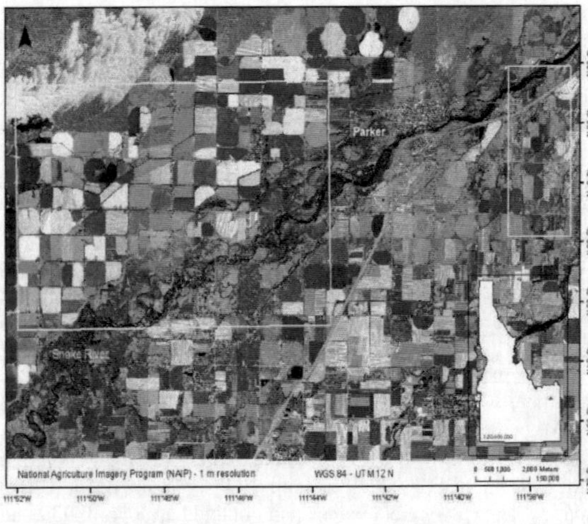

Figure 1. Idaho study area showing field locations near Parker, Idaho. (WGS 84 - UTM 12N)

Figure 2. Lebanon study area field locations in the Bekaa Valley, Lebanon. (WGS 84 - UTM 36N)

2.2 Satellite Imagery

The satellite imagery used in this work includes Sentinel-2A and PlanetScope data. The USGS (United States Geological Services) Earth Explorer online interface allowed open access to the Sentinel-2A imagery where the data is freely available to the public via the ESA (European Space Agency). The PlanetScope imagery operated by Planet, a non-governmental privately-owned commercial company, was accessible through the Planet Application Program Interface using their Education and Research Program license. At the time of this study, in the summer of 2017, Sentinel-2B imagery was unavailable.

Sentinel-2A is a multispectral imager covering 13 spectral bands (443 nm – 2190 nm) at resolutions of 10-20 and 60 m with a five-day revisit period. The Sentinel–2A imagery was processed using ESA's SNAP software for atmospheric correction using the Sen2Cor Plugin. The PlanetScope imagery has four spectral bands (455 nm – 860 nm [Blue: 455-515, Green: 500-590, Red: 590-670, NIR: 780-860]) at a resolution of 3 m with a daily revisit. The downloaded PlanetScope imagery is the surface reflectance ready product provided by Planet so no further atmospheric correction was needed, however, individual bands had to be extracted.

After extracting red, green and NIR bands, and atmospherically correcting the Sentinel-2A bands (Figure 3), the next step was to calculate the various spectral indices that are described below.

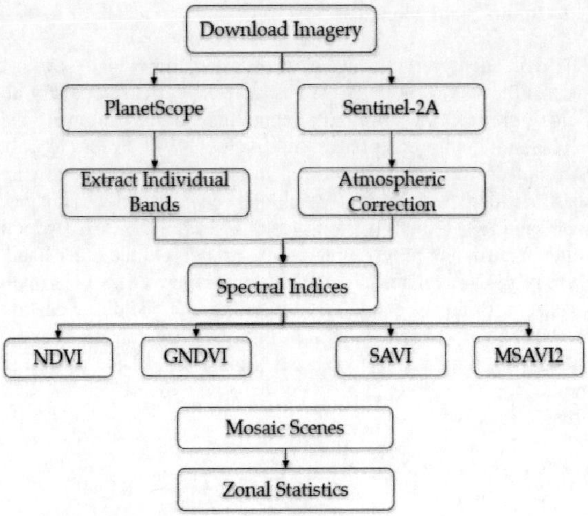

Figure 3. Satellite imagery processing workflow

NDVI (Rouse et al. 1973)

$$\frac{\rho_{NIR} - \rho_{Red}}{\rho_{NIR} + \rho_{Red}} \quad (1)$$

GNDVI (Gitelson, Kaufman, and Merzlyak 1996)

$$\frac{\rho_{NIR} - \rho_{Green}}{\rho_{NIR} + \rho_{Green}} \quad (2)$$

SAVI (Huete 1988)

$$\frac{\rho_{NIR} - \rho_{Red}}{\rho_{NIR} + \rho_{Red} + L} * (1 + \mathrm{L}) \quad (3)$$

MSAVI2 (Qi et al. 1994)

$$\frac{2\rho_{NIR} + 1 - \sqrt{(2\rho_{NIR} + 1)^2 - 8(\rho_{NIR} - \rho_{Red})}}{2} \quad (4)$$

In order to better organize the data, any dates that had multiple scenes for the same date were mosaicked to reduce the number of scenes for processing further. The final step for processing the satellite imagery was running zonal statistics that would be the input for the regression models to predict yield.

2.3 Yield Data

Idaho farmers supported the data analysis by providing information for each field that included planting dates and the potato variety. The farmers provided yield data from the 2017 growing season, which was collected using the GK Technology for Agriculture yield monitor. The sensor attaches to the harvester and records the speed, yield, and pounds per second along with the geographic coordinates every 0.8 meters by 10 meters.

Based on planting date and potato variety, the Idaho fields were sub grouped into three categories (G1, G2, G3) so that fields within the same group have a similar growing timeline which influences the variables for yield forecasting models depending on date and variety. Different planting dates result in offset between models while each potato variety has its own growth cycle (different maturity periods).

2.4 Yield Forecasting

A Python script was written to extract all mean values for each index and bands red, blue and NIR along with the mean yield values for each grid cell making it suitable for running regression analysis to explain yield variability within the fields. The Python code calculated the average values within each grid cell for every date available. This resulted in multiple averaged cells for each index by growth week depending on cloud cover. Exploratory Regression in ArcGIS Pro helped reduce the number of potential explanatory variables for the yield model through several runs. Subsets of the explanatory variables included a selection of the Green, Red and Near Infrared bands while others used the different indices: NDVI, GNDVI, SAVI, and MSAVI2. In addition, a subset included a combination of bands with spectral indices. From each run, the most significant variables became the new subset for the next run. After three exploratory regression runs, the variables that had the highest correlations were used for input into the Ordinary Least Square (OLS) tool. The resulting models were validated using a cross validation leave-one-out approach (Figure 4). In addition, slope values were calculated and averaged within the fields based on the grid cells using the National Elevation Dataset (10m resolution).

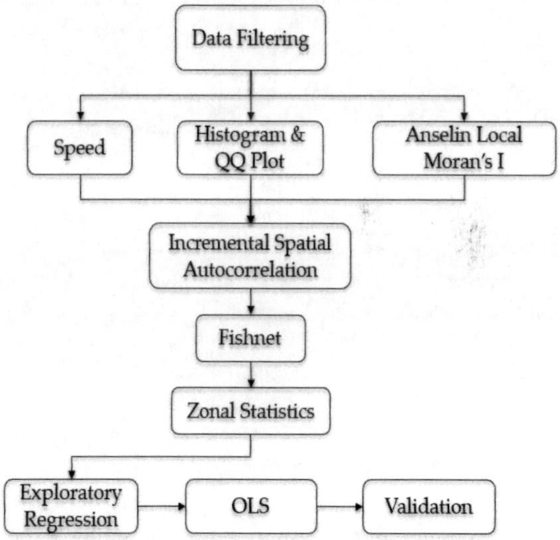

Figure 4. Yield data processing workflow

An incremental spatial autocorrelation approach determined the optimal grid size by examining the corresponding z-score of yield data. This approach utilizes Global Moran's I to determine yield clustering by measuring spatial autocorrelation based on yield points' location and value. An output of this method is the z-score which is the standard deviation returned by Global Moran's I where the peak in z-score corresponds to highest yield clustering density. The result of the incremental spatial autocorrelation method indicated that an 80x80 m grid size was the peak distance for yield value clustering.

3. RESULTS

3.1 Variation of Indices over Growing Season

All five indices used (NDVI, GNDVI, SAVI, and MSAVI2) showed similar variation over the fields throughout the season in both Sentinel-2 and PlanetScope data (Figure 5). However, the PlanetScope imagery gave a more detailed observation about what was happening in the fields along with determining critical dates and stages for the growing season due to its higher temporal resolution. The PlanetScope dataset corresponded to specific dates regarding plant emergence, row closure and full maturity in addition to dates with higher temperatures and increased evapotranspiration rates.

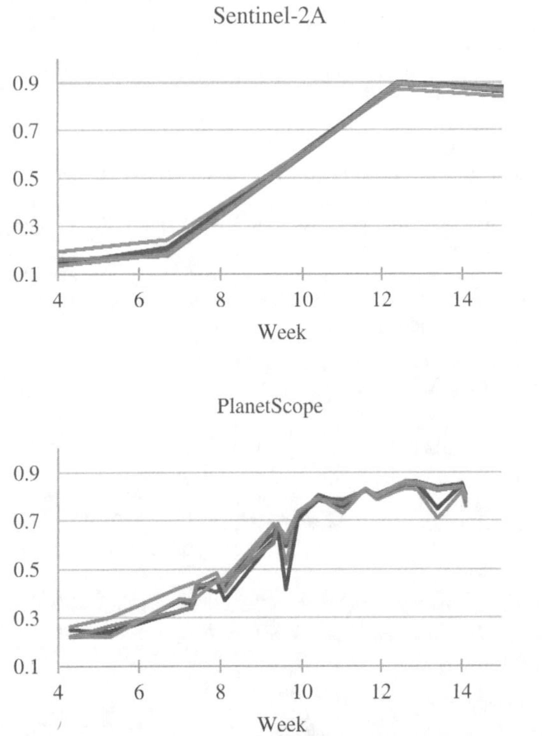

Figure 5. Variation of NDVI over the 2017 growing season Sentinel-2A vs. PlanetScope for Idaho fields

In both the Sentinel and Planet imagery for all the fields, NDVI and SAVI increase gradually during the season to reach a peak value of 0.9 and then decrease to around 0.8 (Figure 5). However, the limitation of Sentinel-2A availability due to cloud cover resulted in fewer data points between critical times over the growing season. PlanetScope imagery gave a more detailed interpretation of the variation between the weeks and the fields. For Lebanon, the cloud free Sentinel-2A dataset consisted of 10 dates, representing only 10 weeks during the growing season. While with PlanetScope there was imagery for 50 available dates for the same time period with an average of 4 or 5 images per week and covered the entire growing season. For Idaho, cloud free Sentinel-2A imagery consisted of 5 dates, representing only 5 growth weeks. For PlanetScope there was 22 available dates for the same time period with minimum of 2 images per week to provide average weekly vegetation response over most of the growing season.

3.2 Idaho Yield Models

The dependent variable, yield, was tested individually against each of the average weekly indices along with field slope. For all fields' groups, the different vegetation indices allowed the explanation and prediction of yield values. With access to only Sentinel-2A (Sentinel-2B was unavailable for the 2017 growing season), there were only five cloud free Sentinel-2A images, representing 5 growth weeks (weeks 1, 3, 7, 11, and 13) which was not sufficient to test a regression model. On the other hand, there were 24 available PlanetScope dates, covering 14 growth weeks, which was enough for building the yield forecasting model. The model calculated R^2 values by growth week over the entire growing season for each index plus slope. The best-fitting model corresponding to growth week was identified for each potato variety.

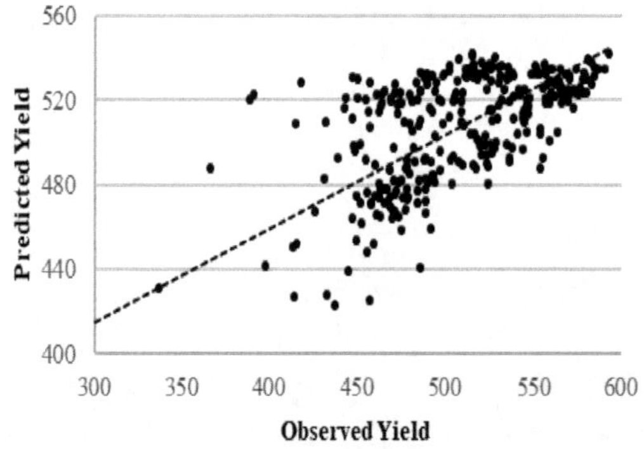

Figure 6. Russet potato fields yield model results

The Russet potato variety yield prediction model (Figure 6) with SAVI response during week 12 as the most significant for yield prediction. The model obtained has an R^2 value of 0.44 and follows the equation of: $y = 0.4435x + 281.6$

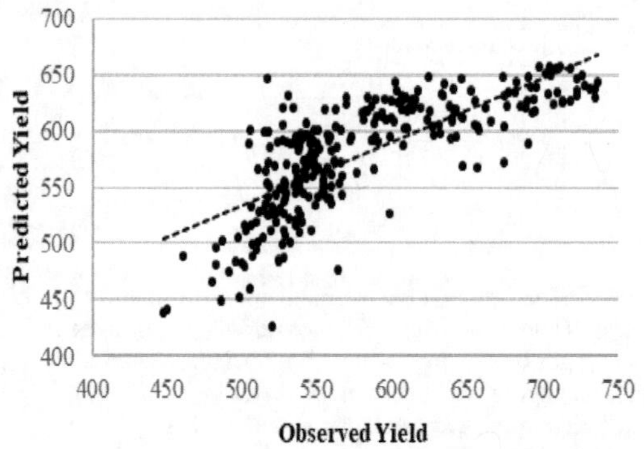

Figure 7. Norkotah potato fields yield model results

The Norkotah potato variety yield prediction model (Figure 7) with SAVI response during week 10 as the most significant for yield prediction. The model obtained has an R^2 value of 0.57 and follows the equation of: $y = 0.570x + 248.8$

3.3 Growth Degree Days and SAVI

Potatoes are an irrigated crop and water is a crucial element in the productivity of the fields. Although irrigation rate was unavailable for this study, we retrieved precipitation and temperature records (Figure 8) to relate to indices' and yield values. A better understanding of these weather variables provided vital information related to critical weeks in the growth stages has the potential to help farmers with irrigation decisions.

In addition to the relationship between SAVI and yield, there was also a relationship between SAVI and GDD (Growth Degree Days). The correlation coefficient between SAVI and averaged weekly GDD (Equation 1) values for the individual fields was between 0.45 and 0.67 (Figure 9).

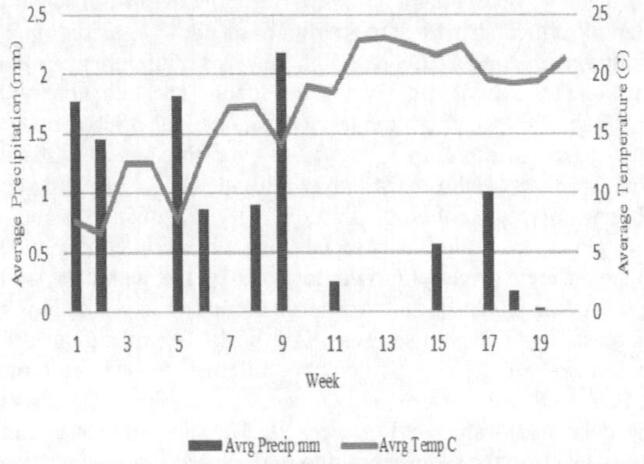

Figure 8. Precipitation and Temperature during the 2017 growing season provided by U.S. Bureau of Reclamation Agrimet weather station.

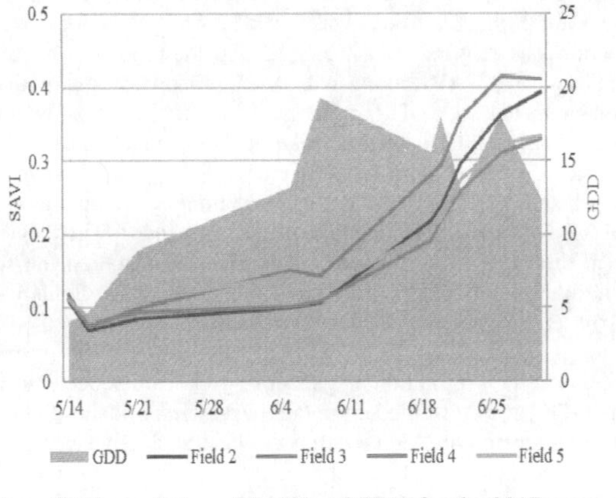

Figure 9. The variation of SAVI and GDD for the 2017 range of critical growth weeks (6 to 13)

Growth Degree Days:

$$(MaxAirTemp+MinAirTemp)/2 - Base\ Temp \quad (6)$$

[Base Temp = 50F]

3.4 Yield Forecasting in Lebanon

The regression model derived from the Idaho data predicted yield for fields in Lebanon. Due to the difference in potato variety, the approach was to calculate yield using the different models. Though the Idaho model used averaged values of the variables within a fishnet cell size of 80m x 80m, the fishnet used to predict yield in the Lebanon data was 32m x 32m. The reason for that choice is the fact that the Lebanese farmer gave an approximation of the yield to have had a range of 3 ton to 5 ton per 1,000 m^2. Hence, a 32m x 32m cell grid is the closest approximation to what the farmer provided.

Figure 10. Predicted yield values in kg/1000m^2 (A) and the corresponding slope values in degrees (B)

The micro topography of the fields affected the predicted yield values due to the variation in slope leading to uneven irrigation. The results obtained from the yield model were verified qualitatively by showing the farmer the yield variation within the fields and confirming with his records. The farmer also mentioned that there are three sprinklers every 100 meters and as the yield values are clustered in a 100-meter neighborhood this could potentially show the positions of the installed sprinklers.

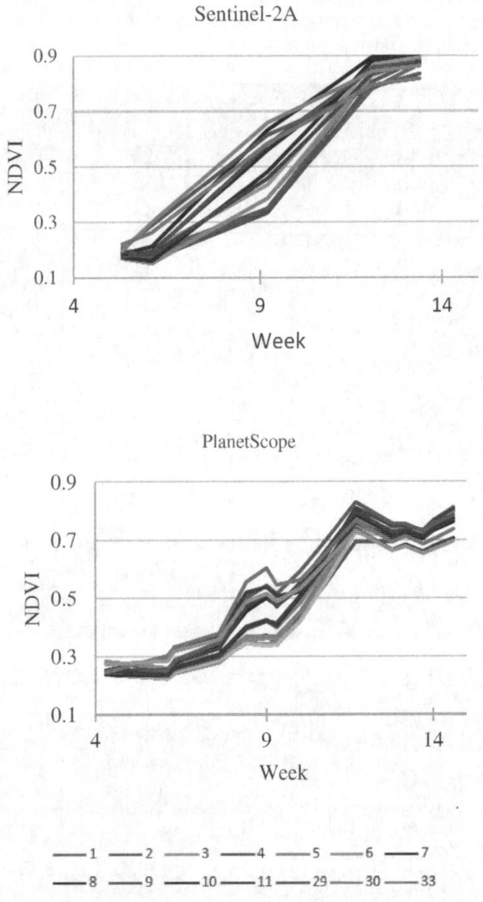

Figure 11. Variation of NDVI over the 2017 growing season Sentinel-2A vs. PlanetScope for Lebanon fields

4. DISCUSSION/CONCLUSION

Over the past five years, new offerings of satellite imagery are providing high spatial and temporal resolution data that advances opportunities for monitoring crops and predicting yield. The high resolution and daily return frequency of PlanetScope allows farmers to apply precision agriculture practices that take in to consideration the variation within the field as opposed to the assumption that variables such as microtopography and soil types within the field are uniform. The higher spatial resolution of PlanetScope is particularly valuable with smaller fields such as those in Lebanon where fields are usually small in size as most of the fields are inherited from father to sons and thus get subdivided when passed from generation to another. With each spectral band delivering specific information about the plants based on the reflectance signature, combining slope data with vegetation index values – particularly SAVI or NDVI daily, growers can rapidly respond to crop health threats. Vegetation indices provide information about the health status of the crop and can help with decision making regarding adding nutrients and water to plants when needed based on the index value. Further, our study demonstrated that SAVI values averaged by week from PlanetScope imagery revealed the growth stage and growth week that was the most critical for impacting potato yield. Though the critical week varied between potato varieties (ranging between week 10 and 12), it corresponded to when the tubers where filling, while the later weeks that corresponded to the crops reaching full maturity had no role in forecasting yield. On the contrary, as the crops reached full row closure and continued to mature before harvesting showed a decrease in the correlation between the indices and yield values.

Our study also highlighted the importance of frequent satellite revisit periods. With only 5 cloud free image dates across the entire 2017 growing season in Idaho with no dates during the critical growth weeks as indicated by PlanetScope and Sentinel 2B data not yet available, Sentinel 2A alone was insufficient to identify critical growth stages or plant response to predict yield. With PlanetScope's daily revisit there was a minimum of 2 cloud free images available per week for Idaho. The Lebanon study area had an even larger number of available PlanetScope images due to more cloud free days (50 over the growing season compared to Idaho's 24) that provided a more frequent observation period of crop variation over the growing season. The graphs from the PlanetScope imagery clearly corresponded to the different stages of growth for potato plants within the study area. As the plants started to emerge around week 4, the indices increased gradually to show a slight peak. The maximum peak within the graphs represented the full maturity of the crops, indicating that they were ready for harvest. A significant drop in the curves also appeared at week 9. This drop indicated high stress levels in the crops related to high recorded temperatures.

Norkotah potato variety crops reached full maturity two weeks earlier than the Russet crops. This was evident by the difference in the critical growth week based on the correlation values and the regression models for the two varieties where the critical week for Norkotah groups was week 10 while the critical week for the Russet group was week 12. All the regression models indicated that SAVI, along with average slope, best explained and predicted yield. The indices' values of the weeks beyond week 12 showed a decreased relationship to yield and thus not considered a good indicator for yield prediction. SAVI proved to be a better indicator of yield over NDVI due to the limitation of NDVI and its relationship with the leaf area index. The variable L in the calculation of SAVI takes into consideration the soil within the fields unlike NDVI and this is crucial in fields with early season soil exposure between plants.

Access to high resolution and accurate yield data collected from the GTK sensor installed on potato harvesters in Idaho (0.8 m x 10 m) coupled with high resolution PlanetScope imagery (3 m) facilitated the development of a yield forecasting model that was suitable to apply in another region (Lebanon) with smaller fields. With agricultural lands being passed down between generations through inheritance, mainly in developing countries such as Lebanon, yield prediction models for smaller fields are a global need and an essential tool towards food security and sustainability. This model can be modified for other crop types at any given geographic location. The key information for adapting this model and applying to other crops is identifying the critical week as it differs between crops and varieties of the same crop as demonstrated for the Norkotah and Russet potato varieties within this research.

ACKNOWLEDGEMENTS

This research received funding from the Fulbright Program by the US Department of State, Planet Education and Research Program by Planet Team, Idaho State Department of Agriculture and U.S. Geological Survey under Grant/Cooperative Agreement No. G18AP00077.

We would like to thank the Geosciences Department at Idaho State University for their assistance and support. Special thanks to the farmers who contributed to this work by providing yield data and information about their fields and growing practices.

REFERENCES

Aditya Shastry, H.A. Sanjay and E. Bhanusree. 2017. "Prediction of Crop Yield Using Regression Techniques." *International Journal of Soft Computing* 12(2): 96–102.

Al-Gaadi, Khalid A. et al. 2016. "Prediction of Potato Crop Yield Using Precision Agriculture Techniques." *PLoS ONE* 11(9): 1–16. http://dx.doi.org/10.1371/journal.pone.0162219.

Burke, Marshall, and David B. Lobell. 2017. "Satellite-Based Assessment of Yield Variation and Its Determinants in Smallholder African Systems." *Proceedings of the National Academy of Sciences* 114(9): 2189–94. http://www.pnas.org/lookup/doi/10.1073/pnas.1616919114.

Gitelson, Anatoly A., Yoram J. Kaufman, and Mark N. Merzlyak. 1996. "Use of a Green Channel in Remote Sensing of Global Vegetation from EOS-MODIS." *Remote Sensing*.

Hatoum, Talal. 2005. "الزراعة في لبنان 2005." *مديرية الدراسات والتنسيق*: 23 http://www.agriculture.gov.lb/Arabic/DataAndAgriStatistics/AgriInLebanon/Documents/Enq_2005.pdf (January 11, 2017).

Huete, A. R. 1988. "A Soil-Adjusted Vegetation Index (SAVI)." *Remote Sensing of Environment* 25(3): 295–309.

Lewin, Paul et al. 2011. "The Role of Agricultural Processing in Idaho' s Economy: Status and Potential."

Qi, J et al. 1994. "A Modified Soil Adjusted Vegetation Index." 126: 119–26.

Rembold, Felix, Clement Atzberger, Igor Savin, and Oscar Rojas. 2013. "Using Low Resolution Satellite Imagery for Yield Prediction and Yield Anomaly Detection." *Remote Sensing* 5(4): 1704–33.

Rouse, J. W., R. H. Hass, J.A. Schell, and D.W. Deering. 1973. "Monitoring Vegetation Systems in the Great Plains with ERTS." *Third Earth Resources Technology Satellite (ERTS) symposium* 1: 309–17.

S.K., McFeeters. 1996. "The Use of the Normalized Difference Water Index (NDWI) in the Delineation of Open Water Features." *International Journal of Remote Sensing* 17(7): 1425–32.

Sellam, V, and E Poovammal. 2016. "Prediction of Crop Yield Using Regression Analysis." *Indian Journal of Science and Technology* 9(October): 1–5.

Song, Renzhong et al. 2016. "Evaluation of Landsat 8 Time Series Image Stacks for Predicting Yield and Yield Components of Winter Wheat." *IEEE*: 6300–6303.

United Nations Food and Agriculture Organization. 2017. *The Future of Food and Agriculture – Trends and Challenges.* Rome.

Zhang, Qin. 2016. *Precision Agriculture Technology for Crop Farming*. Prosser, Washington: Taylor & Francis Group.

ASSESSMENT OF WATERSHED RESOURCES FOR SUSTAINABLE AGRICULTURAL DEVELOPMENT: A CASE OF DEVELOPING AN OPERATIONAL METHODOLOGY UNDER INDIAN CONDITIONS THROUGH GEOSPATIAL TECHNOLOGIES

K. Balasubramani*

School of Earth Sciences, Central University of Tamil Nadu, Thiruvarur, India - geobalas@cutn.ac.in

KEY WORDS: Watershed, Land and Water Resources, Potential-Utilisation Index, SDG, Geospatial Technologies

ABSTRACT:

The watershed based integrated approach is considered as a more efficient and appropriate approach for resource appraisal and implementation of various sustainable development measures. The present study intends to demonstrate the application of geospatial technologies in preparation of comprehensive as well as an operational framework under Indian conditions to evaluate land and water resources patterns through a case study of Andipatti watershed, located in Theni district in the State of Tamil Nadu (India). The study used widely acclaimed various empirical and quantitative models to characterise the watershed and to understand the spatial pattern of potential-utilisation levels of the watershed resources. The study used mostly remotely sensed data for understanding surface characteristics and location enabled in-situ data such as soil, rainfall and groundwater data for understanding the sub-surface characteristics of the study area. All these datasets were integrated in a GIS environment to asses soil productivity, soil erosion, surface runoff, groundwater potential, groundwater quality, crop-land suitability, groundwater suitability for agriculture, determination of potential-utilisation index and to draw land use planning strategies. Since the methodology developed in the study is mostly depends on remotely sensed data, only minimal efforts are required for any agencies to prepare a similar systematic sustainable agricultural development plan.

1. INTRODUCTION

1.1 Background

Over the last 60 years, land and water resources have been severely exploited to meet the rapidly rising demands for food and fibre through input-intensive, mechanised and intensive irrigation based agricultural practices. Currently, agriculture uses 11 per cent of the world's land surface for crop production and accounts for 70 per cent of all water withdrawn from aquifers, streams and lakes (FAO, 2016). All the agricultural lands used for cropping and livestock rearing are more susceptible to degradation than non-agricultural lands. The world's agricultural production has grown between 2.5 and 3 times in the last 60 years, at the same time the cultivated area has grown only by 12 per cent. More than 40 per cent of increase in food and fibre production achieved only by doubling the irrigated areas of the world (FAO, 2011). Now, the water shortage, soil depletion and groundwater quality deterioration are most common problems in these irrigated agricultural areas (Nachtergaele et al., 2011).

The processes of degradation are largely determined by land use change, economic investment, access to markets, availability of infrastructure and farmers' accessibility to land that allows them to produce at maximum capacity (Bai et al, 2008). Behind these obvious direct human causes, there are other deeply rooted drivers such as population pressure, poverty, poor governance and institutional framework and lack of awareness, which leads to resources degradation (Nachtergaele et al., 2011). Besides these anthropogenic impacts, biophysical factors including soil properties, climatic characteristics, topography and vegetation also sometimes interact among themselves to yield a high rate of resource degradation (Kiage, 2013).

The condition of land and water resources and their role in supporting socio-economic development is becoming more and more critical in the arid and semi-arid lands of tropical regions. These areas are highly sensitive to human interference that resulted in a severe depletion of its natural resources due to multiple interactions of natural and anthropogenic factors (Abu-Sharar, 2006). The vulnerability is mainly due to high population density, vagaries of the monsoon and associated floods and droughts, high rates of soil erosion, increasing groundwater exploitation, haphazard use of fertilizers, low productivity, high post-harvest yield losses, long-term climate changes, lack of environmental awareness and fragile ecosystems. Agricultural production of semi-arid lands is increases steadily but this growth has had negative side-effects on land and water resources by poorly adapted production systems which remain a serious threat to future food security (Bindraban et al., 2012).

There are various approaches being adopted to improve agricultural production systems. Despite differences, most of the approaches focus on increasing agricultural production without depleting or over-exploiting the natural resources (Verhagen et al., 2017). The main objective of sustainable agricultural development is thus to integrate people's coexistence with potentials of natural resources over the long-term so that the production from farming lands and supporting services of ecosystems are ensured (Liniger et al., 2011). However, systematic assessments for sustainable agricultural development practices in largest tropical countries like India are

* Corresponding author

scarce (Balasubramani, 2018). With this background, the present study attempts to assess land and water resources in a watershed scale by integrating merits of global methodologies and applications of geospatial technologies towards sustainable agriculture development.

1.2 Need for the Study

In India, land and water resources degradation poses a considerable challenge to agricultural growth and food security. An analysis attempted by leading agencies of the nation reveals that 96.40 mha area of the country i.e. 29.3 per cent of the total geographic area of the country is undergoing a process of land degradation (ISRO, 2016). About 60 per cent of land in the country is rainfed with low productive capacity, leading to high inter-annual fluctuations in agricultural output (Mythili and Goedecke, 2016). To overcome the problems of food security and to meet out the status of self-sufficient, the country has introduced intensive farming practices in the early 1970's in the form the Green Revolution which later leads to higher rate of soil degradation, water shortage, groundwater quality deterioration and other environmental implications. Now, these problems are the foremost challenges to policymakers to balance the multiple goals of poverty eradication, food security and sustainable development. Current agricultural practices in the country are neither economically sound nor environmentally sustainable. Although India has attained self-sufficiency in food staples, the productivity of its farms is still poor mainly due to very small (less than 2 hectares) average size of land holdings and poor water resource management (Narayan, 2017). Most of the agricultural policies have primarily benefited large-scale farmers with productive land and access to water, bypassing the majority of small-scale producers, who are still in a trap of poverty, vulnerability and climatic uncertainty (FAO, 2011). In order to have sustenance of agriculture, it is necessary that the country should adopt a systematic resource assessment and management (Pathak, 2017). Many international agencies/organisations agreed that systematic action at national level is an indispensable basis for sustainable agricultural development. Hence an operational strategies needs to be developed at multiple levels within the country considering unique characteristics of the nation to address certain common issues of agriculture (Huddleston, 1990).

There is a general conscience that land and water resources should never be seen as products to be exploited, instead, they should be viewed as valuable assets and must be wisely managed (Balasubramani, 2016). Due to its complex and multi-faceted nature, degradation of land and water resources cannot be easily monitored or managed. There should be a proper system in a place to assess systematically the natural resources through the principles of sustainable development. Since watersheds forms a natural as well as an ideal unit for assessing land and water resources, a basin or watershed based resources assessment would be an effective approach for holistic development of agriculture, forestry and allied activities and lessening the negative impacts of agricultural land use in India. Hence, the present study attempts to formulate a framework to assess land and water resources in a watershed scale. The study demonstrates the application of geospatial technologies for assessing land and water resources potentials as well as utilizing patterns by taking a case study area. It also attempts to outlines strategies that need to be taken by decision-makers in a bid to pursue the objectives of sustainable agricultural development for the case study area.

1.3 Choice of Study Area

Andipatti watershed, part of Vaigai river basin in the State of Tamil Nadu is considered as case study area. The economics of the watershed is primarily depending on agriculture and related activities and reflecting most of the major characteristics of India. The farmers of the watershed are poor and they are highly dependent on land and water resources. Most of the agricultural fields are less than one hectare but supports more than 10,000 marginal farmers. Even though the watershed falls under semi-arid climate with an annual average rainfall of 790 mm, about 40 per cent of the watershed area falls under irrigated agriculture (Balasubramani, 2018). Due to inappropriate management practices and intense crop production, the watershed faces acute problem of land and water resources degradation. Thus, Andipatti watershed is selected as case study area to demonstrate the application of geospatial technologies in the preparation of comprehensive framework to evaluate land and water resources patterns for better management strategies. However, the proposed methodological framework is applicable to all agricultural dominant rainfed watersheds of the country.

2. METHODOLOGY

2.1 Physical Characteristics

The physical characteristics of watershed were studied mainly through geospatial technologies. The survey of India (SOI) toposheets (58 F/12 SW & 58 G/9 NW), Geology map published by Geological Survey of India (GSI), Soil database generated by Tamil Nadu Agricultural University (TNAU) and National Bureau of Soil Survey and Land Use Planning (NBSS&LUP), climate and groundwater data compiled by State Surface and Ground Water Board, demography data published by Census of India and village agricultural database compiled by Directorate of Economics and Statistics forms basis for preparation of base layers. All these datasets projected into common scale (1:25,000) and projection system (UTM) and stored as geodatabase using GIS software. The IRS LISS-III & IV, CARTOSAT-1 and LANDSAT TM satellite images were used to interpret and generation of landuse / land cover, geomorphology and slope maps of the watershed. The technical details of data sources and interpretation are detailed in (Balasubramani, 2016).

2.2 Micro-watersheds

The watershed is further delineated into 44 micro-watersheds with the help of ArcSWAT tool. The size, shape and areal aspects of micro-watersheds were extracted from automated/semi-automated processes in ArcGIS and stored in geodatabase for detailed morphometric analysis of the watershed. The indices used for morphometric analysis and the results are detailed in (Balasubramani et al., 2019).

2.3 Soil Productivity

The soil productive capacity can be evaluated directly or indirectly. An assessment of soil productivity in terms of quantitative parameters is of paramount significance for proper land use planning and management. The best-known parametric system for rating the productivity of soil is the Storie Index (O'Geen et al., 2008 and Verheye, 2011). The surface and sub-surface soil information including soil series, texture, permeability, depth and different chemical properties along with topographic information were used to understand the soil productivity of the watershed. The factors and their ratings to compute soil productivity (Storie index) are detailed in (Balasubramani, 2016).

2.4 Soil Erosion and Surface Runoff

Since watershed forms a natural boundary to focus on all the effects of downhill runoff, a systematic assessment of soil erosion and surface runoff within each micro-watershed would provide reliable information to draw strategies for sustainable development of agriculture. The dominant model applied worldwide to estimate the soil erosion is Revised Universal Soil Loss Equation (RUSLE) which is reasonably good in estimating soil erosion (Renard et al., 2011). The method of estimation and spatial distribution of soil erosion in the case study area are detailed in (Balasubramani et al., 2015). Similarly, NRCS-Curve Number (CN) approach provides a simple method for runoff estimation that uses empirical formulas based on rainfall, land use and soil type (Rao et al., 2010). The methodology adopted for estimating runoff of the watershed is detailed in (Bharathi, Balasubramani, 2015) and (Balasubramani, 2018).

2.5 Groundwater Potential and Level

Groundwater resources play a major role in the social and economic growth of the watershed. The problems of over-exploitation of groundwater resources are already reported for the case study area (CGWB, 2008). Groundwater potential of the study area was assessed using resistivity, well yield and dug well data as demonstrated in (Balasubramani, 2019a). Groundwater level is also an important variable which indicate indirectly the groundwater potential zones. The seasonal groundwater levels (pre-monsoon and post-monsoon) collected from 34 open wells across the study area and long-term groundwater level data collected from the State and Central Groundwater boards were used to analyse the spatio-temporal variations in the groundwater. The methods and results of the analysis are described in (Balasubramani, 2019a).

2.6 Land Capability and Irrigability

Land capability and irrigability classifications are systematic approaches to evaluate the land characteristics for agricultural purposes (Sehgal, 2012). Some soils are suitable for crops, others for pastures and some others may only be suitable for forestry. Likewise some soils are suitable for extensive irrigation and some poses serious limitations. Therefore, grouping of soil units into capability and irrigability classes serve as a guide to assess the suitability of land. The factors that used to determine the capability of a soil unit in the present study are topography (slope), soil erosion, drainability, permeability, infiltration rate, surface texture, soil depth, electrical conductivity, sodicity, CEC, base saturation and organic matter content. The criteria adopted for both capability and irrigability classifications are explained in (Balasubramani, 2016).

2.7 Crop-Land Suitability

Land suitability evaluation plays an important role in sustainable agricultural development of a region (Davidson, 2002). The data collection, delineating land units and suitability evaluation are now largely facilitated by geospatial technologies. Keeping FAO (2007) guidelines, the thematic layers such as topography (slope), soil series (soil properties), soil erosion (RUSLE), generalised land use / land cover were intersected in GIS to prepare 787 homogenous land units. These units used to describe land qualities and characteristics as well as used to compare with the requirements of major crops. The qualities and characteristics of each land unit were compared with crop requirements and rated either as highly suitable with slight limitations (S1), moderately suitable with moderate limitations (S2), marginally suitable with severe limitations (S3) or not suitable (N). The major crops considered for crop-land suitability analysis are paddy, maize, sorghum/bajra, ragi, pulses, cashew, groundnut, cotton, sugarcane, vegetables, coconut, mango and banana. These crops were selected based on cropping pattern, farmer's preferences and agro-ecological conditions of the study area. The details of land units, land qualities, crop selection and crop requirements are presented elaborately in (Balasubramani, 2016).

2.8 Groundwater Suitability

Groundwater is the main source of irrigation in the watershed. The suitability of groundwater for irrigation purpose can be examined using concentration of chloride (Cl), magnesium ratio (MR), sodium adsorption ratio (SAR), residual sodium carbonate (RSC), soluble sodium percentage (SSP), permeability index (PI) and a method formulated by the US Salinity Laboratory (USSL). The adopted equations and spatial distribution of the quality parameters are detailed in (Balasubramani, 2019a).

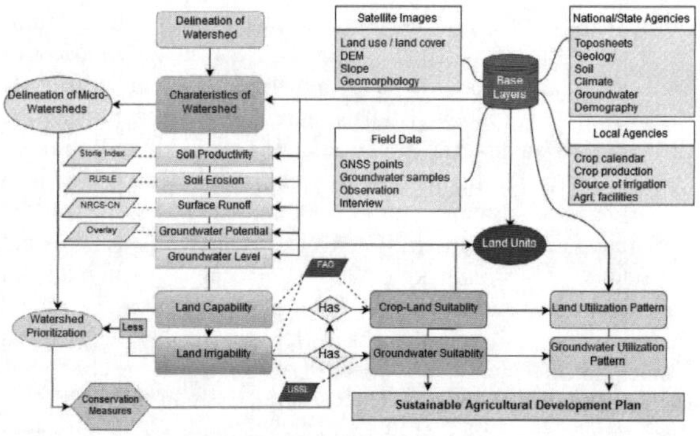

Figure 1 Workflow of developing an operational methodology for sustainable agricultural development

2.9 Resources Utilization Pattern

An analysis on resources potential and utilisation level in micro-level unit is essential for effective watershed management. In this study, land use / land cover pattern and soil productive capacity forms the basis for understanding the land potential-utilisation pattern. Similarly, groundwater level, groundwater quality, percentage of irrigated land, concentration of wet crops, land potential-utilisation pattern and erosion characteristics were used to determine the groundwater potential-utilisation pattern. Both the utilisation patterns were computed in a micro-watershed scale as detailed in (Balasubramani, 2016 and Balasubramani, 2019a). This kind of assessment helpful to the planners not only for understanding the utilisation level but also for regulating the present land and water utilizations scenario.

2.10 Watershed Prioritization

Delineation of micro-units within the watershed and prioritization of those units are essential steps in the integrated watershed management. Each micro-watershed has unique characteristics and limitations. While suggesting soil and water conservation measures, it is important to prioritize the micro-watersheds based on the characteristics and limitations in a systematic manner. With the help of geospatial technologies and multi-criteria decision-making approach, the present study prioritized micro-watersheds on the basis of morphometric parameters, soil erosion and surface runoff rates. The scheme of prioritization and method to calculate priority index are detailed in (Balasubramani, 2019).

The overall workflow of developing an operational methodology through geospatial technologies for sustainable agricultural development under Indian conditions is presented in Figure 1.

3. RESULTS AND RECOMMENDATIONS

The land and water resources degradation are prevalent in many forms throughout the country. It is realised widely now that the consumption of natural resources is higher than their natural regeneration capacity. This has affected the carrying capacity of the ecosystems and created an ecological imbalance, especially in the semi-arid regions. To increase productivity of agriculture in these regions, watershed based approaches are highly recommended because it found effective through a number of case studies. Further, the watershed approach is a system-based integrated approach that facilitates the holistic development of agriculture, forestry and allied activities in the watershed. By interfacing remote sensing, GIS and computation models, different management scenarios could be generated at watershed scale which would help the planners to assess the feasibility of various alternatives before selecting the one that would be most suitable. Thus, the study tries to understand various aspects of suitability of land and water resources for agricultural and other allied uses.

By comparing resource potentials and existing land use patterns, it is observed that about 22 per cent of fallow / wastelands of the watershed have excellent to good land productive capacity, but such lands are not utilised for productive purposes. Though the extreme northern watersheds found with shallow water table and good to very good groundwater quantity, the chemical properties of the watershed are not suitable for irrigation. Both quantity and quality of the groundwater are not supporting agriculture in the watersheds of central and west-central part of the study area, but the utilisation level is very high and about 2/3 of the agricultural crops are water intensive. Unless and otherwise the area under intensively irrigated crops is changed to less intensively irrigated crops or dry crops, the land and water resources of the watershed cannot be saved from complete ruination.

In an area where land potential is marginal and water is a precious commodity, then structural conservation measures and optimization of land use are the most essential requirements for sustainable development. Based on land suitability, the groundwater constraints and socio-economic conditions of the watershed, sustainable land use plan is proposed for the study area (Figure 2). The conservation measures need to be taken up is discussed elaborately in (Balasubramani, 2016). If these suggested measures are adopted promptly, it would not only help to reduce resource degradation, but also reduce the severity of drought and increase the agricultural productivity of the watershed to a considerable extent. Further, the application of geospatial technologies should be encouraged in every dimensions of watershed planning because these culminations of technologies facilitate accurate resource assessments and provide sound platform for monitoring and managing the watershed resources and for achieving the sustainable development goals. The methodology of the case study can be further calibrated and could be used for optimizing utilization of land and water resources at different scales and levels.

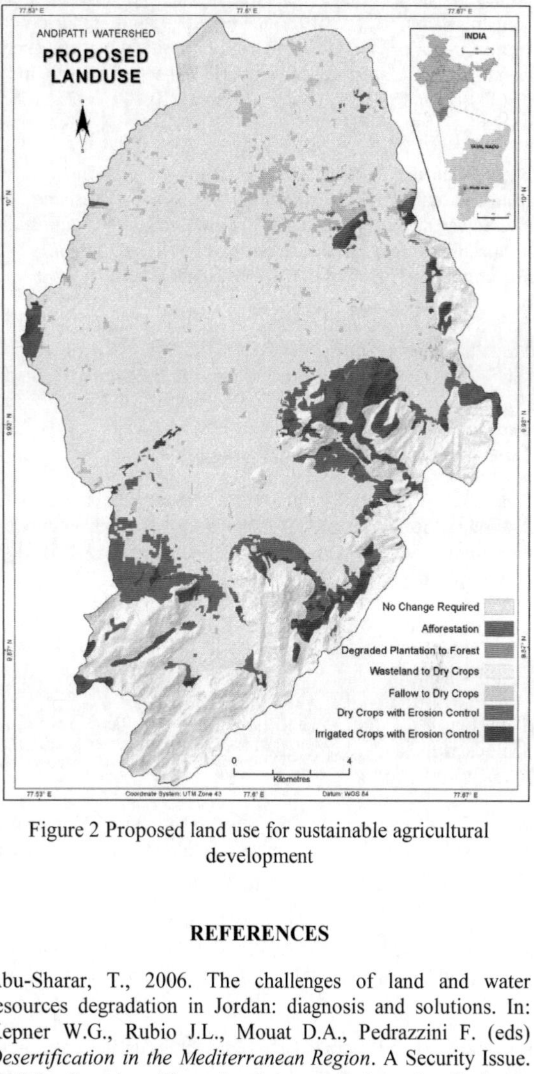

Figure 2 Proposed land use for sustainable agricultural development

REFERENCES

Abu-Sharar, T., 2006. The challenges of land and water resources degradation in Jordan: diagnosis and solutions. In: Kepner W.G., Rubio J.L., Mouat D.A., Pedrazzini F. (eds) *Desertification in the Mediterranean Region*. A Security Issue. NATO Security Through Science Series, 3, Springer, Dordrecht.

Bai, Z.G., Dent, D.L., Olsson, L., Schaepman, M.E., 2008. Proxy global assessment of land degradation. *Soil Use Management*, 24:223–234.

Balasubramani, K., Veena, M., Kumaraswamy, K., Saravanabavan, V., 2015. Estimation of soil erosion in a semi-arid watershed of Tamil Nadu (India) using revised universal soil loss equation (rusle) model through GIS. *Model. Earth Syst. Environ*. 1:10. https://doi.org/10.1007/s40808-015-0015-4

Balasubramani, K, 2016. Application of geospatial technologies in land and water resources evaluation for sustainable development of Andipatti watershed, Tamil Nadu, India. Ph.D. Thesis, Madurai Kamaraj University, Madurai. https://doi.org/10.13140/RG.2.2.12302.97606

Balasubramani, K., 2018. Physical resources assessment in a semi-arid watershed: An integrated methodology for sustainable land use planning. *ISPRS Journal of Photogrammetry and Remote Sensing* 142:358–379. https://doi.org/10.1016/j.isprsjprs.2018.03.008

Balasubramani, K., Gomathi, M., Bhaskaran, G., Kumaraswamy, K., 2019. GIS-based spatial multi-criteria approach for characterization and prioritization of micro-watersheds: a case study of semi-arid watershed, South India. *Appl Geomat* 11:289–307. https://doi.org/10.1007/s12518-019-00261-y

Balasubramani, K., Rutharvel Murthy, K., Gomathi, M., Kumaraswamy, K., 2019a. Integrated assessment of groundwater resources in a semi-arid watershed of South India: implications for irrigated agriculture. *GeoJournal*, 1-23 https://doi.org/10.1007/s10708-019-10050-0

Bharathi, G.P., Balasubramani, K. 2015. Rainfall - Runoff Modelling using Soil Conservation Service – Curve Number (SCS-CN) Method – A Case Study of Ungauged Andipatti Watershed, Tamil Nadu, India. In: Saravanabavan, V. (ed) *Geospatial Technologies for Resources Evaluation and Management*, Jayalakshmi Publications, Madurai, 59-67.

Bindraban, P.S. Marijnvan der V., LimingYe, et al., 2012. Assessing the impact of soil degradation on food production. *Current Opinion in Environmental Sustainability*, 4(5): 478-488. https://doi.org/10.1016/j.cosust.2012.09.015

CGWB, 2008. District groundwater brochure, Theni District, Tamil Nadu. Technical report series, Central Ground Water Board, Chennai.

Davidson, D., 2002. The assessment of land resources: achievements and new challenges. *Australian Geographical Studies*, 40:109-128.

FAO, 2007. Land evaluation towards a revised framework. Land and water discussion paper 6. Food and Agricultural Organization, Rome.

FAO, 2011. The state of the world's land and water resources for food and agriculture - Managing systems at risk. Food and Agricultural Organization, Rome. http://www.fao.org/docrep/017/i1688e/i1688e.pdf (15 August 2019)

FAO, 2016. Land degradation assessment in drylands - Manual for local level assessment of land degradation and sustainable land management Part-1. Food and Agricultural Organization, Rome. http://www.fao.org/ publications/card/en/c/40875549-ecc8-4388-b944-26edc9b58272 (15 August 2019)

Huddleston, B., 1990. FAO's Overall Approach and Methodology for Formulating National Food Security Programmes in Developing Countries. *IDS Bulletin*, 21:72–80. https://doi.org/10.1111/j.1759-5436.1990.mp21003009.x

ISRO, 2016. Desertification and Land Degradation Atlas of India. Space Applications Centre, Ahmedabad. https://vedas.sac.gov.in/vedas/downloads/atlas/DSM/Desertification_ Atlas_2016_SAC_ISRO.pdf

Kiage, L.M., 2013. Perspectives on the assumed causes of land degradation in the rangelands of Sub-Saharan Africa. *Progress in Physical Geography: Earth and Environment*, 37(5):664 – 684. https://doi.org/10.1177/0309133313492543

Liniger, H.P., Mekdaschi Studer, R., Hauert, C., Gurtner, M., 2011. Sustainable land management in practice – guidelines and best practices for sub-saharan Africa. Food and Agricultural Organization, Rome. http://www.fao.org/docrep/014/i1861e/i1861e.pdf (15 August 2019)

Mythili, G., Goedecke, J., 2016. Economics of land degradation in India. In: Nkonya, E. et al. (Eds), *Economics of Land Degradation and Improvement – A Global Assessment for Sustainable Development*. Springer, Heidelberg. https://doi.org/10.1007/978-3-319-19168-3_15

Nachtergaele, F. Biancalani, R., Petri, M., 2011. Land degradation. SOLAW Background Thematic Report 3, Food and Agricultural Organization, Rome. http://www.fao.org/nr/solaw/thematic-reports/en/ (15 August 2019)

Narayan, S., 2017. Indian agriculture: a long way to go. In: Pant, H. et al., (Eds) *Emerging Trends in Agricultural, Environmental and Rural Developmental Challenges and Solution: An Overview*, Society of Biological Sciences and Rural Development, New Jhunsi, 26-42.

O'Geen, A.T., Southard, R.J., Southard, S.B., 2008. A revised storie index for use with digital soils information. ANR Publication 8335, Agricultural and Natural Resources Communication Services, University of California. California.

Pathak, S.C., 2017. Management of natural resources for sustainable agricultural development. Emerging Trends in Agricultural. In: Pant, H. et al., (Eds) *Environmental and Rural Developmental Challenges and Solution: An Overview*, Society of Biological Sciences and Rural Development, New Jhunsi, 1-3.

Rao, K.N., Narendra, K., Latha, P.S., 2010. An integrated study of geospatial information technologies for surface runoff estimation in an agricultural watershed, India. *Journal of Indian Society of Remote Sensing*, 38:255-267.

Renard, K.G., Yoder, D.C., Lightle, D.T., Dabney, S.M., 2011. Universal soil loss equation and revised universal soil loss equation. In: Morgan, R.P.C. Nearing, M.A. (Eds), Handbook of Erosion Modeling, Blackwell Publishing Ltd., England, 137–167.

Sehgal, J. 2012. *A text book of pedology concepts and applications.* Kalyani Publishers, New Delhi, 342-388.

Verhagen, J., Greet, B., Christy van, B., Simone, V., 2017. Approaches aiming at sustainable agricultural production. Wageningen Research, Report WPR-677, Wageningen http://dx.doi.org/10.18174/416937

Verheye, W., 2011. Land evaluation systems other than the FAO system. *Land use, Land cover and Soil Sciences*, 2:3-7.

Revised November 2019

REALTIME PLANETARY-SCALE DATACUBE FUSION

P. Baumann [1]

[1] Jacobs University, Bremen, Germany – p.baumann@jacobs-university.de

Commission VI, WG VI/4

KEY WORDS: datacube, fusion, distributed processing, query planning, rasdaman

1. INTRODUCTION

The datacube model has rapidly gained acceptance as a cornerstone for analysis-ready data, and also the corresponding service model which is more powerful, but easier to use than existing API-based interfaces. Mature and widely adopted datacube standards show the way how datacube functionality can be presented to clients, be they human or m2m connections. Specifically, the OGC Coverage Implementation Schema (CIS) data model and the Web Coverage Service (WCS) service model suite define a framework adopted by the main open-source as well as proprietary tools, including MapServer, GeoServer, GDAL, QGIS, ArcGIS, and python/OWSlib.

While the basic, most widely used functionality includes access, possibly extraction, and reformatting of a data sub(set) for download, processing and in particular data fusion represent complex, resource-intensive challenges. In the WCS suite this is addressed through a concept of modularity where WCS Core (which is mandatory for any WCS to implement) offers the GetCoverage request to accomplish "give me part of this spatio-temporal coverage, in my favourite format" and (optionally implementable) extensions add further functionality facets, up to the spatio-temporal datacube analytics language, Web Coverage Processing Service (WCPS) (Baumann, 2010).

For example, computing the NVDI over Europe on the 1st of July 2018 using Sentinel data stored in a datacube with the same name, and returning the result as a NetCDF file, can be achieved with the following WCPS query:

```
for $c in (Sentinel)
return
 encode(
  (((float) $c.nir – $c.red) / ((float) $c.nir + $c.red))
  [ Lat(35:70), Long(10:40), time("2018-07-01") ],
  "image/tiff"
 )
```

As WCPS allows combination of datasets ("joins" in database terminology) the question arises: what if datacubes to be combined reside on different computers (cloud scenario), or even in different data centers (federation scenario)? A suboptimal implementation obviously could cause massively degraded performance.

In this contribution we present federation methods implemented in the rasdaman datacube engine which is accepted technology leader, standards driver, and reference implementation for datacubes.

2. RASDAMAN ARCHITECTURE

The rasdaman engine resembles a fully-fledged Array Database System which utilizes a declarative query language enhancing standard SQL with high-level array operators; in fact, ISO in 2018 has adopted this language as an extension to the SQL language standard (ISO, 2019). This datacube language is domain agnostic and can serve as well medical imagery, cosmologic simulation results, and the like. A semantic layer on top of it offers geo semantics, i.e., it knows about spatial and temporal coordinates and, hence, also about regular and irregular grids. This semantics is offered via OGC WCS, WCPS, and WMS interfaces so that a plethora of clients can utilize massive datacubes managed by rasdaman. All such requests are translated to array SQL queries internally.

In the server, such incoming queries are optimized, parallelized, and ultimately mapped to datacubes that are tiled on disk (or tape). As opposed to standard regular tiling (i.e., equi-sized tiles) rasdaman allows for arbitrary tiling guided by several strategies (Baumann et al., 2010). Tiling remains invisible to the applications (and, hence, in the queries), it is a tuning parameter for the administrator. Tiles of the same datacube can even sit on different machines.

The rasdaman server is multi-parallel per se, allowing any number of simultaneous parallel requests (inter-query parallelization) as well as parallel execution of incoming queries (intra-query parallelization), see (Dumitru, 2014).

Datacube operations are usually triggered via standardized web services, such as WCS, WMS and WCPS. However, these services are only the interface to the client performing the operation, while the processing itself is handled at a lower level which enables optimized, efficient execution. In rasdaman, the geo layer intercepting the web requests representing datacube operations translates them into array database queries, which are then handled by rasdaman's query processing engine.

The engine analyses each incoming query and compiles it into machine code. During that process, a series of optimization steps is applied, one of which is distribution.

Such federations are being done routinely meantime (Baumann, 2017). In the EarthServer project, Petabyte-scale datacubes have been federated across ECMWF/UK and NCI/Australia. Another real-life use-case of the optimization is given in the federation which was set up between the German Sentinel hub, CODE-DE, established in the BigDataCube project (BigDataCube project team, 2019), and the Alfred Wegener Institute for Polar and Marine Research. Among others, temperature datacubes are available at CODE-DE, and SeaIce datacubes are available at AWI. In one of the use-cases, scientists need to compute the sea ice distribution at different temperature intervals. The WCPS query that achieves that is presented below. The datacubes involved contain data at different resolutions and in different coordinate systems, which gets adjusted in the course of query processing. Most recently, FCU/Taiwain has been federated with CODE-DE.

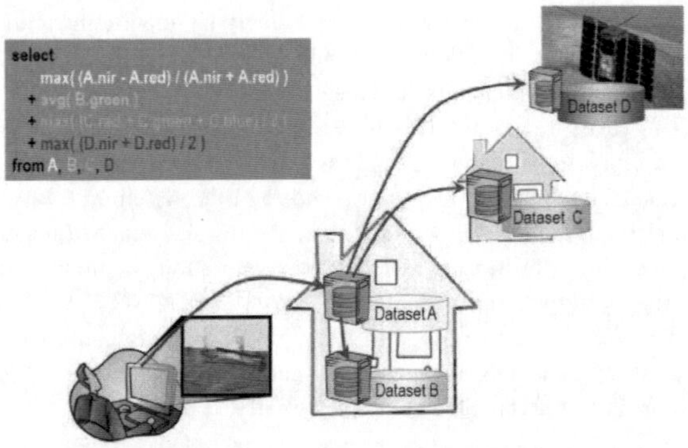

Figure 1. Sample query splitting and distribution in rasdaman

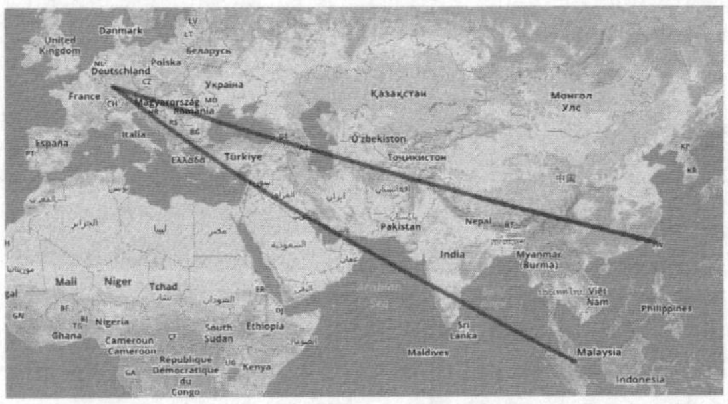

Figure 2. FCU/CODE-DE federation query path: query sent to CODE-DE from Singapore, subquery generated for FCU

3. CONCLUSION

In summary, rasdaman allows federated processing of declarative queries whereby complete location transparency is given: any federation member can receive the query, and the federation will dynamically orchestrate each query individually for optimized processing.

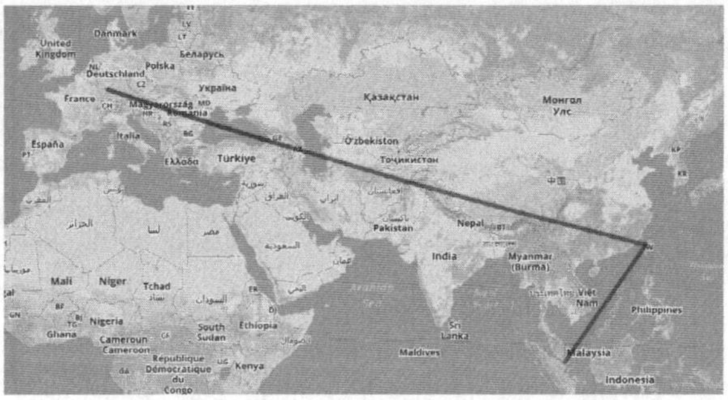

Figure 3. FCU/CODE-DE federation query path: query sent to FCU from Singapore, subquery generated for CODE-DE

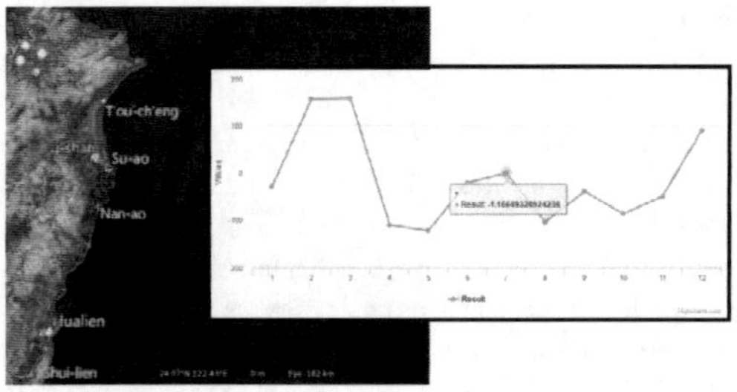

Figure 4. FCU/CODE-DE federation query path: query sent to FCU from Singapore, subquery generated for CODE-DE

ACKNOWLEDGEMENTS

This work is being supported by H2020 LandSupport, H2020 EOSC-hub, and German BMWi BigDataCube.

REFERENCES

Baumann, P., 2010: The OGC Web Coverage Processing Service (WCPS) Standard. *Geoinformatica*, 14(4)2010, pp 447-479

Baumann, P., et al. 2010: Putting Pixels in Place: A Storage Layout Language for Scientific Data. *Proc. IEEE ICDM Workshop on Spatial and Spatiotemporal Data Mining*, Sydney, Australia, pp. 194 - 201

Dumitru, A. et al., 2014: Exploring cloud opportunities from an array database perspective. *Proc. ACM SIGMOD DanaC*, Snowbird, USA, pp. 1 - 4

ISO, 2019: Information technology — Database languages — SQL — Part 15: Multi-Dimensional Arrays. ISO IS 9075-15:2019

N.n., 2019: BigDataCube. http://www.bigdatacube.org/

P. Baumann, A.P. Rossi,et al., 2017: Fostering Cross-Disciplinary Earth Science Through Datacube Analytics. In. P.P. Mathieu, C. Aubrecht (eds.): Earth Observation Open Science and Innovation - Changing the World One Pixel at a Time, International Space Science Institute (ISSI), pp. 91 - 119

ESTIMATION OF CHL-A CONCENTRATION IN LAGUNA LAKE USING SENTINEL-3 OLCI IMAGES

Ariel C. Blanco[1,2]*, Amihan Manuel[2], Rey Jalbuena[2], Kristina Ticman[2], Jommer M. Medin[1], Edgardo Gubatanga[2], Ace Santos[2], Raymund Sta. Ana[2], Eugene Herrera[3], Kazuo Nadaoka[4]

[1]Training Center for Applied Geodesy and Photogrammetry, University of the Philippines, Diliman, Quezon City 1101, Philippines
[2]Department of Geodetic Engineering, University of the Philippines, Diliman, Quezon City 1101, Philippines
[3]Institute of Civil Engineering, University of the Philippines, Diliman, Quezon City 1101, Philippines
[4]Tokyo Institute of Technology, Tokyo, Japan
*acblanco@up.edu.ph

KEY WORDS: regression analysis, inland water, water quality, remote sensing, phytoplankton

ABSTRACT:

The use of Sentinel-3 Ocean and Land Color Instrument (OLCI) images in estimating chlorophyll-a (total and class-differentiated)a concentration is promising owing to Sentinel-3's 21 bands. This was investigated for the case of Laguna de Bay (or Laguna Lake), Philippines. Field surveys were conducted on 13-17 November 2018 using FluoroProbe, a submersible fluorimeter capable of quantifying concentrations of spectral classes of microalgae. These were regressed with reflectance data obtained from 10-day composite Sentinel-3 reflectance images as well as ten empirical algorithms (indices) for OLCI. Compared to band reflectance, the 10 indices yielded stronger correlations, especially with R665/R709, R674/R709, and (1/R665-1/R709)xR754 with the following respective correlation values: -0.623, -0.646, and 0.628. Multiple regression results indicates that 48% of the variability of total chl-a concentration is explained by five explanatory (reflectance) variables (R412, R443, R560, R681, and R754) with RMSE of 2.814 ug/l. In contrast, the two indices R674/R754 and (1/R665-1/R709)xR754 accounted for about 46% of the variability of total chl-a concentration with RMSE of 2.475 ug/l. For diatoms and bluegreen microalgae, R560/R665 and (1/R665-1/R709)xR754 constitute the models with R^2 of 0.21 and 0.435, and RMSE of 2.516 and 2.163 ug/l, respectively. Green microalgal concentration is jointly described by three indices: R560/R665, R674/R754, and R709-R754, with R^2=0.182 and RMSE=1.219 ug/l. From cryptophytes, the model comprising of R560/R665, (1/R665-1/R709)xR754, and R709-R754 produced an R^2=0.289 and RMSE=0.767 ug/l. It can be said that the empirical algorithms can be used for Sentinel-3 OLCI data providing acceptable estimations of total and spectral class-differentiated chl-a concentration.

1. INTRODUCTION

Chlorophyll-a is one of the standard water quality parameters monitored in bodies of water. Typically, chl-a concentration is measured by water sampling and laboratory analysis and/or through in situ measurement using water quality checker/instrument at designated sampling stations. There are however time-consuming, labor-intensive, and costly (Schaeffer et al, 2013). With limited number of sampling stations, it would be difficult to describe the temporal and spatial variability of large scale lake phenomena (Dörnhöfer and Oppelt, 2016). Due to uneven distribution of sampling points and differences in time of sampling/field measurement, these may result in unreliable water quality assessment (Bresciani et al., 2011).

Many studies have utilized remote sensing to assess water quality, describing its spatial and temporal variability. Chlorophyll-a, turbidity, suspended particulate matter, and colored dissolved organic matter (CDOM) are the parameters commonly mapped and monitored using satellite images, particularly those freely available including Landsat images (Chen and Feng, 2016, Manzo et al,. 2015, Palmer et al., 2015). Recent studies have focused on the use of Sentinel-2 and Sentinel-3 data (Toming et al., 2017, Blix et al., 2018, Petus et al., 2019). Toming et al. (2017) retrieved water reflectance, inherent optical properties (IOPs), and water quality parameters including chl-a, total suspended matter (TSM) and CDOM in the Baltic Sea using standard Case-2 Regional/Coast Colour (C2RCC) processing chain. Blix et al. (2018) used and evaluated Sentinel-3 OLCI L2 products, namely, chl-a, CDOM, and total suspended matter (TSM) to monitor an optically highly complex shallow Lake Balaton. These and other studies estimated total chl-a and not the chl-a associated with different classes of microalgae. The estimation of the latter is important as abundance and proportions/percentages of microalgal classes signify or indicate occurrence of particular environmental phenomena and issues.

Laguna de Bay is a multi-use productive lake in the Philippines devoted to aquaculture, water supply, and transportation of people and goods. The lake supplies significant percentage of milkfish and tilapia sold in Metro Manila and provinces surrounding the lake. Laguna Lake is already being tapped for domestic water supply with water concessionaires abstracting hundreds of million liters per day. The lake is eutrophic and turns green in color when microalgae blooms occur as a result of high nutrient inputs from surrounding watersheds and abundant sunlight. Excessive algae may result in fish kills and affect water supply from Laguna Lake. Abundance of algae in terms of chlorophyll (green pigment) must be mapped and monitored.

This study investigates the applicability and performance of regression models generated from Sentinel-3 OLCI reflectance data and indices in estimating chl-a concentration. The abundance of phytoplankton is estimated not only through total chl-a but also the spectral class-differentiated chl-a concentration.

2. MATERIALS AND METHODS

2.1 Study Area

Laguna de Bay (more commonly called Laguna Lake) is the largest inland waterbody in the Philippines with a surface area of approximately 900 sq. km. (Laguna Lake Development Authority, 2017). The lake is shallow with an average depth of 2.5 m only (Santos-Borja and Nepomuceno, 2003). Primary uses of the lake include aquaculture, navigation, and water source for domestic use, irrigation, and power generation.

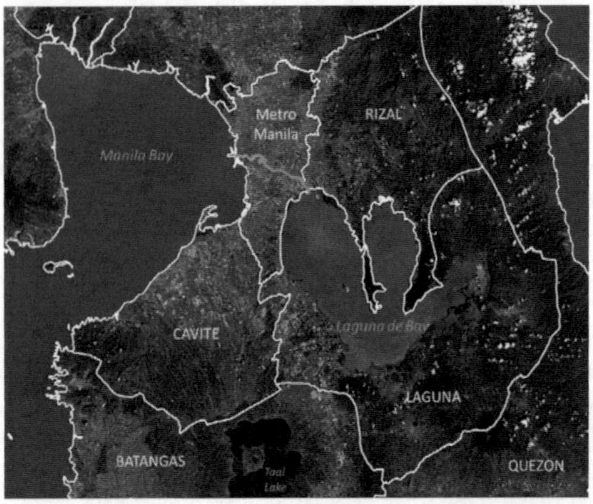

Figure 1. Laguna de Bay (Laguna Lake) and surrounding land areas. The lake is connected to Manila Bay via the Pasig River

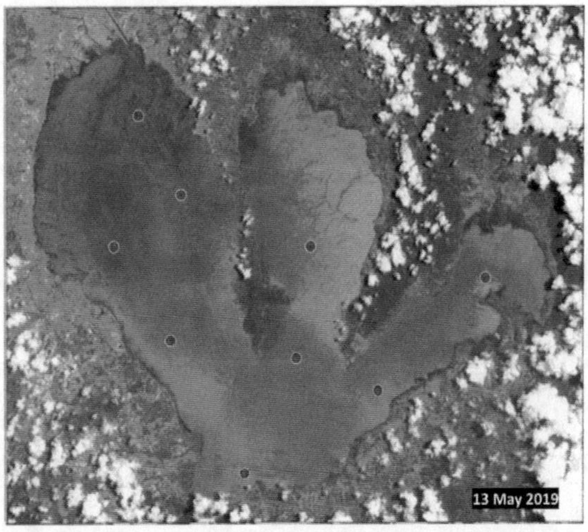

Figure 2. Landsat 8 image (acquired on 13 May 2019) showing the spatial variation of "greenness" considered as indication of varying levels of abundance of phytoplankton. The red circles indicate the water quality monitoring stations of the Laguna Lake Development Authority

As shown in Figure 1, Laguna Lake is connected to Manila Bay through the Pasig River, which serves as the only outlet of the lake. During dry season, backflow occurs making the water brackish with the intrusion of saline water (Santos-Borja and Nepomuceno, 2003). The lake is surrounded by Metro Manila and the Provinces of Laguna, Rizal, Cavite, Batangas, and Quezon. Its watershed has an area of about 2,980 sq. km. (Santos-Borja and Nepomuceno, 2003). Water quality degradation has been observed in Laguna Lake due to intensified exploitation of the its resources and pollution from industrial, household, and agricultural sources (Delos Reyes and Martens, 1994; Tamayo-Zafaralla et al., 2010).

2.2 Field Surveys

Field surveys were conducted in the lake on 13-17 November 2018 using FluoroProbe, a submersible fluorimeter capable of quantifying concentrations of spectral classes of microalgae. FluoroProbe, water quality, and spectral measurements were taken at forty-seven (47) stations in the lake (see Figure 3). In situ water quality measurements of turbidity, chl-a, conductivity, temperature, and pH were conducted using the AAQ multi-parameter water quality instrument.

2.3 Sentinel-3 OLCI Data

OLCI provides the opportunity and capability to monitor aquatic environments utilizing nine (9) spectral bands, which allows to retrieve detailed information about the water quality of various type of waters (Blix et al., 2018). In this study, 10-day composite Sentinel-3 reflectance images were downloaded from the Copernicus Global Land Service (CGLS), which is a component of the Land Monitoring Core Service (LMCS) of the European flagship programme on Earth Observation called Copernicus. CGLS "systematically produces a series of qualified bio-geophysical products on the status and evolution of the land surface, at global scale and at mid to low spatial resolution, complemented by the constitution of long-term time series" (see https://land.copernicus.eu/global/). This study did not consider the Level 2 (L2) products which have been available for the public since July 2017.

2.4 Correlation and Regression Analysis

The chl-a (total and spectral class-differentiated) concentration values were tested for correlation and regressed with reflectance data obtained from 10-day composite Sentinel-3 reflectance images. downloaded from the Copernicus Global Land Service. In addition, ten (10) empirical algorithms (indices) for OLCI based on previous studies were also used in the regression analysis as listed below:

- R560/R665
- R665/R709
- R665/R754
- R674/R709
- R674/R754
- R709/R754
- (1/R665-1/R709)xR754
- (R490 - R665)/R560
- R709 - ((R665 + R754)/2)
- R709-R754

In this study, the reflectance bands In multiple regression, variables with the highest variance inflation factor (VIF) are eliminated one at a time until all variables have VIF less than or equal to 7.5.

3. RESULTS AND DISCUSSION

3.1 Spatial Variation of Chl-a

As shown in Figure 3, chl-a concentrations were high in the central lobe of Laguna Lake. In the west lobe, there are areas where the chl-a levels were also relatively high. What is noticeable from Figure 3 is the dominance of diatoms in most parts of the lake. However, bluegreens were relatively more abundant in the central lobe. Green microalgae were considerably dominant in some areas of the east lobe of the lake.

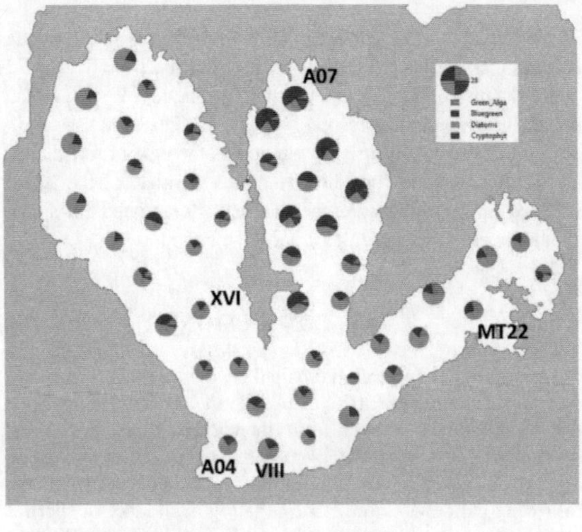

Figure 3. Spatial distribution of chl-a (ug/L) as measured using FluoroProbe. The size of the pie indicates the total chl-a relative to the legend and the colors of the pie sectors indicate the percentage of green microalgae, bluegreen microalgae (cyanobacteria), diatoms, and cryptophytes.

3.2 Spectral Signatures

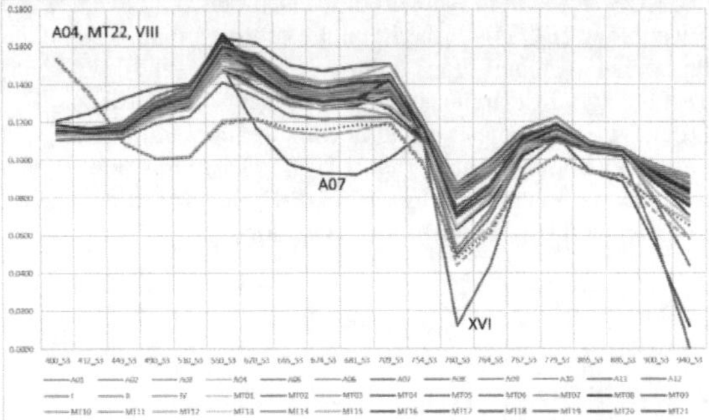

Figure 4. Spectral signatures obtained at the Laguna Lake water quality survey stations shown in Figure 3. Labelled lines are significantly different from the rest.

The spectral signatures obtained for each of the stations in Figure 3 were similar to each except at five (5) stations, namely, A04, MT22, VIII, A07, and XVI (Figure 4). These stations can be removed in the subsequent regression analysis as optical water quality types at these stations were considerably different Considering the majority of the spectral signatures, local peaks can be observed at R560, R709, and R779. Local dips are located at R665, R674, R681, and R760. These peaks and dips might be useful in developing indices for spectral class-differentiated chl-a estimation.

Further examination of the field spectral data at the survey stations using factor analysis yielded similar results. The first two factors F1 and F2 accounted for 83.47% of the data variability. As shown in Figure 5, points representation stations A07, XVI, MT22, A04, and VIII are separated clearly from the cloud of points representing all other stations.

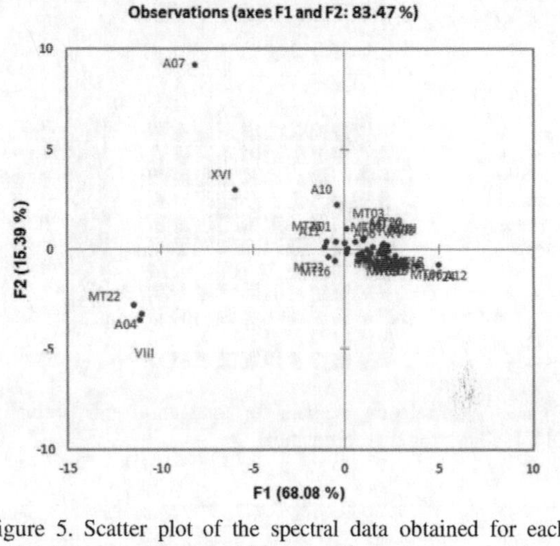

Figure 5. Scatter plot of the spectral data obtained for each station based on the results of factor analysis.

3.3 Correlation Levels

Weak correlations were observed between the microalgal concentrations and the band reflectance data (Figure 6). This points to the need to utilize in combination these bands. However, bands to be included in a model must be carefully chosen as multicollinearity will be a problem since most of the bands are highly intercorrelated positively.

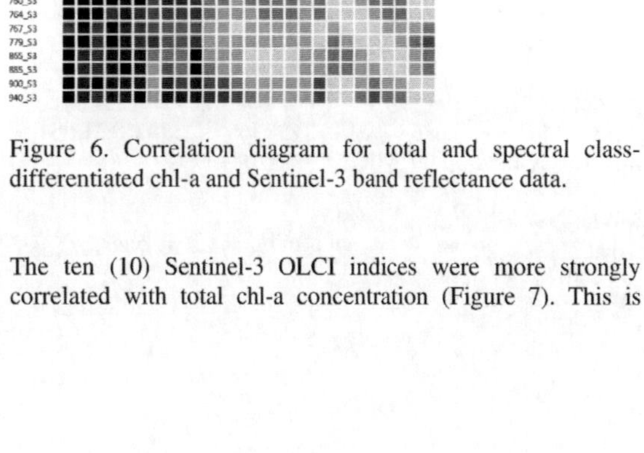

Figure 6. Correlation diagram for total and spectral class-differentiated chl-a and Sentinel-3 band reflectance data.

The ten (10) Sentinel-3 OLCI indices were more strongly correlated with total chl-a concentration (Figure 7). This is especially true for R665/R709, R674/R709, and (1/R665-1/R709)xR754 with the following respective correlation values: -0.623, -0.646, and 0.628. The correlation of these indices with the class-differentiated microalgal concentration is shown in Table 1. These are generally much lower compared to that of total chl-a, indicating the apportioning chl-a contributions to microlagal classes is difficult. It should be noted that per microalgal class, at least two of the indices turned out to be significant at alpha = 0.05.

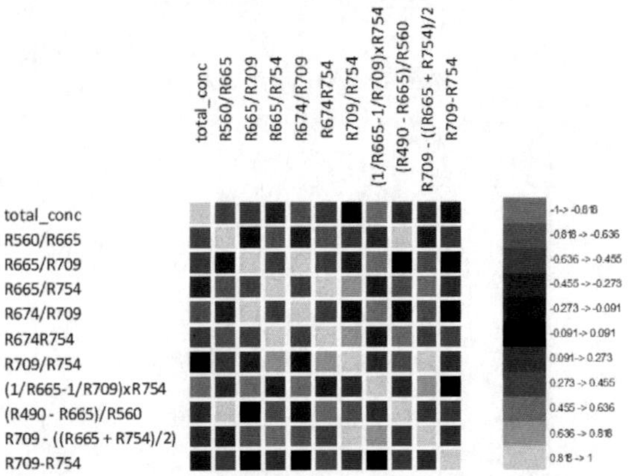

Figure 7. Correlation diagram for total chl-a and Sentinel-3 OLCI indices for chl-a estimation.

Variables	Green_Alga	Bluegreen	Diatoms	Cryptophyt	Yellow_sub	total_conc
Green_Alga	**1**	-0.213	-0.167	-0.178	-0.087	0.044
Bluegreen	-0.213	**1**	**-0.406**	**0.909**	0.227	**0.620**
Diatoms	-0.167	**-0.406**	**1**	**-0.425**	**-0.804**	**0.364**
Cryptophyt	-0.178	**0.909**	**-0.425**	**1**	0.189	**0.567**
Yellow_sub	-0.087	0.227	**-0.804**	0.189	**1**	**-0.507**
total_conc	0.044	**0.620**	**0.364**	**0.567**	**-0.507**	**1**
R560/R665	**-0.345**	**0.472**	-0.281	**0.421**	0.247	0.110
R665/R709	0.105	**-0.444**	-0.272	-0.264	**0.376**	**-0.631**
R665/R754	-0.036	**-0.311**	-0.089	-0.204	0.197	**-0.404**
R674/R709	0.117	**-0.459**	-0.260	-0.280	**0.365**	**-0.633**
R674/R754	-0.003	**-0.381**	-0.105	-0.254	0.221	**-0.475**
R709/R754	-0.145	0.210	0.205	0.108	-0.217	**0.321**
(1/R665-1/R709)xR754	-0.113	**0.444**	0.268	0.262	**-0.366**	**0.623**
(R490 - R665)/R560	**-0.367**	**0.390**	**-0.365**	**0.394**	**0.349**	-0.047
R709 - ((R665 + R754)/2)	-0.118	**0.353**	0.261	0.203	**-0.328**	**0.525**
R709-R754	-0.132	-0.112	0.227	-0.255	-0.089	-0.016

Values in bold are different from 0 with a significance level alpha=0.05

Table 1. Correlation of total and class-differentiated microalgal chl-a concentration with selected indices

3.4 Total Chl-a Concentration

Multiple regression results indicate that 48% of the variability of total chl-a concentration is explained by five explanatory (reflectance) variables (R412, R443, R560, R681, and R754) with RMSE of 2.814 ug/l. In contrast, the two indices R674/R754 and (1/R665-1/R709)xR754 accounted for about 46% of the variability of total chl-a concentration with RMSE of 2.475 ug/l. The equations for these two total chl-a models are given below.

Model 1:
Total chl-a = 20.086 – 213.721*R560 – 314.078*R681 + 409.807*R709 + 304.925*R764 – 176.081*R940

Model 2:
Total chl-a = 29.968 – 14.302*R674/R754 + 62.278*(1/R665-1/R709)xR754

3.5 Spectral Class-differentiated Chl-a

For diatoms and bluegreen microalgae, R560/R665 and (1/R665-1/R709)xR754 constitute the models with R^2 of 0.21 and 0.435, and RMSE of 2.516 and 2.163 ug/l, respectively. Green microalgal concentration is jointly described by three indices: R560/R665, R674/R754, and R709-R754. This yielded an R^2=0.182 and RMSE=1.219 ug/l. For cryptophytes, the model comprising of R560/R665, (1/R665-1/R709)xR754, and R709-R754 produced an R^2=0.289 and RMSE=0.767 ug/l.

The relatively lower goodness of fit indicates that estimating spectral class-differentiated chl-a concentration from Sentinel-3 OLCI data is challenging but promising. The low R^2 values can be attributed to the low resolution of OLCI data. The use of 10-day composite data is also a limitation as the date of reflectance data acquisition may not directly match the date of the actual field surveys. Furthermore, more class-differentiated chl-a data should be collected to further improve the model.

4. CONCLUSIONS

Concentration of spectral microalgal classes can be estimated to a certain degree of accuracy using Sentinel-3 OLCI reflectance data and indices derived from them. Based on this study, bluegreens and cryptophytes are better estimated compared to diatoms and greens. The preliminary models showed that the following indices are useful of estimating abundance of spectral class-differentiated microalgae: R560/R66, (1/R665 – 1/R709) x R754, and R674/R754. This study is the first attempt to

ACKNOWLEDGEMENTS

This research was done as part of the Multi-platform and Cross-sensor Water Quality Monitoring (MCWQM) Project. The Project was implemented by the University of the Philippines Training Center for Applied Geodesy and Photogrammetry (TCAGP), through the support of the Department of Science and Technology (DOST) of the Republic of the Philippines and the Philippine Council for Industry, Energy, and Emerging Technology Research and Development (PCIEERD). The Project was also implemented in collaboration with the Laguna Lake Development Authority (LLDA). Dissemination of this research is supported by the UP-DOST-SEI Engineering Research and Development for Technology (ERDT) Program.

REFERENCES

Blix, K., Pálffy, K., Tóth, V.R. and Eltoft, T. (2018). Remote Sensing of Water Quality Parameters over Lake Balaton by Using Sentinel-3 OLCI. Water 2018, 10, 1428; doi:10.3390/w10101428

Bresciani, M., Stroppiana, D., Odermatt, D., Morabito G., Giardino, C. (2011). Assessing remotely sensed chlorophyll-a for the implementation of the Water Framework Directive in European perialpine lakes. Science of The Total Environment 409(17):3038-91. doi.org/10.1016/j.scitotenv.2011.05.001.

Chen, X. and Feng, L. (2016). Remote Sensing of Lakes' Water Environment. Comprehensive Remote Sensing, pp.249-277.

Delos Reyes, M. and Martens, R. (1994). Geoecology of Laguna de Bay, Philippines I. Techno-commercial impact on the trophic level structure of the Laguna de Bay aquatic ecosystem 1968–1980. Ecological Modelling, 75-76, pp.497-509.

Dörnhöfer, K. and Oppelt, N. (2016). Remote sensing for lake research and monitoring – Recent advances. Ecological Indicators, 64, pp.105-122.

Laguna Lake Development Authority. (2017). 2017 Annual Report.

Manzo, C., Bresciani, M., Giardino, C., Braga, F. and Bassani, C. (2015). Sensitivity analysis of a bio-optical model for Italian lakes focused on Landsat-8, Sentinel-2 and Sentinel-3. European Journal of Remote Sensing, 48(1), pp.17-32.

Palmer, S., Kutser, T. and Hunter, P. (2015). Remote sensing of inland waters: Challenges, progress and future directions. Remote Sensing of Environment, 157, pp.1-8.

Petus C., Waterhouse J., Lewis S., Vacher M., Tracey D., Devlin M. (2019) A flood of information: Using Sentinel-3 water colour products to assure continuity in the monitoring of water quality trends in the Great Barrier Reef (Australia). J Environ Manage. 2019 Oct 15;248:109255. doi: 10.1016/j.jenvman.2019.07.026. Epub 2019 Jul 26.

Tamayo-Zafaralla, M., Santos, R., Orozco, R. and Elegado, G. (2002). The ecological status of Lake Laguna de Bay, Philippines. Aquatic Ecosystem Health & Management, 5(2), pp.127-138.

Toming, K, Kutser, T., Uiboupin, R., Arikas. A., Vahter, K., and Paavel, B. (2017) Mapping Water Quality Parameters with Sentinel-3 Ocean and Land Colour Instrument Imagery in the Baltic Sea. Remote Sensing, 9, 1070; doi:10.3390/rs9101070

Santos-Borja, A., Nepomuceno, D. S. (2003). Experience and Lessons Learned Brief for Laguna de Bay, Philippines. Lake Basin Management Initiative ILEC/LakeNet Regional Workshop for Asia: Sharing Experience and Lessons Learned in Lake Basin Management.

Schaeffer, B., Schaeffer, K., Keith, D., Lunetta, R., Conmy, R. and Gould, R. (2013). Barriers to adopting satellite remote sensing for water quality management. International Journal of Remote Sensing, 34(21), pp.7534-7544.

IDENTIFYING HURRICANE IMPACTS ON BARBUDA USING CITIZEN SCIENCE GROUND OBSERVATIONS, DRONE PHOTOGRAPHY AND SATELLITE IMAGERY

Rebecca Boger[1] , Russanne Low[2], Peder Nelson[3]

[1] Brooklyn College, City University of New York
[2] Institute for Global Environmental Strategies, Arlington
[3] College of Earth, Ocean, Atmospheric Sciences, Oregon State University

KEY WORDS: Barbuda, land cover, citizen science, hurricanes

ABSTRACT:

Barbuda is a low-lying leeward island in the West Indies. In September 2017 Hurricane Irma battered the island with 185 mph winds. Damage to housing and infrastructure was so extensive that all 1800 residents of the island were immediately evacuated post hurricane to the neighboring island of Antigua. Our research is focused on documenting and analyzing the landscape scale changes that took place on the island as a result of the hurricane using a nested methodology. Ground level observations were recorded using a beta version of the Global Learning and Observation to Benefit the Environment (GLOBE) Observer Land Cover tool. Aerial photographs provide medium scale landscape resolution. Satellite images were processed to provide NDVI values. We describe how nested data at several spatial scales enable detailed description of Hurricane Irma's impact across the landscape of this 62 sq. mile island. Here we are connecting observations at different spatial scales, using citizen science observations, aerial drone photography and satellite imagery to document and analyze hurricane impacts on the island of Barbuda, eastern Caribbean.

1. INTRODUCTION

This project employs both remotely-sensed and ground observational data in an effort to describe and quantify the environmental impacts of Hurricane Irma on the island of Barbuda in the West Indies. In September 2017, category 5 Hurricane Irma directly hit Barbuda and devastated the built and natural environments. The main human habitation zone of the island is the village of Codrington, located on the eastern side of Codrington Bay. Damage to Barbuda's infrastructure was so extensive that the 1800 residents were under mandatory evacuation to the neighboring island of Antigua immediately following the extreme weather event. By early 2019, when this fieldwork took place, about half of the residents had returned and are now rebuilding their lives on the island (Boger, Perdikaris, and Rivera-Collazo, in Press).

Pre- and post- hurricane scientific visualizations of Barbuda using Normalized Difference Vegetation Index (NDVI) data provide a quantitative comparison of damage to vegetation zones (Figure 1). Damage resulted from both the storm surge that submerged the low-lying areas to the west and north, as well as high winds that uprooted vegetation and stripped off leaves. The storm surge from Hurricane Irma is estimated at over 2.4 meters (Cangialosi et al., 2018), which would have caused an inundation of about a third of the island (Figure 1).

Quantifying the impact of the hurricane event on Barbuda's landscape is an important critical step guiding restoration of ecosystem communities. Periodic natural disturbance events such as hurricanes contribute to the characteristic vegetation and land cover of the island. Coastal habitats are the critical first line of defense from storm damage. Barbuda is fringed by mangroves, known to significantly reduce both annual and catastrophic damages from storms through flood protection (Narayan et al., 2020).

Satellite data documenting land cover is available for both pre- and post- Hurricane Irma, but the classification land cover products is a challenge on small islands such as Barbuda, which display unique vegetation structures and highly fragmented vegetation complexes. Existing global land cover land use (LCLU) products have spatial resolutions ranging from 30 m – 4 km, resulting in highly heterogenous pixels averaging the spectral response of multiple land cover types (Révillion et al., 2019).

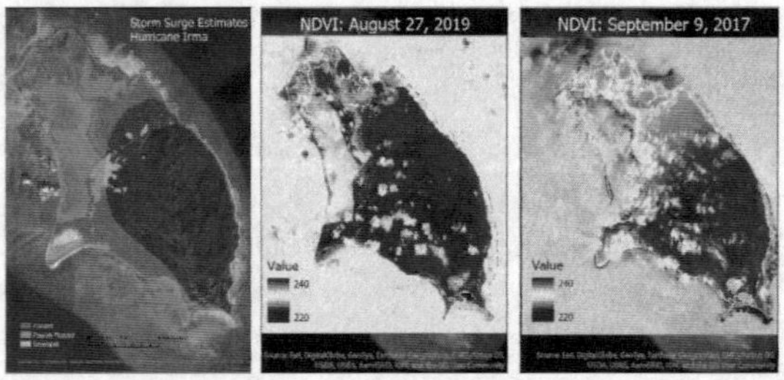

Figure 1: A (left): Hurricane Irma storm surge model. Areas in orange are within the storm surge area. The area in yellow, Palmetto Point, was likely flooded during the event. A value of 2.64 m was initially used as the height of terrain impacted by the storm surge and refined by ground observations and local informants. B and C: NDVI maps of Barbuda contrasting before (B) and after (C) Hurricane Irma hit in September 2017.

Data collected by citizen scientists is gaining recognition as potential and important source of data for research (Bonnie et al. 2014; Follett and Strezov, 2015; McKinley et al., 2017). The National Aeronautics and Space Administration's (NASA) GLOBE Observer (observer.globe.gov) released the Land Cover app as part of the suite of tools available for use on the mobile GLOBE Observer app. Designed for use by citizen scientists, it enables the capture of photo validation data to accompany land cover classification in the field. A secondary goal of this project was to employ the mobile app in a research setting and make any recommendations for a future version so it best serves the needs of the land cover research community. Barbuda provides a variety of unique landscapes and environments where the methodologies and categories employed in the app can be applied and tested, to ensure the tool is appropriate for use in all land cover conditions.

To proceed with land cover classification of Barbuda's terrain, this project initiated descriptive research documenting land cover characteristics. This report describes outcome of fieldwork conducted January 4-18, 2019, where aerial photography and ground observation of land cover were collected simultaneously for eventual application to development of a regional a tool to interpret satellite land cover data products.

Guiding questions used to develop the research design include:

- Can ground observations of land cover be used to support classification of drone imagery and remotely sensed land cover observations from space?
- Can ground observations of storm surge or hurricane damage be identified in the drone imagery and remotely sensed land cover data collected from space?
- Are the features and data collection capabilities of the GLOBE Observer land cover tool sufficient for deployment in a scientific field research context.
- How well do the reported observations using the GLOBE Observer Land Cover tool correspond to published maps and descriptions of Barbuda vegetation data?

2. SITE LOCATION

Barbuda is a low-lying coral island of 161 square km in the Caribbean located midway along the island chain known as the Lesser Antilles. Barbuda is part of the nation of Antigua and Barbuda and is about 50 km northeast of its sister island (Figure 2). The highest elevations rise to 40 meters and are located toward the east in an area called the Highlands. There is a network of historic wells built during the Colonial period in the 1700 and 1800s that are fed by either springs or surface aquifers. The limestone interior of the island contains numerous sinkholes and caves that contain fresh water throughout the year. The soil is generally thin and nutrient-poor. The coastline is almost completely surrounded by fringing and barrier reefs teeming with extensive marine resources, some of the richest in the region (Perdikaris and Hejtmanek 2014). The mangroves at the Codrington lagoon to the northwest are the home of one of the largest breeding colonies (after the Galapagos) of the Frigate birds (*Fregata magnificens*).

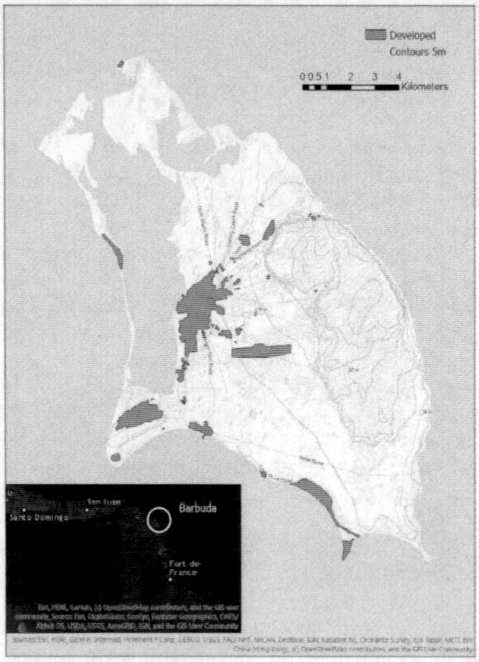

Figure 2: Location of Barbuda along the chain of islands that form the Lesser Antilles. The highest elevations on Barbuda are along the eastern side.

Barbuda has a tropical climate with wet and dry seasons and an average temperature in Codrington is 26.7 °C. Figure 3 shows the monthly average precipitation[1]; the driest months are January – March and the wettest months are August – November. The average annual rainfall is 924 mm.

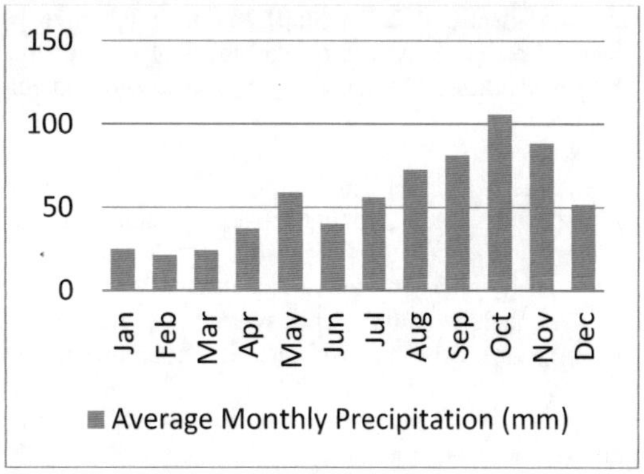

Figure 3. Average monthly temperatures for Codrington, Barbuda. Source climate-data.org.

The Köppen- Geiger climate classification classifies Barbuda as an Aw, tropical savanna climate with non-

[1] Data downloaded from WorldWeatherOnline.com

seasonal or dry-winter characteristics[2]. The surrounding ocean waters averages around 27°C with little variation. Barbuda is located within the Atlantic hurricane belt. Most hurricanes occur between August and October, with September being the month with the most hurricanes[3].

3. METHODS

3.1 Field work: GLOBE Observer is NASA's data collection and sharing mobile app developed for citizen science participation in Earth system observations. Ground observations and voucher photographs recorded by citizen scientists using this app are collected for use to augment the interpretation of data remotely sensed from space. The Land Cover tool provides a protocol for data collection. Users identify the centroid of their observational field. The app automatically identifies the latitude, longitude, date and time of observation. Observations are made in the four cardinal directions from the center point, and voucher photographs are taken to record the land cover observed in each direction. Up and down photographs provide information about the atmospheric conditions, vegetation canopy, as well as detail about low lying ground cover. A series of questions about the surface conditions (wet/dry) and phenology provide additional information.

To make an observation and photograph, the user is instructed to tilt the mobile device slightly downward so that approximately 50 m of the ground is observed in the photograph. A built-in compass function enables the user to point their mobile device in the cardinal directions without additional instruments.

Once photographs have been uploaded the device, there is a series of pull-down menus that enable the user to distinguish different vegetation growth forms (such as forests, wetlands, grasslands, barren) as well as some common land use features (such as structures, pavement, croplands, pastures, athletic fields, cemeteries) (Figure 4). The landcover features are identified in the app, and then the user is prompted to estimate the percentage cover of each of the cover types identified. Both ground cover and canopy cover are estimated.

The GLOBE Observer tool returns a land cover classification based on percent cover by applying the Modified UNESCO Classification. (MUC). This classification system is physiogmonic-structural in character and does not include cultural landcover types (i.e. messicol vegetation), nor other land cover classes that are needed to support classification of remotely sensed land cover data from satellites, so other categories have been added to the Land Cover tool to support those needs.

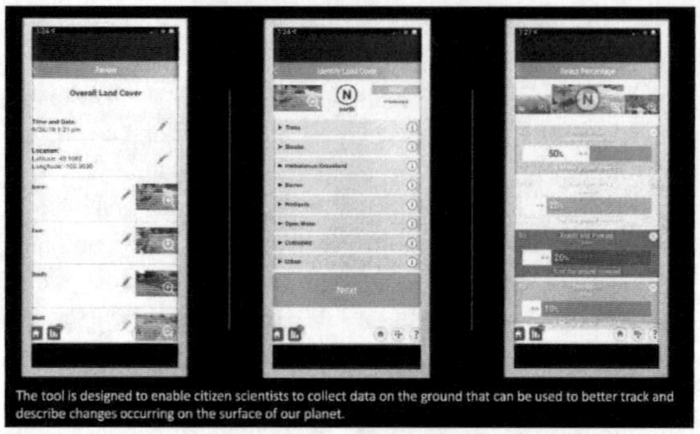

Figure 4. Screenshots of the GLOBE Observer Landcover tool. Left: voucher photographs of landcover for each cardinal direction. Middle: landcover elements available for selection by app user. Right: Slides enable user to estimate landcover in 10% increments.

One limitation of the GLOBE Observer app is that its data upload function is only available in countries where the government has agreed to share data with the GLOBE Program. Since the Nation of Antigua and Barbuda is not a GLOBE member country, Esri Survey123[4] web-based tools were used to recreate the data fields within the GLOBE Observer (GO) Land Cover app, and exported data as a comma delimited file which was then brought into Excel for further analysis.

3.2 Drone Photography: A 3D Robotics Solo drone with a GoPro Hero 3 camera was flown at a 50-meter altitude. The open source Mission Planner[5] was used to create the flight plans. Aerial photos were collected with 80% transect overlap and 80% consecutive photo overlap. Aerial drone photography was collected throughout the island at many locations where the land cover data were being collected. In this way, there was a coordination of map and side views of the vegetation cover estimates.

3.3 Computer Analysis: Although data can be directly brought into ArcGIS Pro, the data was first brought into Excel to edit the data and create summaries for total vegetative cover. The points with latitude and longitude values were brought into ArcGIS Pro. The individual photos taken with the drone were mosaiced with AgiSoft[6]. The orthrectified mosaiced image was then brought into ArcGIS Pro and overlayed with the point Survey123 data.

A search for Sentinel 2 (S2), and Landsat 8 (L8) imagery with less than 20% cloud cover per scene was done through GloVis[7] for the time period of one month before or after the time of field collection. For S2 the best imagery was collected on February 1, 2019 and for L8, the best one was collected on December 4, 2018. Although there is nearly two months difference between the two dates, they are both within a similar period of rainfall and plant growth. The hurricane season is over and the dry season has not started in

[2] Climate-Data.org: https://en.climate-data.org/north-america/antigua-and-barbuda-165/#example0
[3] NOAA historical hurricane tracks: https://coast.noaa.gov/hurricanes/

[4] https://survey123.arcgis.com/
[5] http://ardupilot.org/planner/
[6] https://www.agisoft.com/
[7] https://glovis.usgs.gov/

full. The months of December and January have relatively moderate rainfall. So, although all three sets of imagery (Sentinel 2, Landsat 8, and drone) were not collected at exactly the same time, the vegetative cover should be similar to explore a nested analysis.

In ArcGIS Pro, composited images were created for both the L8 and S2 datasets. Bands 1-7 with 30-meter spatial resolution were composited for L8 while bands 2, 3, 4, and 8 with 10-meter resolution were composited for S2. In this way, NDVI images using the visible and Near Infrared bands could be created for 30-meter (L8) and 10-meter (S2) resolutions.

Fifty-meter buffers were created around the survey points to create an area around the citizen scientist on the ground for which the citizen scientist is asked to collect data about (e.g., percentages of vegetative ground and canopy cover). The buffers were then used to clip out the NDVI data and the drone imagery. Percent vegetative cover can be estimated from the drone imagery and compared to the estimates collected by the citizen scientist. NDVI data can be compared between the satellite imagery at the coarser and finer resolutions, as well as compared to the drone.

4. RESULTS

Land cover data using the Survey123 app was collected at 61 100-meter quadrants around the island. After data processing, 9 quadrants with incomplete data were removed from the study to ensure data quality, leaving 52 quadrants available for analysis (Figure 5). Drone photography were collected at 17 sites.

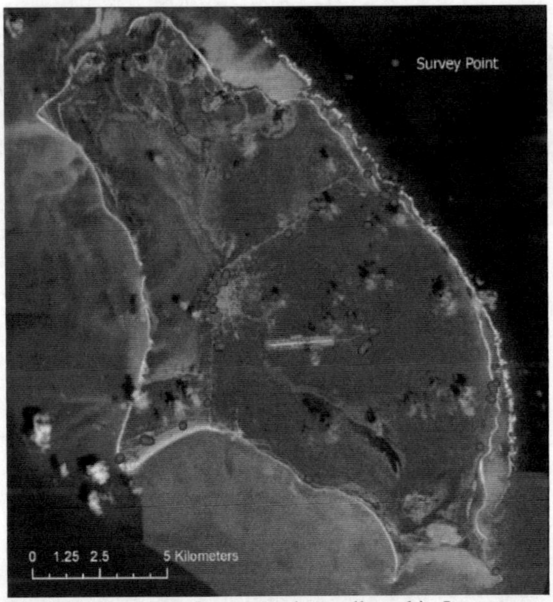

Figure 5: Location of survey points collected in January 2019.

To illustrate the process of comparison between citizen scientist collected data and satellite and drone imagery, one site called Pie Crust will be shown. Figure 6 shows the two satellite datasets and the drone photography. The citizen scientist stands in the center and a 50-meter radius buffer shown in the red circle. The graduated green tones for the NDVI imagery goes from light to dark green, low to higher NDVI values. The Sentinel 2 data captures the variability of the landscape shown in the drone photography. This area is characterized by patchy shrub vegetation with exposed limestone bedrock and reddish-brown soil. There are large amounts of dead trees and branches that are remnants from Hurricane Irma (Figure 7).

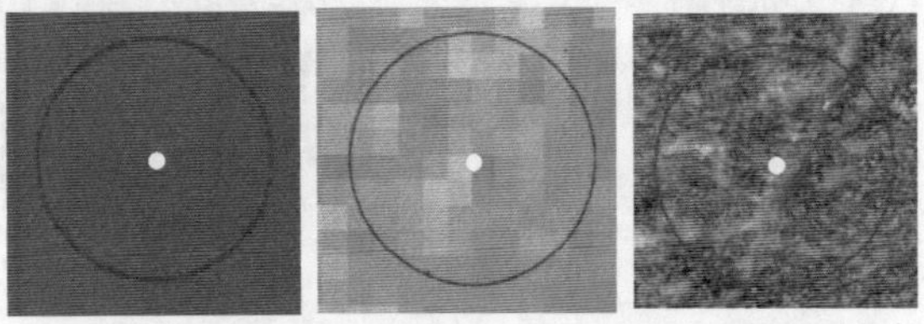

Figure 6: Pie Crust site comparing NDVI for the Landsat imagery collected on December 4, 2018 at 30 meter resolution to the left, Sentinel 2 imagery collected on February 1, 2019 at 10 meter resolution, and the drone imagery collected on January 8, 2019 with a submeter resolution. NDVI images showed more contrast among individual pixels than either true-color or NIR false-color composites, especially for Sentinel 2 data. The diameter of the circles is 100 meters.

Figure 7: Close-up of the drone photography to show the exposed limestone, soil, dead trees, and green shrubs.

Figure 8 shows the drone photography in the center with the photos collected using the Survey 123 app going North, East, South, and West. The picture in the lower left shows the ground where the observer was standing. At this particular site, vegetation cover was estimated each in the four directions and then averaged to give one value characterizing the entire 100-meter quadrant (or circle for the buffer analysis). A visual comparison indicates that there is a fairly good match between the ground estimates and the drone photography. Other sites with different vegetation coverage also seem to have a reasonable match.

Figure 8: Comparison of the map view of the drone imagery (middle mosaiced photo) and coincident ground photos looking down (photo lower left) and in the cardinal directions (N, E, S, W). The circle in the drone image shows a circle with 50-meter radius (100-meter diameter).

To see if we can quantify the vegetation coverage for the drone photography, we could then classify the pixels into two categories, vegetation and bare. Figure 9 shows the results for an unsupervised classification that originally had ten classes that were then grouped into two, bare (soil and rock) and vegetation of all types.

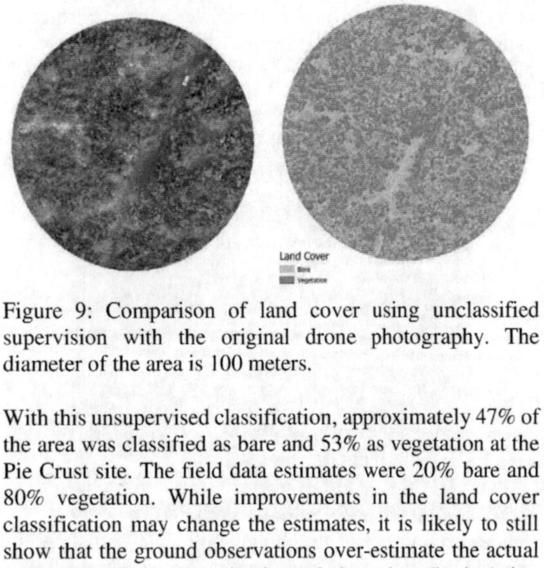

Figure 9: Comparison of land cover using unclassified supervision with the original drone photography. The diameter of the area is 100 meters.

With this unsupervised classification, approximately 47% of the area was classified as bare and 53% as vegetation at the Pie Crust site. The field data estimates were 20% bare and 80% vegetation. While improvements in the land cover classification may change the estimates, it is likely to still show that the ground observations over-estimate the actual vegetation cover. Examination of the other 51 is being undertaken to classify the land cover in the drone photography and satellite data and compare ground observations with the imagery.

5. DISCUSSION AND CONCLUSION

The Barbuda fieldwork provided an opportunity to test a new citizen science land cover methodology in the field in conjunction with a research project. During fieldwork, some issues encountered in the study relate to the protocol used to create estimates of land cover that had not been encountered in the terrain where the app was first tested. For the dense evergreen bushland and thickets in Barbuda, it is not possible to obtain photos with a clear 50m view that can be used to solely document land cover accurately. We list some of the limitations we experienced here:

1. Except for in the highlands, the land surface has low relief, and a 50-m photographs, documenting the vegetation cover for use in validation are impossible to obtain, except in salt barrens, pasture, and low lying xeric shrublands.
2. Mangroves are not easily classified in the MUC system which conflates physiogmonic-structural attributes with land use categories (urban), and with ecological zones (wetlands). Mangroves can be described both in broad-leaved evergreen tall shrub and wetland categories.
3. For dense shrubland and dense forest vegetation, options to describe smaller quadrants would improve reproducibility of results. A 100-m quadrant was rarely possible to describe accurately in the shrub thicket vegetation found on Barbuda, because it was not possible to obtain a sightline more than a few meters. This is even true when overall vegetation cover is 50% or less. The patchiness of the vegetation leads to thick patches interspersed with bare areas.
4. Subjective estimates of cover will always be prone to observer error because no two people will estimate cover identically. The protocol instructs users to make photos, so that the data can be verified by others. However, as a rapid assessment tool, these estimates may give an indication of the land cover components from this ground level.
5. In this environment, estimation of land cover class percentages, conducted by examining 50-m distances proved difficult to reproduce from photo voucher specimens, and difficult to do in the field. For users with little-to-no experience in vegetation classification, data may not be especially reliable.

The difficulty associated with user estimates of percentage cover is well-known and has resulted in estimation tools such as the Domin Scale, which simplifies % cover into a series of categories to increase inter-rater reliability (Adds et al. 1999). Tests are planned to compare visual % land cover assessments at 10 m, 30 m, and 100 m quadrants to determine whether citizen scientist evaluation of smaller quadrants produces outcomes with higher accuracy and if decreasing the size of sampling quadrants has potential significance for data applications in the classification and ground validation of remotely-sensed data.

6. FUTURE RESEARCH

The exploratory research here has potentially great significance, both in the ultimate creation of a land cover classification system honed for use in Caribbean island nations, but also in providing potential test data for newly developed tools now being tested in other regions. Révillion et al. (2019) recently released an object-based classification process in use interpreting terrain on small islands in the Indian Ocean, which is returning results with an accuracy of 86%. Continued development of land cover classification tools that work on small islands will require extensive field work and analysis, and both ground observations collected by both scientists and citizen scientists and drone photography will contribute to this goal.

The research discussed here is a component of a multiyear sustainability project conducted in collaboration with the Barbuda Archaeological Research Center, an NGO providing scientific and logistical support for visiting research scientists, located in Codrington, Barbuda.

7. REFERENCES

Adds, J. and Larkcom, E., 1999. *Tools, techniques and assessment in biology: a course guide for students and teachers*. Nelson Thornes.

Boger, R., Perdikaris,S., and Rivera-Collazo, I., 2019. Cultural heritage and local ecological knowledge under threat: Two Caribbean examples from Barbuda and Puerto Rico, *Journal of Anthropology and Archaeology*, 7(2): 1-14.

Bonney, R., Shirk, J.L., Phillips, T.B., Wiggins, A., Ballard, H.L., Miller-Rushing, A.J. and Parrish, J.K., 2014. Next steps for citizen science. *Science*, *343*(6178), pp.1436-1437.

Cangialosi, J.P., Latto, A.S. and Berg, R., 2018. Hurricane Irma (AL112017). *National Hurricane Center Tropical Cyclone Report.*

Follett, R. and Strezov, V., 2015. An analysis of citizen science based research: usage and publication patterns. *PloS one*, *10*(11), p.e0143687.

Küchler, A.W. and Zonneveld, I.S., 2012. *Vegetation mapping* (Vol. 10). Springer Science & Business Media.

McKinley, D.C., Miller-Rushing, A.J., Ballard, H.L., Bonney, R., Brown, H., Cook-Patton, S.C., Evans, D.M., French, R.A., Parrish, J.K., Phillips, T.B. and Ryan, S.F., 2017. Citizen science can improve conservation science, natural resource management, and environmental protection. *Biological Conservation*, *208*, pp.15-28.

Narayan, S., Thomas, XC., Matthewman, J., Shepard, C., Geselbrach, L, Nzerem, K., Beck, M., 2020. Valuing the Flood Risk Reduction Benefits of Florida's Mangroves. UC Santa Cruz and the Nature Conservancy. Downloaded from https://www.rms.com/assets/blt681abe87cb1cb8d6/Mangrove_Report_digital_FINAL.pdf.

Révillion, C., Attoumane, A. and Herbreteau, V., 2019. Homisland-IO: Homogeneous Land Use/Land Cover over the Small Islands of the Indian Ocean. *Data*, *4*(2), p.82 -92.

Santiago-Valentin, E., Olmstead, R.G., 2004.. Historical Biogeography of Caribbean Plants: Introduction to Current Knowledge and Possibilities from a Phylogenetic Perspective. *Taxon*, vol. 53(2): 299–319.

Wulder, M. A., Coops, N. C., Roy, D. P, White, J. C. and Hermosilla, T., 2018. Land cover 2.0, *International Journal of Remote Sensing*, 39:12, 4254-4284, DOI: 10.1080/01431161.2018.1452075

INTRODUCING AUTOMATIC SATELLITE IMAGE PROCESSING INTO LAND COVER MAPPING BY PHOTO-INTERPRETATION OF AIRBORNE DATA

H. Costa [1, 2], P. Benevides [1], F. Marcelino [1], M. Caetano [1, 2] *

[1] Direção-Geral do Território, 1099-052 Lisbon - (hcosta, pbenevides, fmarcelino, mario.caetano)@dgterritorio.pt
[2] NOVA Information Management School (NOVA IMS), Universidade Nova de Lisboa, Campus de Campolide, 1070-312 Lisbon, Portugal

KEY WORDS: Time series, Sentinel, Landsat, NDVI, Change detection, classification, crops, COS

ABSTRACT:

A series of five land cover maps, widely known as COS (*Carta de Uso e Ocupação do Solo*), have been produced since 1990 for mainland Portugal. Previous to 2015, all maps were produced through photo-interpretation of orthophotos. Land cover and land use changes were detected through comparison of previous and recent orthophotos, which were used for map updating, thereby producing a new map. The remaining areas of no change were preserved across the maps for consistency. Despite the value of the maps produced, the method is very time-consuming and limited to the single-date reference of the orthophotos. From 2015 onwards, a new approach was adopted for map production. Photo-interpretation of orthophoto maps is still the basis of mapping, but assisted by products derived from satellite data. The goals are three-fold: (i) cut time production, (ii) increase map accuracy, and (iii) further detail the nomenclature. The last map published (COS 2015) benefited from change detection and classification analyses of Landsat data, namely for guiding the photo-interpretation in forest, shrublands, and mapping annual agriculture. Time production and map error have been reduced comparing to previous maps. The new 2018 map, currently in production, further explores this approach. Landsat 8 time series of 2015-2018 are used for change detection in vegetation based on NDVI differencing, thresholding and clustering. Sentinel-2 time series of 2017-2018 are used to classify Autumn/Winter crops and Spring/Summer crops based on NDVI temporal profiles and classification rules. Benefits and pitfalls of the new mapping approach are presented and discussed.

1. INTRODUCTION

The importance of land cover and land use (LCLU) has long been recognized, such as in environmental sciences and policy, and LCLU mapping is nowadays a well-established activity all over the world at a diversity of scales. In Portugal, the national mapping agency, *Direção-Geral do Território* (DGT), produces and publishes a LCLU map for the continental territory, widely known as COS (*Carta de Uso e Ocupação do Solo*) (Figure 1). This map was first produced for the year of 1990 and updated for 1995, 2007, 2010, and 2015. Currently, COS for 2018 is under production and its publication is expected by the end of 2019.

COS is used nationally and internationally by a large spectrum of users for numerous purposes, including landscape planning, decision-making, reporting obligations, research, education, and business. The uptake of COS is related to the fact that it is made available freely through an open data policy, and is the most detailed product of mainland Portugal in terms of thematic content and spatial representation.

The minimum mapping unit (MMU) of COS is 1 hectare (Figure 1), and the LCLU nomenclature is based on that of CORINE Land Cover (CLC), which constitutes a reference LCLU product at European level (Büttner, 2014). The nomenclature of COS 2015 includes 48 classes aggregated to 9 main classes: Artificial land, Agriculture, Pastures, Agro-forestry, Forest, Shrublands, Bare soil and sparse vegetation, Wetlands, and Water.

Figure 1. Subset of COS 2015 over the city of Coruche (38°57'30,05"N; 8°31'36,013"W). Background: false colour orthophoto of 2015.

COS has been produced through photo-interpretation of orthophotos (spatial resolution of 50 cm and better). LCLU changes are normally detected through comparison of previous and recent orthophotos. The changes detected are mapped and superimposed on the outdated map, thereby producing a new map. The remaining areas of no change are preserved across the maps for consistency.

The traditional methods used in map production holds several problems. Mostly, map production is very time-consuming, expensive, and orthophotos provide information only related to the specific date of the flight campaigns (while multi-temporal data are valuable for accurate photo interpretation of some classes, such as annual crops).

* Corresponding author

From 2015 onwards, map production is evolving to include (semi)-automatic methods of image analysis applied to Landsat and Sentinel-2 data. Therefore, while photo-interpretation of orthophotos is still the basis of mapping, additional products derived from multi-temporal satellite data are used with three main objectives: (i) cut time production, (ii) further detail the nomenclature, and (iii) increase map accuracy.

COS 2015 already benefited from change detection analysis of Landsat multi-temporal data, namely inter-annual differencing of Normalized Difference Vegetation Index (NDVI). This was used for guiding the photo-interpretation on detection of changes between 2010 and 2015, and classifying Autumn/Winter crops and Spring/Summer crops. Here, however, the production of the new 2018 map is presented, namely the (semi)-automatic methods used to analyze satellite data.

First, a change detection method based on inter-annual time series of Landsat 8 data is described. This method applies thresholding and k-means clustering on image differencing of NDVI data of 2015-2018 for detecting changes in forest and shrublands. Then, a classification method based on intra-annual Sentinel-2 data is presented for annual croplands. This method uses expert knowledge and statistics extracted from the intra-annual time series to distinguish between Autumn/Winter crops and Spring/Summer crops.

2. COS PRODUCTION

COS 2018 is under production and its publications is expected by the end of 2019. It will be a map of polygons released in vector format with 1 hectare MMU. The technical specifications of COS 2018 (Table 1) are similar to the precedent map of 2015, except the number and name of some classes. The maps are, however, compatible as all classes have direct correspondence among maps. The differences between nomenclatures are not discussed here, but they result from collaborative work between DGT and other Portuguese public institutions, so COS can serve them better.

Reference year	1990, 1995, 2007, 2010, 2015 (and 2018)
Format	Vector (polygons)
MMU (ha)	1 ha
Minimum distance between lines (m)	20 m
Base data	Orthorectified digital aerial images
Nomenclature	Hierarchical classification system
Production method	Visual interpretation
Geometric accuracy	≥ 5,5m
Thematic Accuracy target	≥ 85%
Coordinate reference system	ETRS89/PT-TM06

Table 1. Technical specifications of COS

The production of COS 2018 is founded on visual interpretation of orthophotos and assisted and complemented with novel methods of image analysis. First, photo-interpreters are assisted with auxiliary information produced with Landsat 8 data to find changes in forest and shrublands. The auxiliary information indicates where vegetation loss occurred between 2015 and 2018, which is indicative of changes, such as new constructions or agricultural fields. Photo-interpreters use these auxiliary data as alerts deserving careful inspection. The benefit of these alerts is that photo-interpreters focus their attention on sites of probable change, hence reducing time wasted in inspecting areas where vegetation change was not observed from space. Because forest and shrubland cover 51% (2015) of mainland Portugal, the time needed to detect and map changes across a large area is reduced considerably.

Second, photo-interpretation is complemented with analysis of Sentinel-2 data to distinguish two main types of annual crops: Autumn/Winter crops and Spring/Summer crops. These two classes have a strong seasonal variation in terms of vegetation cover and status, which a single-date orthophoto misses to capture. On the contrary, the temporal resolution of Sentinel-2 is large and tracks the vegetation cycles typical of annual agriculture. Therefore, the photo-interpreters are responsible for delimiting the borders of annual agricultural fields without making a distinction on the type of annual crops. Later, Sentinel-2 data is classified on either Autumn/Winter crops or Spring/Summer crops across the agricultural fields identified previously. The relative abundance of each of the crop type found in the polygons classified as annual crops is used as an attribute. The satellite data overcomes the temporal limitation of the orthophotos, making it possible to map two classes with otherwise insufficient accuracy.

3. METHODS

3.1 Detection of vegetation loss

Comparison of remotely sensed data acquired on different dates is a simple but effective approach for change detection. Some techniques based on this approach, often called differencing or layer arithmetic (Tewkesbury et al., 2015; Zhu, 2017), calculate descriptive statistics for the difference of vegetation indices between two points in time. Then, the descriptive statistics are analysed to discriminate change from no-change. This approach was used here to analyse differences of NDVI (Tucker, 1979) across 2015-2018.

The analysis of NDVI differencing is summarized in Figure 2. All available images (Surface Reflectance Level-2 Data Products) with cloud cover < 50% and covering mainland Portugal (Figure 3a) during summer of 2015-2018 were downloaded from EarthExplorer. Phenological differences between the years are expected to be small as only images of summer are used. All images were processed according to the specifications described in the Landsat 8 Land Surface Reflectance Product Guide. Specifically, pixels affected by clouds, shadow and with invalid values were reclassified as "no data".

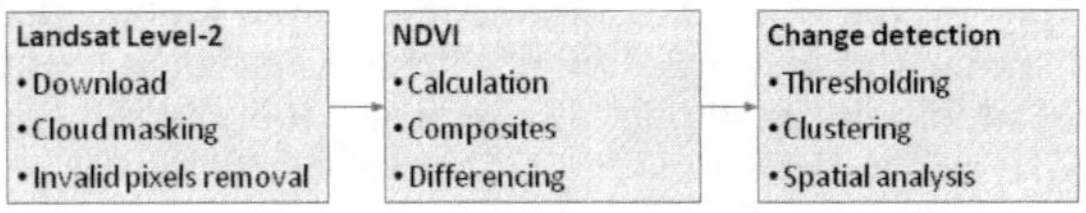

Figure 2. Methods used detecting vegetation loss.

The NDVI was calculated from each processed image. Then, images were combined to produce composites without gaps ("no data"). Therefore, images of the same year (e.g. 2015) were combined to produce one composite per year. Each pixel was given the minimum NDVI value among all images of the same year.

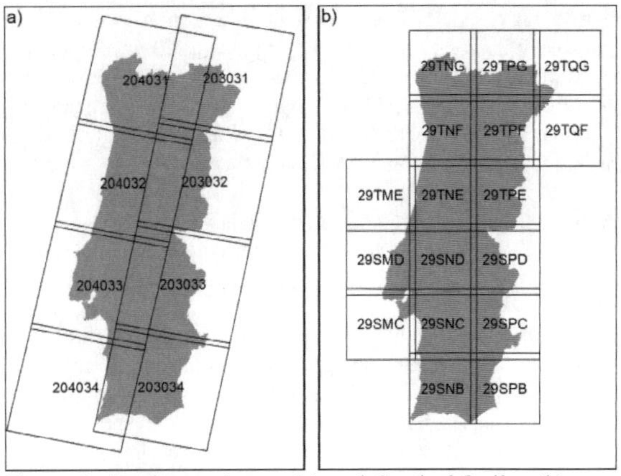

Figure 3. Landsat pathrows (a) and Sentinel-2 tiles (b) over mainland Portugal.

The minimum operator was selected for the composites production because it selects the NDVI associated with potential vegetation loss. For example, if three images are available from June, July and August, and a clear cut of forest occurs in August, the composite will select the NDVI associated with change (the smallest NDVI of August). Other operators such as the mean would select a value unrepresentative of change. There is one exception thought. For the first year of the analysis (2015), the maximum NDVI was selected for the composite because 2015 is the baseline of comparison. If the minimum NDVI was selected, any change occurring in 2015 would go unnoticed while comparing it with 2016, in which the NDVI will certainly remain small. Using the maximum NDVI of 2015 ensures that the removals of vegetation of 2015 are detected as compared to 2016.

Differences of NDVI between the years were then calculated and analysed based on the assumption that spectral changes among years are due to many reasons (e.g., vegetation health), but only substantial changes reflect land cover changes. Therefore, the difference of NDVI between two years should follow a normal distribution as most of the differences observed are small (close to zero) and only few marked differences are caused by land cover change (Jin et al., 2013; Lu et al., 2004; Pu et al., 2008).

The NDVI differences were submitted to two analyses: thresholding and clustering. The former relies on a threshold of difference. That is, if the difference of NDVI exceeds the threshold, a land cover change is expected to have occurred in the pixel. Here, a threshold was defined empirically and corresponded to -1.5 standard deviations of the average NDVI difference (Pu et al., 2008). The second analysis was k-means clustering (Hartigan and Wong, 1979). Specifically, the NDVI difference was classified in 10 clusters, and those characterized by a median difference of < -0.25 were flagged as change. Because the analyses are based on relative comparisons between pixels, they are sensitive to the range of NDVI as a function of the land cover class. Therefore, calculations were done independently by strata, which were eucalyptus, pines, other forest species, and shrubs, all of them as mapped in 2015.

All pixels associated with change, irrespective of type of analysis (thresholding or clustering) and strata, were merged together in a single layer to represent the cumulative changes identified across the whole period on analysis in forest and shrubland. Then changes were submitted to spatial analysis. Patches of change were enlarged and contracted by one pixel, and only patches >= 5 pixels (4.5 hectares) were retained. This step removed noise and extremely small changes. Finally, the changes were converted to vector format.

The accuracy of the method was assessed in two different ways. First, a set of polygons produced in advanced for COS by the photo-interpreters without the help of satellite data were taken as reference data. These polygons were labelled as either "change" or "no-change". The first case corresponds to changes to be mapped in COS 2018, and the latter corresponds to area of no change between 2015 and 2018. Commission and omission error were estimated simply by counting the relative number of reference polygons that spatially intercepted the alerts.

Second, the accuracy of the method was assessed based on a sample of alerts collected across the Landsat pathrow 204032 (Figure 3a), which were labelled as either "change" or "no-change" by visual interpretation of orthophotos and satellite data. In this case, "change" was interpreted as all obvious spectral changes visible in the satellite data associated with vegetation loss, regardless if they should correspond to a thematic change in COS. Commission error was estimated simply by counting the relative number of alerts associated with no vegetation loss. No omission error was calculated.

3.2 Classification of crops

Autumn/Winter crops and Spring/Summer crops normally follow a different annual cycle, which satellites can track from space. The analysis of temporal profiles of vegetation indices can help detect phenological traits of crops such as vegetation growth and time of maturation. Temporal analysis of vegetation indices are increasingly used for crop monitoring (Belgiu and Csillik, 2018; Defourny et al., 2019).

The analysis performed is summarized in Figure 4. All Sentinel-2 level-2 (L2A) data available with cloud cover < 50%, covering mainland Portugal (Figure 3b), and acquired between the 1st October 2017 to 30th September 2018 were downloaded from the French Theia Land Data Centre (THEIA). The period analysed corresponds to the agricultural year relevant for 2018. The level 2A data distributed by THEIA corrects the level-1 data for atmospheric and slope effects using MAJA (Baetens et al., 2019). Clouds and their shadows were converted to "no data". All bands were saved with 10 m pixel size without resampling methods applied.

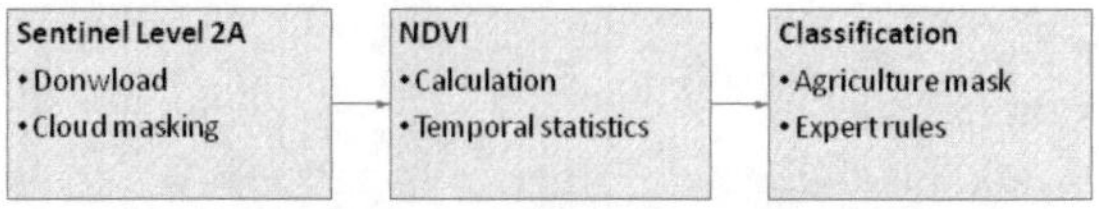

Figure 4. Methods used for crop classification.

The NDVI was calculated from each processed image, and the entire time series was analysed with the estimation of a set of statistical indicators. For example, extracting the date of the maximum NDVI can be important to identify different types of crops. Some statistics take into account the entire agricultural year, while other statistics were extracted independently for each quarter of the agricultural year.

The statistics extracted were then used together with rules predefined with expert knowledge to make the particular distinction between Autumn/Winter crops and Spring/Summer crops. A pixel was classified as Autumn/Winter crop if the maximum annual amplitude of NDVI was >0.4, the mean annual value of NDVI was > 0.2 and the mean NDVI observed in the third year quarter (April to June) was greater than that of the fourth year quarter (July to September). Spring/Summer crops were classified if the difference between the maximum NDVI of the fourth year quarter and the minimum NDVI of the third quarter was > 0.4, combined with the mean annual NDVI value being > 0.3.

The rules were assessed by inspecting a large sample of NDVI temporal profiles grouped by Autumn/Winter and Spring/Summer crops, such as the ones presented in Figure 5. Here, we can visualize that the maximum NDVI for Autumn/Winter crops is observed at the end of the winter and the respective maximum for the Spring/Summer crops at the end of Summer. Large amplitude values between the minimum and the maximum value throughout the year are also noticeable for both types of crop.

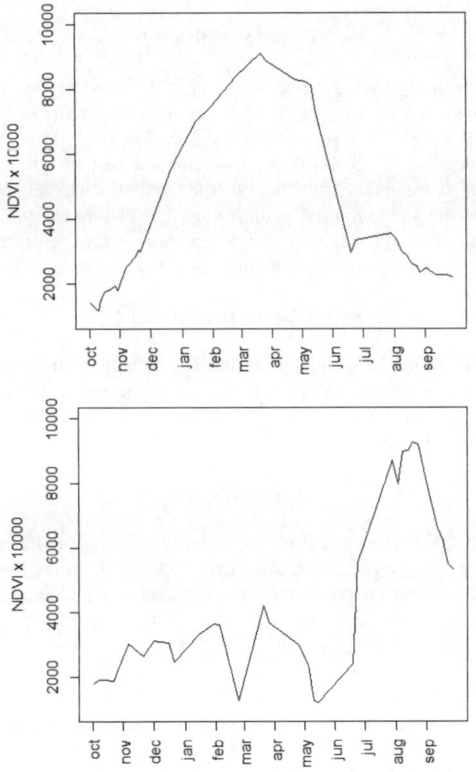

Figure 5. Examples of typical NDVI profiles observed for Autumn/Winter crops (top) and Spring/Summer crops (bottom).

The classification was applied only to pixels that fall inside polygons identified as annual crops in the visual interpretation of orthophotos performed in COS production. Therefore, the rules used are under revision while the geometry on COS is not closed yet for the entire country. Finally, the polygons will be enriched with the relative abundance of each of the crop type as an attribute.

The accuracy of the method was assessed by comparing the 2018 crop map produced to the Land Parcel Identification System (LPIS) of the *Instituto de Financiamento da Agricultura e Pescas* (IFAP) in tile T29SND (Figure 3b). LPIS is a geographical data set used in the framework of the Common Agricultural Policy (CAP) of the European Union for the administration and control of payments to farmers. The agreement between the classification and LPIS was calculated, corresponding to the area of the LPIS's polygons intersected by the crop map of the same class. Commission error was also calculated, corresponding to the area of the LPIS's polygons intersected by the crop map of a different class.

4. RESULTS

4.1 Change detection

The change detection method produced 148793 polygons used as alerts for vegetation loss (Figure 6). When considering the polygons produced in advanced for COS as reference (first accuracy assessment), around 77% of the "change" polygons and 49% of the "no-change" polygons overlap an "alert". This may be interpreted as a crude estimation of 33% omission error and 51% commission error. The second accuracy assessment undertaken only with alerts in pathrow 204032 found 85% of the cases associated with vegetation loss, which corresponds to 15% commission error.

Commission and omission error were observed to occur mostly in small areas of land cover change. Therefore, when using the 148793 alerts produced operationally in the production of COS, they were divided by size and given to the photo-interpreters as different layers. Alerts >2 ha were first inspected, than alerts from 1 to 2 ha, and finally <1 ha.

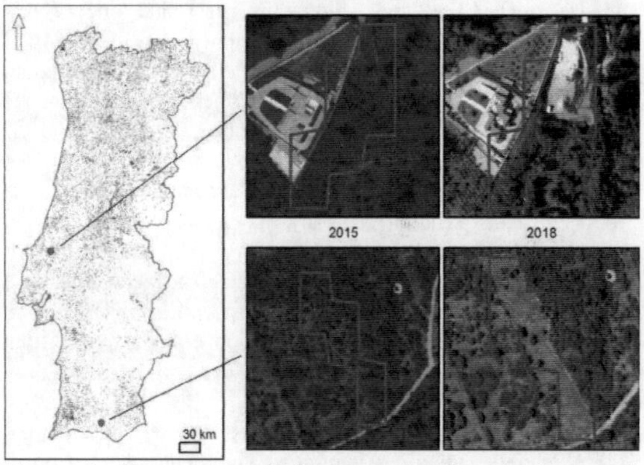

Figure 6. Alerts produced for vegetation loss in forest and shrubland. Two examples of change between 2015 and 2018 are shown for central and south Portugal.

4.2 Classification

The intra-annual statistics extracted from Sentinel-2 dataset combined with the set of predefined were able to produce two classification maps in raster format, one of each for Autumn/Winter and Spring/Summer crops (Figure 7).

The classification of Autumn/Winter crops covers 91% of the area represented by the polygons of the LPIS of the same class (9% omission error), and 39% of the area identified as Spring/Summer crop (commission error). On the other hand, the classification of Spring/Summer crops reaches 94% of the area covered by the same class in LPIS (6% omission error) with 24% commission error.

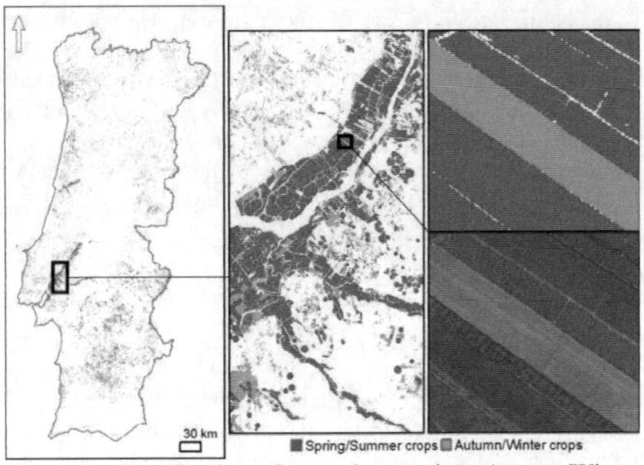

Figure 7. Classification of annual crops into Autumn/Winter crops and Spring/Summer crops.

5. DISCUSSION

The change detection method produced numerous alerts for false changes in vegetation cover, which are expressed by large commission error (51%) in the first accuracy assessment based on COS polygons. However, this was expected because the method was implemented to minimise omission error. Commission error is preferred over omission error as the former has no negative impact on the accuracy of COS. For example, forest clear-cuts occurring as part of the normal cycle of forest management are detected as the NDVI drops sharply, but no change is mapped as land use remains forest, and a new forest patch is expected to follow. The only inconvenient of commission error is that photo-interpreters inspect areas of no thematic change, wasting some time and hence diminishing the effectiveness of the change detection method.

However, it is still important to assess the accuracy of the alerts produced. The second accuracy assessment performed in pathrow 204032 shows a larger accuracy as commission error was only around 15%. The difference between the two assessments is that the later considered spectral changes such as forest clear-cuts as change in vegetation cover (even if that change should not be mapped).

Omission error, on the contrary, impact negatively on mapping as real changes may go unnoticed and hence not mapped. Omission error was larger than desired (33%) and should be reduced. This may be achieved, for example, by improving the implementation of thresholding and clustering analyses, which were conducted sometimes based on empirical decisions on parameterization (e.g., number of clusters in k-means). Exploring change detection methods other than these based on NDVI differencing is also recommended (e.g. Verbesselt et al., 2010).

Despite the large omission error of the alerts, the real omission error of COS 2018 across forest and shrublands will be certainly smaller than 33%. This is because the first accuracy assessment performed was a simple spatial overlap operation between the alerts and the reference data, which maps thematic changes only. Furthermore, the photo-interpreters see beyond the spatial extent of the alerts, and hence are able to spot changes not flagged automatically.

With regard to the classification of crop types, classification problems are expressed mainly by large commission errors. This is caused by the fact that the same polygon of LPIS could be considered as both Autumn/Winter crop and Spring/Summer crop in the accuracy assessment, as long as intersecting the crop map of these classes. This was regarded as appropriate to simulate the cases in which COS includes different crop types in the same polygon, such as in complex cultivation patterns. Therefore, a single polygon can include Autumn/Winter and Spring/Summer crops, whose abundance is to be included in the polygons attributes.

Note that these classification rules are a preliminary approach to the technique and will be improved. The rules were initially derived analysing the typical temporal profile signature observed for these types of crops (Figure 5), and thus they disregard less common cases such as the occurrence of multiple crops during the crop year. Therefore, alternative rules are likely to be implemented to address specific cases. In addition, semi-automatic approaches based on supervised classification should be tested and adopted.

6. WAY AHEAD

The methods discussed will be further developed beyond 2018 to satisfy the production of the future mapping model of COS. The latter will be a map of polygons similar to previous maps, but enriched with attributes that characterize the polygons. While the geometry of the map should continue to be defined by visual interpretation of orthophotos, the attributes will be produced from raster maps through automatic change detection and classification of satellite time series data.

ACKNOWLEDGEMENTS

The work has been funded by the Portuguese Foundation for Science and Technology (FCT), under projects IPSTERS (DSAIPA/AI/0100/2018), foRESTER (PCIF/SSI/0102/2017), and SCAPE FIRE (PCIF/MOS/0046/2017).

REFERENCES

Baetens, L., Desjardins, C., Hagolle, O., 2019. Validation of copernicus Sentinel-2 cloud masks obtained from MAJA, Sen2Cor, and FMask processors using reference cloud masks generated with a supervised active learning procedure. *Remote Sens.* 11, 1–25. doi:10.3390/rs11040433

Belgiu, M., Csillik, O., 2018. Sentinel-2 cropland mapping using pixel-based and object-based time-weighted dynamic time warping analysis. *Remote Sens. Environ.* 204, 509–523. doi:10.1016/j.rse.2017.10.005

Büttner, G., 2014. CORINE Land Cover and Land Cover Change Products, in: Manakos, I., Braun, M. (Eds.), *Land Use and Land Cover Mapping in Europe: Practices & Trends. Springer Netherlands, Dordrecht*, pp. 55–74. doi:10.1007/978-94-007-7969-3_5

Defourny, P., Bontemps, S., Bellemans, N., Cara, C., Dedieu, G., Guzzonato, E., Hagolle, O., Inglada, J., Nicola, L., Rabaute, T., Savinaud, M., Udroiu, C., Valero, S., Bégué, A., Dejoux, J.F., El Harti, A., Ezzahar, J., Kussul, N., Labbassi, K., Lebourgeois, V., Miao, Z., Newby, T., Nyamugama, A., Salh, N., Shelestov, A., Simonneaux, V., Traore, P.S., Traore, S.S., Koetz, B., 2019. Near real-time agriculture monitoring at national scale at parcel resolution: Performance assessment of the Sen2-Agri automated system in various cropping systems around the world. *Remote Sens. Environ.* 221, 551–568. doi:10.1016/j.rse.2018.11.007

Hartigan, J.A., Wong, M.A., 1979. Algorithm AS 136: A K-Means Clustering Algorithm. *Appl. Stat.* 28, 100–108.

Jin, S., Yang, L., Danielson, P., Homer, C., Fry, J., Xian, G., 2013. A comprehensive change detection method for updating the National Land Cover Database to circa 2011. *Remote Sens. Environ.* 132, 159–175. doi:http://dx.doi.org/10.1016/j.rse.2013.01.012

Lu, D., Mausel, P., Brondízio, E., Moran, E.F., 2004. Change detection techniques. *Int. J. Remote Sens.* 25, 2365–2401. doi:10.1080/0143116031000139863

Pu, R., Gong, P., Tian, Y., Miao, X., Carruthers, R.I., Anderson, G.L., 2008. Using classification and NDVI differencing methods for monitoring sparse vegetation coverage: a case study of saltcedar in Nevada, USA. *Int. J. Remote Sens.* 29, 3987–4011. doi:10.1080/01431160801908095

Tewkesbury, A.P., Comber, A.J., Tate, N.J., Lamb, A., Fisher, P.F., 2015. A critical synthesis of remotely sensed optical image change detection techniques. *Remote Sens. Environ.* 160, 1–14. doi:10.1016/j.rse.2015.01.006

Tucker, C.J., 1979. Red and photographic infrared linear combinations for monitoring vegetation, *Remote Sens. Environ.* doi:10.1016/0034-4257(79)90013-0

Verbesselt, J., Hyndman, R., Zeileis, A., Culvenor, D., 2010. Phenological change detection while accounting for abrupt and gradual trends in satellite image time series. *Remote Sens. Environ.* 114, 2970–2980. doi:10.1016/j.rse.2010.08.003

Zhu, Z., 2017. Change detection using landsat time series: A review of frequencies, preprocessing, algorithms, and applications. *ISPRS J. Photogramm. Remote Sens.* 130, 370–384. doi:10.1016/j.isprsjprs.2017.06.013

AGRICULTURALLY CONSISTENT MAPPING OF SMALLHOLDER FARMING SYSTEMS USING REMOTE SENSING AND SPATIAL MODELLING

A. Crespin--Boucaud[1,2,*], V. Lebourgeois[1,2], D. Lo Seen[1,2], M. Castets[1,2], A. Bégué[1,2]

[1] CIRAD, UMR TETIS, F-34398 Montpellier, France – (arthur.crespin-boucaud, valentine.lebourgeois, danny.lo_seen, agnes.begue)@cirad.fr
[2] TETIS, Univ Montpellier, AgroParisTech, CIRAD, CNRS, IRSTEA, Montpellier, France.

KEY WORDS: Landscape, Crop type mapping, Cropping practices, Sentinel-2, Spatio-temporal modelling, Smallholder agriculture

ABSTRACT:

Smallholder agriculture provides 90 % of primary food production in developing countries. Its mapping is thus a key element for national food security. Remote sensing is widely used for crop mapping, but it is underperforming for smallholder agriculture due to several constraints like small field size, fragmented landscape, highly variable cropping practices or cloudy conditions. In this study, we developed an original approach combining remote sensing and spatial modelling to improve crop type mapping in complex agricultural landscapes. The spatial dynamics are modelled using Ocelet, a domain-specific language based on interaction graphs. The method combines high spatial resolution satellite imagery (Spot 6/7, to characterize the landscape structure through image segmentation), high revisit frequency time series (Sentinel 2, Landsat 8, to monitor the land dynamic processes), and spatiotemporal rules (STrules, to express the strategies and practices of local farmers). The method includes three steps. First, each crop type is defined by a set of general STrules from which a model-based map of crop distribution probability is obtained. Second, a preliminary crop type map is produced using satellite image processing based on a combined Random Forest (RF) and Object Based Image Analysis (OBIA) classification scheme, after which each geographical object is labelled with the class membership probabilities. Finally, the STrules are applied in the model to identify objects with classes locally incompatible with known farming constraints and strategies. The result is a map of the spatial distribution of crop type mapping errors (omission or commission) that are subsequently corrected through the joint use of spatiotemporal rules and RF class membership probabilities. Combining remote sensing and spatial modelling thus provides a viable way to better characterize and monitor complex agricultural systems.

1. INTRODUCTION

The ESA (European Spatial Agency) Sentinel 2 (S2) mission provides free imagery with an unprecedented combination of high spatial (10-60m), temporal (5 days) and spectral (13 bands) resolutions, having promising applications in land use/cover characterization and crop monitoring (Drusch et al., 2012). A study carried out on 12 contrasted agricultural sites around the world showed that image time series acquired at a decametric resolution are adapted to field crop characterization for the majority of intensive agricultural systems (Inglada et al., 2015). However, methods usually used in northern countries are not efficient in the context of smallholder tropical agriculture. In such complex systems, the small size of fields and the landscape fragmentation are significant constraints to crop characterization using remote sensing. To overcome these limitations, studies proposed OBIA (Objects Based Image Analysis) approaches to improve crop identification (Peña-Barragán et al. 2011), or combined with high-resolution time series, very high resolution imagery and auxiliary spatialized data (i.e. Digital Elevation Model; Lebourgeois et al., 2017). However, limitations remain for a crop characterization in smallholder agriculture areas.

To improve current approaches, several studies combine remote sensing with other disciplines. For example, the number of papers where remote sensing is coupled with neural networks or deep learning is increasing. They integrate satellite data, and environmental or socio-economic data (called hereafter context data) as inputs into the network, in order to extract essential variables for each class. However, these researches have mainly focused on cultivated areas with relatively large fields, a large number of ground truth points, and have not, or only slightly, been applied to smallholder agricultural areas.

Only a few studies focused on the combination of remote sensing and expert rules or spatial modelling. Bailly et al. (2018) highlight that taking into account the spatio-temporal structures of fields improves small objects classification. Two main approaches can be used for modelling agro-systems. The first one consists in using expert knowledge to create crop rotation models (e.g. Dury et al. 2012). To do so, it is necessary to understand the economic, social and environmental stakes of these agroecosystems and therefore to rely on experts in these disciplines. The second consists in using machine learning methods (such as deep learning and neural network) to free from the limitations of expert knowledge. The direct integration of context data into the classifier allows the latter to learn automatically from the ground truth data. Nevertheless, this approach is particularly suitable to address one issue at a time, such as for crop rotation in Bailly et al. (2018) and with a large and diversified dataset.

To go beyond traditional approaches in remote sensing, this research aims at investigating the complementarity of remote sensing and spatio-temporal modelling to study:

- How can spatial modelling help converting remote sensing data into more reliable thematic products?
- How can remote sensing provide evidence of the landscape processes expressed as formalized knowledge in models?

Such an approach is expected to be useful for a better understanding of complex fragmented and heterogeneous

* Corresponding author

landscapes, such as those shaped by smallholder tropical agriculture in developing countries. To illustrate our method, we describe its application for the characterization of smallholder agriculture in Madagascar Highlands. This agricultural landscape is an interesting case study because of its high complexity that led to mapping difficulties when remote sensing approaches were used alone, as shown in Inglada et al., (2015) and Lebourgeois et al. (2017).

The approach is divided into three main parts: the first one concerns remote sensing data processing (producing a first land use map), the second one focuses on the spatio-temporal modelling (based on the implementation of spatio-temporal rules expressing farmers' strategies and practices), and the last one on the combination of remote sensing and spatio-temporal modelling (for obtaining a more agriculturally consistent map).

2. STUDY AREA AND DATA

2.1 Study Area

Our study area is 20 km x 30 km in size and is part of the JECAM network (Joint Experiment for Crop Assessment and Monitoring, www.jecam.org). It is located, close to Antsirabe, the administrative center of Vakinankaratra (coordinates: 19°31'56" to 19°51'32"S, 46°57'24" to 47°13'14"E) in Madagascar Highlands. This area is characterized by complex heterogeneous and predominantly agricultural landscapes. Within the cropland, the complexity is mainly due to the very low average area cultivated per household (ranging from 0.1 to 0.5 ha, according to Sourisseau et al. (2016)). There is also a large variety of crop types with different crop calendars, and the practice of intercropping is widespread. Moreover, the landscape is fragmented by the presence of natural vegetation between the fields that is phenologically synchronized with the crops. The primary growing season occurs during the rainy season, from November to April, constraining the acquisition of optical cloud-free satellite images. The crops are diversified, but rice and maize dominate. They are cultivated under rainfed (in slopes and plateaus) or irrigated conditions (mainly lowland rice paddies).

2.2 Data

For this study, we used several types of data: (i) both high spatial resolution and high revisit frequency time series satellite data, (ii) ground truth data, (iii) agro-environmental and socio-economic knowledge data on the farmers' strategies (for the spatio-temporal rules definition), and (iv) auxiliary data (for the rules implementation).

2.2.1 Remote sensing data and pre-processing: The remote sensing dataset corresponds to (i) one very high spatial resolution SPOT 6 image, and (ii) a high spatial resolution (10m-30m) time series of 37 Sentinel 2 and 9 Landsat 8 images, for the period between October 1st, 2017 and April 29th, 2018, corresponding to the main growing season.

The SPOT 6 image was acquired on February 14th 2017 around the peak of the growing season. The image was ortho-rectified, converted to Top Of Atmosphere reflectance and pansharpened using the open-source Orfeo ToolBox (OTB; Grizonnet et al., 2019) to obtain multispectral images at 1.5 m spatial resolution. Sentinel-2 image time series (Donadieu, L'Helguen, 2016) were acquired from THEIA website (`https://theia.cnes.fr`). These images are provided in surface reflectance, and with a cloud mask, thanks to the MAJA (MACCS ATCOR Joint Algorithm) processing chain (Hagolle et al., 2017). Then, additional preprocessing steps were performed using the MORINGA chain (Gaetano et al., 2019). The functions used in MORINGA include: a) the automatic resampling (with bicubic interpolation) of S2 20 m bands to 10 m resolution, b) a multi-temporal gap-filling (Inglada et al., 2017), and c) a co-registration of S2 and L8 time series on the SPOT 6 image used as a reference.

Data	Spatial resolution / scale	Temporal resolution	Data source
Satellite data			
SPOT 6	1.5	Once	CNES / Airbus Defense & Space
Sentinel 2	10 m	Every 5 days	ESA
Landsat 8	30 m	Every 16 days	NASA / USGS
Auxiliary data			
Altitude	1 arc second	Once	SRTM v3.0 Global 1 arc second
Slope	1 arc second	Once	Computed from SRTM v3.0 Global
Landform (TPI)	1 arc second	Unique	Produced from SRTM v3.0 Global
Pedology	Area at Madagascar scale	Unique	Riquier (1968)
Hydrology	1 arc second	Unique	Hydrological modelling (GRASS GIS)
Fokontany	Municipality scale	Unique	Personal communication

Table 1. Satellite and auxiliary data used in the study.

2.2.2 *In situ* data on land use: A ground truth database on actual land use was built during field surveys around the peak growing season (beginning of February to end of March 2018). GPS waypoints were recorded following an opportunistic sampling approach. Waypoints were chosen according to their accessibility (but some parts of the area could not be reached due to heavy rains affecting the road conditions), and the size of the object (at least 20 * 20 m, so that they can be identified on satellite images). We recorded land uses according to a modified version of the JECAM nomenclature (see http://jecam.org/wp-content/uploads/2018/10/JECAM_Guidelines_for_Field_Data_Collection_v1_0.pdf) to better take into account the specificities of tropical agriculture. In the case of intercropping, we recorded the main and the second crop in the field. Once the waypoints were acquired, the boundaries of each field were digitized on the SPOT 6 image, and the class labels (and other attributes) were joined to the polygon database. Additional non-cropland polygons were also added from photo-interpretation of the SPOT 6 image for easily recognizable classes such as built-up surfaces, water bodies, wetlands, mineral soils, or natural forests. At the end, 2170 polygons were obtained (1759 crop and 511 non-crop).

2.2.3 Knowledge on farmer strategies and practices: Our approach aimed at integrating agronomic, environmental or socio-economic expert knowledge that express farmer strategies and practices. To do so, three sources of information were considered: (i) documents (research papers, books, reports, technical documents, territorial analysis, etc.) describing the agricultural functioning of the region, (ii) interviews of a set of experts with different backgrounds (agronomy, geography or socio-economy), (iii) semi-directive interviews of farmers, local officials or persons involved in rural development. These last interviews were conducted at the Fokontany (municipality/village) scale using a questionnaire. The questions of the surveys concerned the local agricultural practices (crop calendars, cropping constraints and spatial distribution), and the socio-economic context that has an impact on farmer strategies.

2.2.4 Auxiliary Data: These include existing spatial information such as altitude, slope and crop calendars. The full list of auxiliary data used is given in Table. The topographic position index (TPI) was used to generate the local and global topography using a multi-scale grid from the digital elevation model (Weiss, 2001) . The hydrological network was modelled with GRASS GIS (GRASS Development Team, 2017) to obtain an active network for the rainy season.

3. METHODOLOGY

The overall approach involves remote sensing data processing, spatio-temporal modelling and their combination (Figure 1). It is divided into five steps:

(i) A preliminary RS-based crop type map is produced using satellite image processing based on a combined OBIA and Random Forest (RF) classification scheme. Each object is labelled with its corresponding class membership probabilities, i.e. the probability that the object pertains to each of the classes (hereafter referred to as CMVPsat, for class membership vector of probabilities based on satellite image processing).

(ii) Each crop type is defined by a set of general spatio-temporal rules (STrules) that are formalized from the analysis of knowledge gathered on farmer strategies and practices.

(iii) The STrules are implemented in a modelling platform to produce a model-based map of crop distribution probability (CMVPrule for STrules-based class membership probabilities) of each object.

(iv) Possible misclassified objects are detected following specific rules.

(v) The membership probability of the selected objects is recalculated (to obtain a corrected class membership vector probabilities, CMVPcorr) to produce a corrected and more agriculturally consistent land use map (Figure 1).

3.1 Remote sensing

The preliminary land use map was produced from remote sensing. This involved the following steps: (i) segmentation of the very high spatial resolution (VHSR) image into homogeneous objects, (ii) extraction of spectral and textural features from satellite images and other products (DEM), and (iii) training and validation of a Random Forest classifier which was then applied to the whole study area. For this study, all steps were performed using the MORINGA processing chain (Gaetano et al. 2019), which uses OTB and Gdal functions driven by Python scripts.

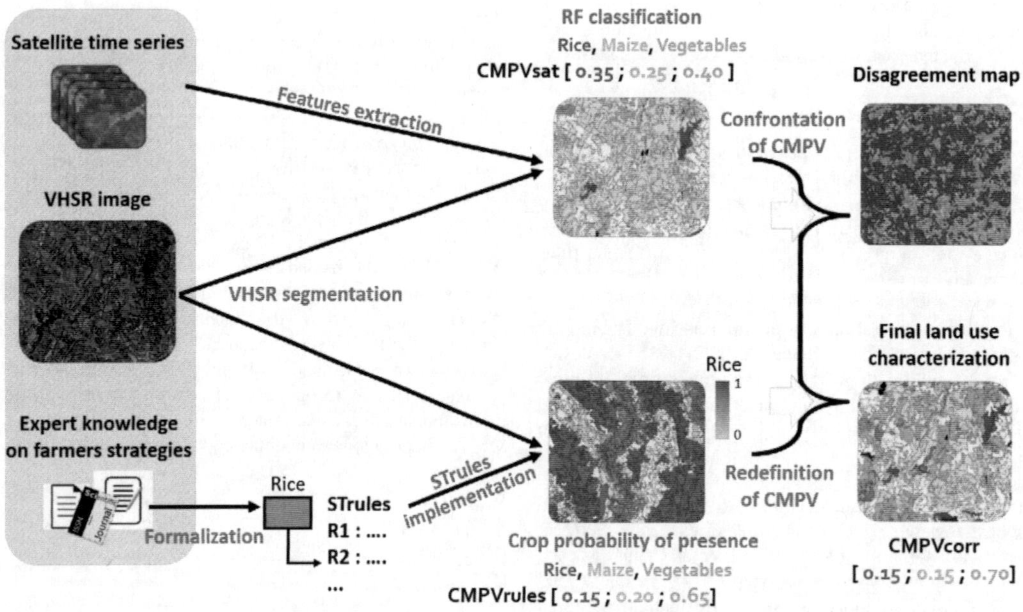

Figure 1: Framework of the method. The processes are presented in red color. CMVPsat corresponds to class membership vector probabilities obtained after random forest (RF) classification of satellite images. CMVPrules refers to the same vector obtained after spatio-temporal rules (STrules) implementation in the model. Finally, CMVPcorr presents the redefinition of the class membership vector probabilities from the two previous ones using fusion methods.

3.1.1 OBIA: According to the OBIA approach (Blaschke et al., 2014), the segmentation of the VHSR image aimed at producing objects delineating the boundaries of agricultural fields, or at least groups of homogeneous fields to characterize the landscape structure. This segmentation determines the landscape structure of the study area, i.e. the boundaries of the "entities" that will be used during the complete workflow. The Baatz and Schäpe segmentation algorithm (Baatz and Schäpe, 2000) implemented in OTB was used with the following parameters: 0.7 for the homogeneity criteria, 0.5 for the spectral homogeneity and the threshold was fixed at 200.

3.1.2 Feature extraction: Satellite features were extracted for each entity of the ground truth database to build a learning database. This step involved the extraction of the following features: (i) reflectance from the Sentinel-2 and Landsat-8 time series (471 features), (ii) spectral indices from the same time series (267 features), (iii) textural indices from the Spot 6 panchromatic image (6 features), and (iv) altitude and slope from the digital elevation model (2 features). Those features were chosen according to a previous study by Lebourgeois et al. (2017). Concerning the spectral indices, we computed the most common ones: NDVI (Rouse et al., 1974), NDWI (Gao, 1996), MNDWI (Xu, 2006), MNDVI (Jurgens, 1997), Brightness Index, and RNDVI (Jorge et al., 2019) for each image of Sentinel-2 and Landsat 8. Haralick textures (Haralick et al. 1973) were computed using mean and correlation for internal textures and contrast as edge texture. Two windows were chosen (7 pixels and 14 pixels) to traduce the diversity of fields organisation???. SRTM-30m ("SRTM V3.0, 1 arcsec") (NASA JPL, 2014) was used to extract altitude and to compute slope.

3.1.3 Classification: In the final step of the remote sensing approach, the objective was to obtain a land use map and to extract and a class membership vector of probabilities for each object (CMVPsat). This vector is useful for understanding the probabilities distribution between classes, and in the following steps, for redefining class membership probabilities for selected objects. The Random forest (RF) algorithm (Breiman, 2001) is widely used in remote sensing (e.g. Lebourgeois et al., 2017; Debats et al., 2016) and has been shown to outperform other classifiers in benchmark studies (e.g. Fernández-Delgado et al., 2014; Inglada et al., 2015). The RF classifier was used to produce a land-use classification based on the learning database at the end of the growing season. The main output, the class membership vector of probabilities (CMVPsat), was extracted for each object.

3.2 Spatio-temporal modelling

The model was built with the Ocelet Modelling Platform (OMP, www.ocelet.org). OMP allows the spatial modelling of a system and their dynamics (Degenne and Lo Seen, 2016) using both raster and vector data. Ocelet is based on the concept of interaction graphs to describe how entities composing a system (e.g. fields, rivers, weather, municipality) are inter-connected. Each entity is characterized by a number of properties that can evolve through time with the application of interaction functions.

3.2.1 Formalization and selection of STrules: The knowledge of farmer strategies and practices gathered during literature review and interviews was formalized so that crop type can be described by a set of general spatio-temporal rules (STrules). The spatial rules can concern distances or intersection with specific entities (such as roads, villages, markets and hydrographic network) or zones (such as soil map units, altitude and slope classes). The temporal rules can be linked to the cropping calendars or rainfall regime during the growing season, or the previous crop rotation on the same field. The spatio-temporal rules for each crop type were selected according to the following criteria: (i) the rule should be representative of the whole study area, (ii) the rule should be frequently mentioned by the persons interviewed or in the documents, (iii) the rule must be consistent with agronomic knowledge and observable in the field, and (iv) the data must be available for the rule to be applied. A crop type can be defined by exclusive rules (e.g. maximum distance from a point) or optional rules (e.g. generally in fertile soil). The assignment of an entity to a given class must comply with all exclusive rules and part (or none) of the optional rules.

3.2.2 Implementation: In the model, the landscape structure of the system is based on the segmentation produced during step 3.1.1, which provide the boundaries of the entities. For each crop, rules are then implemented in the OMP at the entity level to produce a map giving the probability of presence of the corresponding crop in the study area. From this step onwards, the application of all the rules will define a class membership vector of probabilities for each entity, based on formalized knowledge on farmer strategies (referred to as CMVPrule for STrules-based membership probabilities).

3.3 Redefinition of class membership probabilities

Traditional evaluation methods used in remote sensing to validate a classification are based on the computation of indices such as overall accuracy, Kappa, F-score (Labatut, Cherifi, 2012) or the quantity and allocation disagreement (Pontius, Millones, 2011). These methods can be derived from confusion matrices, giving the number of well-classified and misclassified ground truth occurrences per class. These methods allow evaluating the results of a given classifier, but only ground truth information. Other entities are not evaluated. This step proposes a spatial evaluation of the remote sensing classification with respect to farmer strategies (expressed through STrules in the model), and to redefine class membership probabilities to obtain more consistent agricultural maps.

When comparing the CMVPrule to CMVPsat for the whole study area, a map can be generated highlighting areas or entities where the rules disagree, and for which corrections are required. For a given entity, in order to decide which of the two probability vectors is more reliable, other information such as the number of satellite acquisition for an object during the growing season, the set of expert rules concerned, or the confidence value obtained during the random forest vote are used. Then, to redefine class membership, various fusion methods can be used such as fuzzy logic, average between CMVPsat and CMVPrule, or rules weighting.

4. RESULTS

In the following sections, we give examples of STrules used and present three main outputs of the approach. Based on the application of STrules for each entity in the model, probabilities of presence of each crop type were mapped. This map was combined with the map produced during the remote sensing step to create a disagreement map. The final output of the method was a new land use map with an associated class membership probability for each object.

4.1 Spatio-temporal rules

Literature review and interviews allowed determining a set of STrules for each crop. The type of dominant rules can differ between crops. For example, some crops are strongly dependent on economic rules (e.g. good connection to markets) while other crops are more constrained to a given biophysical environment or linked to a specific crop calendar.

Examples of different exclusive and optional rules are given below for three crop classes of the study area:

Market gardening:
Exclusive rules:
R1. Less than 30 minutes by truck or less than 15 km from markets (the presence of market gardening is strongly linked to the presence of markets).
R2. Distance from villages less than 150 m (to avoid steals because market gardening is a high-value crop).
Optional rules:
R3. The field is usually in fertile soil (sandy soil, recent alluvial soil).
R4. The field could be irrigated or close to a watercourse.
R5. The slope can be between 0% and 7%.

Irrigated rice:
Exclusive rules:
R1. The field must be in an irrigated lowland or in the riverbed.
R2. The parcel must be close to a watercourse.
R3. The crop calendar spreads from 11/21 to 12/20 (for transplantation phase) and harvest occurs from 04/20 to 05/10.
Optional rules:
R4. The field is frequently in a hydromorphic soil.

Maize:
Exclusive rules:
R1. The field is located in poor soil.
R2. The crop calendar is 10/20 - 12/30 (seeding) and 02/20 - 04/30 (harvest) for an intermediary cycle and 10/20 - 12/10 (seeding) and 03/20 - 05/10 (harvest) for a long cycle (most common).
Optional rules:
R3. The crop is frequently in intercropping with rainfed rice or soya (to be implemented, this rule thus relies on the probabilities of presence of rainfed rice and soya)
R4. Maize is frequently present in the livestock sector, which corresponds to slopes or plateaus.

4.2 Map of probabilities

The maps of probabilities are the result of STrules Implementation in the model.
Figure 2 shows the irrigated rice probability map in the study area. For each crop, the same map of probability of presence can be computed based on STrules as defined during step 3.3.1. For irrigated rice, this map highlights high probabilities of presence around watersheds and lowlands. In the center of the map, the Ambohibary lowland mainly consists in rice and market gardening crops. On plateaus and slopes, the probability of the presence of irrigated rice is low as expressed by STrules.

Maps of probabilities allow to analyze and understand the possible spatial distribution of the crop in the landscape, and to identify the main underlying explanatory factors of this repartition.

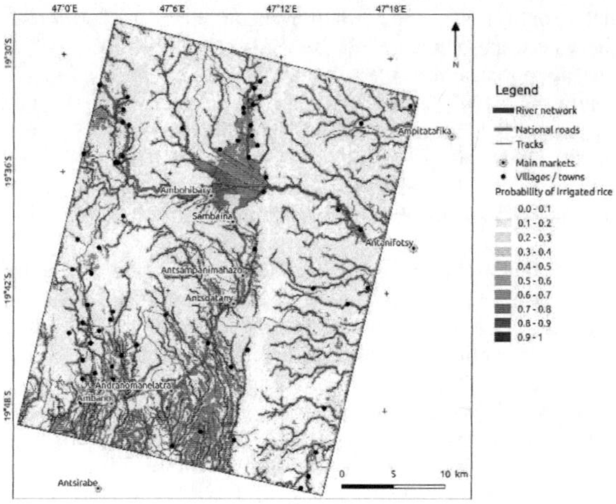

Figure 2. Map of probability for irrigated rice from spatio-temporal modelling (STrules implementation).

4.3 Disagreement map

A comparison between CMVPsat and CMVPrule (see 3.4) allowed producing a disagreement map. This map highlights the entities where remote sensing and STrules seem to be in disagreement. These disagreements mean that either the expert rules or the classifier are not accurate for the concerned entities . Assuming that the STrules have been verified by experts and were found to be consistent, this map thus allows the spatialization of remote sensing image classification errors.

The spatialization of the classification errors can also help to identify production basins or specific zones where remote sensing has more difficulties to discriminate between crop types. These maps support the analysis of the possible factors limiting the success of remote sensing approach in such context. In our study case, the disagreement map presented in Figure 3 shows remote sensing classification errors for maize, irrigated rice, rainfed rice and market gardening.. Majority of irrigated rice is well classified with a predominance in lowlands, but a part of market gardening and other "minor" crops (e.g. cassava, taro, etc.) in lowlands is misclassified. For these minor crops, the weakness of the learning set, leading to an overlearning of the RF classification algorithm can explain the disagreement . In the North-East part of the map, an important area is misclassified. This inaccessible zone had no ground truth data (used in the learning of the RF classification algorithm) to represent its land use variability, and few auxiliary data to support STrules. Consequently, few entities reached an agreement in this zone. A photo-interpretation of this place showed that, except near watercourses, a majority of natural vegetation and rainfed crops were present. In the South-East of the map, another large disagreement area is identified. In this place, ground truth and auxiliary data were present in sufficient quantity. Disagreements are related to the important misclassification of market gardening and orchards with other rainfed crops. An analysis of CMVPsat showed that all these crop types had a very close membership probability. .

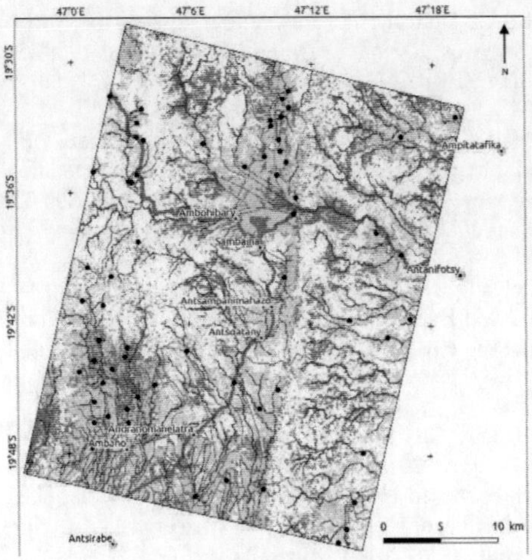

Figure 3. Map showing the disagreements between rule-based (CMVPrule) and remote sensing-based (CMVPsat) class assignment for rice, maize and market gardening.

From this first result, an analysis of the type and characteristics of the entities where a disagreement is found (the type of crop assigned according to CMVPsat and CMVPrule, and values of probability of presence of each crop for the same entity) can be performed to improve the understanding of processes and the characterization of such complex system.

4.4 Corrected map

The main output of the method is a new modified land use map, more consistent with agricultural strategies and practices of the region. An example of a corrected map, produced after the redefinition of probability values by averaging CMVPsat and CMVPrule for each entity in disagreement, is presented in Figure 4. In Madagascar highlands, a first comparison with ground truth data showed that this method allowed better discrimination of rainfed crop types. This crop type is challenging to characterize due to the very small field sizes and the high level of fragmentation of the rainfed cropland area, as shown previously in Lebourgeois et al. (2017).

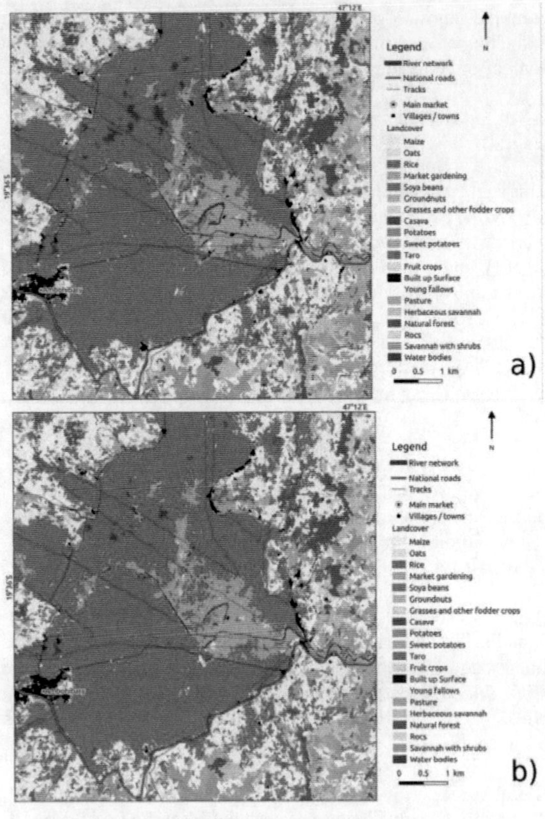

Figure 4. Zoom on land use maps obtained using the remote sensing approach, before (a) and after (b) the redefinition of CMVP values of each entity for which rule-based and remote sensing-based class assignment differs for main crops.

5. CONCLUSION AND OUTLOOKS

This work proposes a new framework in which remote sensing and expert knowledge on farmer practices and strategies are used in a spatio-temporal modelling approach to produce more agriculturally consistent crop type maps in complex agricultural systems. The method has the originality of combining remote sensing and modelling with knowledge from human and social sciences that reflect farmers' behaviour. Such approach can help to improve (i) the understanding of processes driving land use distribution in complex agricultural areas such as in smallholder farming systems and (ii) the accuracy of crop type maps in regions where traditional remote sensing approaches present limitations.

This approach is also being currently tested in the same study area with a new set of satellite images with higher spatial and temporal resolutions (Pléiades instead of SPOT6/7 for the very high spatial coverage, and addition of Venμs data in complement of the Landsat-8 / Sentinel-2 time series), building on the same set of STrules. An analysis of the model sensitivity to the integration of each spatio-temporal rule will also be performed. Beyond its application to the complex smallholder agricultural landscape of Madagascar Highlands, the approach proposes a generic framework that explicitly brings rich thematic expert knowledge into the analysis of satellite time series data. Such a framework can certainly be adapted to other regions of interest or thematic applications (e.g. biodiversity). However, in a broader perspective, this method can be considered as a first attempt to combine knowledge-driven and data-driven approaches for extracting useful information from one of the most abundant sources of geographical data that are earth observation satellites.

ACKNOWLEDGEMENTS

This study was conducted as part of the SAMSAM project funded by CNES, the French Space Agency (TOSCA APR 2017). This work was also supported by the French National Research Agency under the Investments for the Future Program, referred to as ANR-16-CONV-0004. The authors would like to thank Eloise Rasoamalala and Valérie Andoniaina for their contributions to the field surveys, and the researchers of SPAD, the CIRAD platform in partnership on "Highland production systems and sustainability in Madagascar".

REFERENCES

Baatz, M., Schäpe, A., 2000. Multiresolution Segmentation: an optimization approach for high quality multi-scale image segmentation. *Angewandte Geographische Informations-Verarbeitung XII*, 12–23.

Bailly, S., Giordano, S., Landrieu, L., Chehata, N., 2018. Crop-Rotation Structured Classification using Multi-Source Sentinel Images and LPIS for Crop Type Mapping. *IGARSS 2018 - 2018 IEEE International Geoscience and Remote Sensing Symposium*, IEEE, Valencia, pp. 1950–1953. https://doi.org/10.1109/IGARSS.2018.8518427

Blaschke, T., Hay, G.J., Kelly, M., Lang, S., Hofmann, P., Addink, E., Queiroz Feitosa, R., van der Meer, F., van der Werff, H., van Coillie, F., Tiede, D., 2014. Geographic Object-Based Image Analysis – Towards a new paradigm. *ISPRS Journal of Photogrammetry and Remote Sensing* 87, 180–191. https://doi.org/10.1016/j.isprsjprs.2013.09.014

Bontemps, S., Arias, M., Cara, C., Dedieu, G., Guzzonato, E., Hagolle, O., Inglada, J., Matton, N., Morin, D., Popescu, R., Rabaute, T., Savinaud, M., Sepulcre, G., Valero, S., Ahmad, I., Bégué, A., Wu, B., De Abelleyra, D., Diarra, A., Dupuy, S., French, A., Ul Hassan Akhtar, I., Kussul, N., Lebourgeois, V., Le Page, M., Newby, T., Savin, I., Verón, S.R., Koetz, B., Defourny, P., 2015. Building a Data Set over 12 Globally Distributed Sites to Support the Development of Agriculture Monitoring Applications with Sentinel-2. *Remote Sensing* 7, 16062–16090. https://doi.org/10.3390/rs71215815

Breiman, L., 2001. Random forests. *Machine learning* 45, 5–32.
Debats, S.R., Luo, D., Estes, L.D., Fuchs, T.J., Caylor, K.K., 2016. A generalized computer vision approach to mapping crop fields in heterogeneous agricultural landscapes. *Remote Sensing of Environment* 179, 210–221. https://doi.org/10.1016/j.rse.2016.03.010

Degenne, P., Lo Seen, D., 2016. Ocelet: Simulating processes of landscape changes using interaction graphs. *SoftwareX* 5, 89–95. https://doi.org/10.1016/j.softx.2016.05.002

Donadieu, J., L'Helguen, C., 2016. SENTINEL-2A L2A Products description (Technical note). CNES, Direction du Centre Spatial de Toulouse.

Drusch, M., Del Bello, U., Carlier, S., Colin, O., Fernandez, V., Gascon, F., Hoersch, B., Isola, C., Laberinti, P., Martimort, P., Meygret, A., Spoto, F., Sy, O., Marchese, F., Bargellini, P., 2012. Sentinel-2: ESA's Optical High-Resolution Mission for GMES Operational Services. *Remote Sensing of Environment* 120, 25–36. https://doi.org/10.1016/j.rse.2011.11.026

Dury, J., Schaller, N., Garcia, F., Reynaud, A., Bergez, J.E., 2012. Models to support cropping plan and crop rotation decisions. A review. *Agronomy for Sustainable Development* 32, 567–580. https://doi.org/10.1007/s13593-011-0037-x

Fernández-Delgado, M., Cernadas, E., Barro, S., Amorim, D., 2014. Do we Need Hundreds of Classifiers to Solve Real World Classification Problems? *Journal of Machine Learning Research* 15, 3133–3181.

Gaetano, R., Dupuy, S., Lebourgeois, V., Le Maire, G., Tran, A., Jolivot, A., Bégué, A., 2019. The MORINGA Processing Chain: Automatic Object-based Land Cover Classification of Tropical Agrosystems using Multi-Sensor Satellite Imagery. ESA Living Planet Symposium (LPS 2019), Milan, Italy.

Gao, B., 1996. NDWI—A normalized difference water index for remote sensing of vegetation liquid water from space. *Remote Sensing of Environment* 58, 257–266. https://doi.org/10.1016/S0034-4257(96)00067-3

GRASS Development Team, 2017. Geographic Resources Analysis Support System (GRASS) Software. Open Source Geospatial Foundation.

Grizonnet, M., Michel, J., Poughon, V., Inglada, J., Savinaud, M., Cresson, R., 2017. Orfeo ToolBox: open source processing of remote sensing images. *Open Geospatial Data, Software and Standards* 2. https://doi.org/10.1186/s40965-017-0031-6

Hagolle, O., Huc, M., Desjardins, C., Auer, S., Richter, R., 2017. MAJA Algorithm Theoretical Basis Document. https://doi.org/10.5281/zenodo.1209633

Haralick, R.M., Shanmugam, K., Dinstein, I., 1973. Textural Features for Image Classification. *IEEE Transactions on Systems, Man, and Cybernetics SMC*-3, 610–621. https://doi.org/10.1109/TSMC.1973.4309314

Inglada, J., Arias, M., Tardy, B., Hagolle, O., Valero, S., Morin, D., Dedieu, G., Sepulcre, G., Bontemps, S., Defourny, P., Koetz, B., 2015. Assessment of an Operational System for Crop Type Map Production Using High Temporal and Spatial Resolution Satellite Optical Imagery. *Remote Sensing* 7, 12356–12379. https://doi.org/10.3390/rs70912356

Jurgens, C., 1997. The modified normalized difference vegetation index (mNDVI) a new index to determine frost damages in agriculture based on Landsat TM data. *International Journal of Remote Sensing* 18, 3583–3594. https://doi.org/10.1080/014311697216810

Labatut, V., Cherifi, H., 2011. Accuracy Measures for the Comparison of Classifiers. *The 5th Inter-national Conference on Information Technology*. Amman, Jordanie.

Lebourgeois, V., Dupuy, S., Vintrou, E., Ameline, M., Butler, S., Bégué, A., 2017. A Combined Random Forest and OBIA Classification Scheme for Mapping Smallholder Agriculture at Different Nomenclature Levels Using Multisource Data (Simulated Sentinel-2 Time Series, VHRS and DEM). *Remote Sensing* 9(3), 259. https://doi.org/10.3390/rs9030259

NASA JPL, 2014. U.S. Releases Enhanced Shuttle Land Elevation Data. NASA Jet Propulsion Laboratory (JPL). http://www.jpl.nasa.gov/news/news.php?release=2014-321 (28 October 2019).

Peña-Barragán, J.M., Ngugi, M.K., Plant, R.E., Six, J., 2011. Object-based crop identification using multiple vegetation indices, textural features and crop phenology. *Remote Sensing of Environment* 115, 1301–1316. https://doi.org/10.1016/j.rse.2011.01.009

Pontius, R.G., Millones, M., 2011. Death to Kappa: birth of quantity disagreement and allocation disagreement for accuracy assessment. *International Journal of Remote Sensing* 32, 4407–4429. https://doi.org/10.1080/01431161.2011.552923

Riquier, J., 1968. Carte pédologique de Madagascar. One mosaic of 3 map sheet. Scale of 1 :1 000 000. ORSTORM.

Rouse, W., Haas, R.H., Schell, J., Deering, D., 1974. Monitionring Vegetation Systems in the Great Plains with ERTS. *Third Earth Resources Technology Satellite-1 Symposium, Technical presentations*, section A, vol. I, National Aeronautics and Space Administration, Washington, DC, 309–317.

Sourisseau, J.-M., Rasolofo, P., Bélières, J.-F., Guengant, J.-P., Ramanitriniony, H.K., Bourgeois, R., Razafimiarantsoa, T.T., Andrianantoandro, V.T., Ramarijaona, M., Burnod, P., Rabeandriamaro, H., Bougnoux, N., 2016. Diagnostic Territorial de la Région du Vakinankaratra à Madagascar (Prospective territoriale sur les dynamiques démographiques et le développement rural en Afrique subsaharienne et à Madagascar). AFD.

Weiss, A.D., 2001, Topographic position and landform analysis. Poster presentation, ESRI User Conference, San Diego, California, USA.

Xu, H., 2006. Modification of normalised difference water index (NDWI) to enhance open water features in remotely sensed imagery. *International Journal of Remote Sensing* 27, 3025–3033. https://doi.org/10.1080/01431160600589179

USING GUYANA'S MONITORING REPORTING & VERIFICATION SYSTEM TO GUIDE NATIONAL FOREST MANAGEMENT & DECISION MAKING

Nasheta Dewnath[1], Pradeepa Bholanath[1], Rosa Rivas Palma[2], Blair Freeman[2], Pete Watt[2]

[1] Guyana Forestry Commission, 1 Water Street, Georgetown, Guyana - ndewnath@gmail.com, project.coordinator@forestry.gov.gy

[2] Indufor Asia Pacific Ltd, Auckland City 1147, New Zealand - rosa.rivaspalma@indufor-ap.com, blair.freeman@indufor-ap.com, pete.watt@indufor-ap.com

KEY WORDS: Guyana, REDD+, Satellite Data, Forest Management, Decision Making

ABSTRACT:

The Guyana Forestry Commission's (GFC) Monitoring, Reporting and Verification System (MRVS) is a combined Geographic Information System (GIS) and field-based monitoring system, which has underpinned the conducting of a historical assessment of forest cover as well as eight national assessments of forest area change to date. The System seeks to provide the basis for measuring verifiable changes in Guyana's forest cover and resultant carbon emissions from Guyana's forests, which will provide the basis for results-based REDD+ compensation in the long-term. With the continuous compilation, analysis and dissemination of MRVS results on a typically annual basis, the GFC envisioned a larger role for this data, in informing national processes such as natural resources policy and management. This resulted in a significant broadening of the application of the MRVS data and products for purposes that are aligned or complementary to national REDD+ objectives and forest policy and management. These broader applications have allowed for a beneficial shift towards the increased use of remote sensing data and scientific reporting to inform forest management, governance and decision making on natural resource management across forested land. This has resulted in a transformation in the nature of data available to inform decision making on forest management and governance, and overall environmental oversight, from predominantly social science data and factors to now incorporating remote sensing and scientific observations and reporting. Primary decision makers are turning to scientific based reporting to determine best approaches for developmental initiatives in Guyana. This study shows how Guyana has demonstrated significant progress in making remote sensing products accessible and useful to policy makers in Guyana.

1. INTRODUCTION

Guyana's forests cover approximately 87.5% of the country and contain an estimated 19.5 billion tons (or Gt) of CO_2 in live and dead biomass pools (GoG, December 2014)[1]. Recognising the value of this extensive natural capital, the Government of Guyana (GoG) has engaged in a national programme to protect and maintain its forests, to reduce global carbon emissions and promote development along a green development pathway.

Guiding the continued protection and conservation of Guyana's forests is the Green State Development Strategy (GSDS), which is built on a vision of "*green, inclusive development based on sound education and social protection, low-carbon resilient development, green and decent jobs, economic opportunities, individual equality, justice, and political empowerment*" (GoG and UNEP, 2017). Key areas of focus within the GSDS are improvements in land use planning, sustainable agriculture, sustainable forest management (encompassing timber harvesting and non-timber forest products), freshwater management and the conservation of coastal ecosystem services.

In this context, this paper describes the use of Guyana's national forest cover monitoring program to inform decision-making on forest management and natural resource management across a broad range of sectors.

1.1 Guyana-Norway Agreement

In November 2009, the Government of Guyana and the Kingdom of Norway entered into a bilateral agreement regarding cooperation on issues related to mitigating climate change. More specifically, it addressed reducing emissions from deforestation and forest degradation (REDD+) in developing countries, biodiversity protection and the enhancement of sustainable, low carbon development (Joint Concept Note, 2012).

While committing up to US$250 million over the five years of the Agreement, Norway's level of support was to be determined by Guyana's performance, measured against two sets of indicators:

- *Indicators of enabling activities*: These were a set of policies and safeguards designed to ensure REDD+ efforts contribute to the achievement of the goals set out in the Agreement. These indicators spoke to arrangements to ensure systematic and transparent multi-stakeholder consultations throughout the process; protection of the rights of Indigenous peoples; ensuring environmental integrity and biodiversity protection; ensuring continuous improvements in forest governance; and providing transparent,

[1] The Reference Level for Guyana's REDD+ Program, December 2014

accountable oversight and governance of the financial support received.

– *REDD-plus Performance Indicators*: A set of forest-based greenhouse gas emissions-related indicators. It was agreed these indicators would gradually be replaced as the monitoring, reporting and verification system became fully operational. The indicators were developed based on conservative estimates while encouraging the development of a more accurate system over time through building national capacities.

Under the Guyana-Norway Agreement, Guyana developed its Monitoring, Reporting and Verification System (hereafter, MRVS) for REDD+. The MRVS is a combined GIS-based and field-based monitoring system developed to track national-level forest change of deforestation and forest degradation. It was developed to be the primary mechanism to measure Guyana's performance against the agreed indicators as set out in the agreement. Furthermore, it was required for the acquisition and analysis of ground-based and satellite data to report on forest cover and change on a predominantly annual basis.

The MRVS allowed for Guyana to conduct a historical assessment of forest cover (with 2009 as the baseline year), and eight national assessments of forest area change to date. National assessments for Years 1-7 (2009-2017) have been completed and the results disseminated to stakeholders; the national assessment for Year 8 (2018) is currently being completed.

The Guyana Forestry Commission (GFC) has led the development and implementation of the MRVS. Its development was guided by extensive engagement and consultation with government agencies, international experts and non-government stakeholders in Phase 1 (2010- 2014); and this broad-based engagement continues through Phase 2 (2015- 2020[2]).

Over the years, significant progress has been made in the development of the MRVS; not only in the building of local capacities to manage and operate the System, but in technologies and techniques used to monitor forest change. The data collected over the period 2009 to 2018 has been stored in a repository that has been used primarily to facilitate reporting to Norway to trigger payments under the agreement.

As the MRVS continues to evolve, so has the vision for the role of the System in Guyana. The GFC sees a broadening of the current role of the System, from annual performance reporting to supporting forest and environmental management, governance and decision making within the natural resources sectors. The MRVS can be used further to report on Guyana's progress in the implementation of international commitments to the United Nations Framework Convention on Climate Change (UNFCCC) as well as supporting the country's implementation of the indicators of the Sustainable Development Goals.

This Paper describes the development of the MRVS, its applications and the approaches made to move the System towards this vision.

1.2 MRVS Phase 1 – Development of the System

The primary focus of Phase 1 of the MRVS was the development of the System. A Roadmap was developed to guide the development of the MRVS over the period 2009 to post 2012. The aim of the Roadmap in Phase 1 was to establish a comprehensive national system for monitoring, reporting and verifying forest carbon emissions resulting from deforestation and forest degradation in the country.

The Terms of Reference for Developing Capacities for a national Monitoring, Reporting and Verification System to support REDD+ participation of Guyana Background, Capacity Assessment and Roadmap (2009)[3] set out the key building blocks for the development of the MRVS. This sought first to identify the requirements for the System, which are based on the internationally accepted principles and guidance at the international level, as specified by the IPCC Good Practice Guidelines and Guidance for reporting.

The Phase 1 Roadmap further sought to develop the local capacities to implement MRV through the development of a detailed plan to establish and sustain MRV capacities within the country. A detailed capacity assessment was conducted on the state of local capacities in keeping with the requirements for a MRVS; and this informed the development of a capacity building roadmap[4].

Over the period 2009 and 2014, the GFC was able to take progressive steps to build a full MRVS in three different phases: (i) National Strategy Formulation, (ii) Country Readiness Phase and (iii) Implementation. The result was a sustainable, efficient national mechanism and an accompanying institutional framework with competencies for MRVS at different levels. Through the implementation of Phase 1, significant achievements were made in conducting a national forest monitoring program and MRVS. The System has served to report on the country's performance against the "REDD+ Interim Indicators" agreed upon between the Governments of Guyana and Norway.

It also enabled Guyana to make significant contributions to the global dialogues on REDD+. In the context of evolving guidance from the UNFCCC forums regarding MRVS and methodologies, Guyana has been able to demonstrate significant progress in the piloting of methodologies and data collection techniques. Guyana has been able to share lessons and take stock of those lessons for the future as well as working closely with the international community and high-level stakeholders.

1.3 MRVS Phase 2 – Consolidating and expanding capacities for national REDD+ monitoring

The activities planned for implementation during the Phase 2 of the MRVS (2015 – 2019) were designed to support the establishment and maintenance of a world-class MRVS as a key component of Guyana's national

---

[2] The 2019 assessment will be completed in 2020.

[3] Terms of Reference for Developing Capacities for a national Monitoring, Reporting and Verification System to support REDD+ participation of Guyana Background, Capacity Assessment and Roadmap, November 13, 2009.

[4] Ibid.

REDD+ programme. This System will provide the basis for measuring verifiable changes in Guyana's forest cover and the resultant carbon emissions as an underpinning for results-based REDD+ compensation in the long-term.

The objective of the Phase 2 Roadmap[5] is to consolidate and expand capacities for national REDD+ monitoring and MRVS. It has sought to support Guyana in meeting the evolving reporting requirements from the UNFCCC while continuing to fulfil additional international reporting requirements, as well as support Guyana in further developing forest monitoring as a tool for REDD+ implementation. The Phase 2 Roadmap sought to enable Guyana to fulfil its REDD+ objectives to:

- Underpin and stimulate strategies and priorities for REDD+ implementation
- Track performance of REDD+ activities and their impacts (carbon & non-carbon)
- Continue to support the building of capacity for MRV implementation at the government and non-government level and other parties that have a role in MRVS related activities.

Three key activities recommended for Phase 2 in order to consolidate and expand capacities included:

a) Consolidate capacities and routine REDD+ monitoring and MRV
b) Develop national forest monitoring as a tool for REDD+ implementation
c) Facilitate knowledge sharing and capacity building

2. MRVS DATA

Over the duration of the Guyana-Norway Agreement (2009- 2014), the focus of the MRVS (Phase 1) was the production of annual performance reports against this framework of Interim Indicators. These performance reports (the MRVS Interim Measures Reports) detailed forest change by activity data and related emissions. Phase 1 of the MRVS produced the following reports (Table 1):

Report	Time period included
Historic assessment	1990 to Sept 2009
Year 1: First annual assessment	1st October 2009 to 30th September 2010
Year 2: Second annual assessment	1st October 2010 to 31st December 2011
Year 3: Third annual assessment	1st January 2012 to 31st December 2012
Year 4: Fourth annual assessment	1st January 2013 to 31st December 2013
Year 5: Fifth annual assessment	1st January 2014 to 31st December 2014

Table 1 -List of reports completed under Guyana Norway Agreement/MRVS Phase 1 (2009 to 2014)

Guyana was able to secure continued support from the Government of Norway for implementation of Years 6-9 (2015- 2019) of the MRVS under Phase 2. This will lead to an additional set of MRVS Interim Measures Reports (Table 2):

Report	Time period included
Year 6: Sixth annual assessment	1st January 2015 to 31st December 2016 (24 months)
Year 7: Seventh annual assessment	1st January 2017 to 31st December 2017
Year 8: Eight annual assessment	1st January 2018 to 31st December 2018
Year 9: Ninth annual assessment	1st January 2019 to 31st December 2019

Table 2 -List of reporting under MRVS Phase 2 (2015-2019)

2.1 Satellite Imagery

Reporting under the MRVS draws on a combination of ground-based and satellite data. As the MRVS has evolved, so has the data that has been used; most notably, the satellite data.

In 2009, as work commenced on the analysis of historical change in forest cover, freely available Landsat imagery at 30 metre resolution was acquired and used. This covered the period 1990-2009. The reporting for Year 1 (2010) utilised a combination of DMC (32 m) and Landsat imagery.

From 2011 to 2014, the data utilised was purchased from RapidEye at 5 metre resolution.

Forest change mapping conducted for 2015 and 2016 (Year 6) acquired a combination of Landsat and Sentinel (10 metre resolution) data. The Sentinel (2A/2B) has been the primary dataset used for mapping of forest change since 2017, supplemented by Landsat.

The use of all of the above-mentioned satellite imagery providers (i.e. Landsat, DMC, RapidEye, Sentinel) has enabled Guyana to track land cover change greater than one hectare in size, and attribute this change to direct drivers (i.e. mining, shifting agriculture etc.).

Another aspect of the MRVS is conducting an Accuracy Assessment. The Accuracy Assessment is used to provide an evaluation of the quality of the GFC's mapping of change across Guyana's forests. Two datasets have been used to inform the Accuracy Assessment:

- High-resolution aerial photographs have been captured using a Cessna-mounted aerial multispectral imaging system. Image resolution ranges from about 25 to 60 cm (variations due to the altitude of the aircraft at the time of capture); and
- PlanetScope images as provided by Planet Labs. The Planet constellation comprises approximately 200 micro-satellites imaging areas at (approximately) 3-5 metre resolution.

[5] Terms of Reference for continuing to Develop Capacities for a National Forest Monitoring and Measurement, Reporting and Verification System to Support REDD+ Participation of Guyana, Roadmap Phase 1 Achievements, Evolving Requirements and Roadmap for Phase 2 Activities, September 19, 2014.

2.2 National Datasets

Along with satellite data, a range of national datasets are incorporated in the MRVS for mapping of forest change. These datasets relate to infrastructure, mining concessions, agricultural leases, land tenure, protected areas etc. These datasets are managed across Government agencies involved in the active management and allocation of land resources in Guyana. The main Government agencies responsible for these functions and an overview of their relevant management datasets are set out in Table 3.

Agency	Role	Datasets
Guyana Forestry Commission (GFC)	Management of forest resources	Resource management related datasets
Guyana Geology and Mines Commission (GGMC)	Management of mining and mineral resources	Mining concessions, active mining areas
Protected Areas Commission (PAC)	Management of Protected Areas System in Guyana	Spatial representations of all protected areas
Guyana Lands and Surveys Commission (GLSC)	Management of land titling and surveying of land	Land tenure, settlement extents and country boundary

Source: Guyana Forestry Commission (2018). Guyana REDD+ Monitoring Reporting & Verification System (MRVS) Year 7 Interim Measures Report, November 2018.

Table 3- Primary agency datasets incorporated in MRVS

3. EXISTING APPLICATIONS OF MRVS DATA

As implementation of the second phase of the MRVS progressed, the data was used to inform several new applications for natural resource management by agencies that worked closely with the GFC. These applications included the identification of high intensity land uses related to forestry and mining activities, national protected areas planning, assessing land degradation, as well as sub-national level MRV (Community MRV) activities. These activities are detailed below.

3.1 Identification of areas of high intensity land use

The Ministry of Natural Resources (MNR), which has oversight of the forestry and mining sectors in Guyana, has utilised the MRVS data to identify areas of relatively high-level activity (termed 'hotspots') to support the prioritisation of monitoring and compliance measures by these agencies. The MNR, as part of its oversight and governance functions, engages with GFC and the Guyana Geology and Mines Commission (GGMC) on the monitoring and compliance measures for these areas.

Though the MRVS data was used to inform this strategic analysis, the routine use of the MRVS data has not yet been mainstreamed within the MNR's management systems. A constraint they have identified in the use of the MRVS data has been the data lag associated with annual reports, in that the reporting conducted by the GFC is retrospective, rather than a near real time monitoring system.

3.2 Analysis of Protected Areas for Nationally Determined Contribution commitments

The MRVS data has been used as part of the national process for identifying additional areas to extend and expand the Protected Areas System in Guyana. In 2015, the President of Guyana announced the country's submission of its Nationally Determined Contribution (NDC) as part of landmark Paris Climate Change Agreement; and subsequently, the President confirmed that this contribution would include the conservation of an additional 2 million hectares through Guyana's National Protected Area System (NPAS) and other effective area-based conservation measures.

To implement this, the GoG and its implementing agencies worked together with non-government organisations (NGOs) including Conservation International (CI) to firstly identify candidate areas and then evaluate the respective benefits and costs ('trade-offs') for each of these areas. Benefits and costs were addressed through a multi-criteria analysis, encompassing environmental, economic and social impacts for each of the sites. In addition, the analysis incorporated the objective of ensuring a geographic distribution across all the regions.

The lead agencies in the process, the Protected Areas Commission (PAC) and CI observed that MRVS data was one of the important spatial datasets used to identify forest cover and areas of forest cover change over time, as part of this comprehensive analysis of candidate areas across Guyana. Key considerations included deforestation rates within or around candidate areas, and 'hotspots' of relatively high intensity use in these areas.

While this comprehensive analysis is largely completed, the President has identified that the PAC is responsible for modifying (enlarging, extending and expanding) the NPAS to ensure that these targets are met. Under this mandate, there is scope for further analysis and ongoing monitoring of the NPAS over time, using MRVS data and other spatial datasets to monitor changes in land use, notably in forested areas.

3.3 Assessment of land degradation on non-forest land areas

The Guyana Lands & Surveys Commission (GLSC) is actively engaged in the development of the MRVS, through regular inter-agency liaison and coordination with the GFC. Through these linkages, GLSC is highly familiar with the MRVS outputs and has sought to use these outputs for a broader range of purposes aligned with its mandate. GLSC acts as the custodian for all public lands, rivers and streams of Guyana and is the National Focal Point Agency of the United Nations Framework Convention to Combat Desertification (UNCCD).

With the acquisition of RapidEye 5 metre resolution data (2012-2014), the GFC has purchased user licences for the MNR, GLSC, the GGMC and the Environmental Protection Agency (EPA). The GLSC has access to the licence of the RapidEye imagery acquired for the MRVS assessment and has received the MRVS outputs based on this imagery. In the implementation of Guyana's *Mainstreaming Sustainable Land Development and*

Management Project (Government of Guyana and FAO, 2017), the GLSC has used the RapidEye imagery to assess land degradation in the non-forest areas defined by the MRVS assessment. However, the GL&SC has pointed out that the imagery has proven less suitable for:

- updating topographic maps that have a scale of 1:50,000 (some map scales are presented in metres, while some are presented in feet);
- geo-referencing of cadastral plans and detailed maps for land use planning.

A further limitation that the GLSC identified with the use of MRVS outputs to date is the resolution of the satellite imagery data available. The GLSC indicated that 2 metre resolution and lower would be required for the fulfilling of the GLSC's mandate.

3.4 Capacity Building for Community MRV

Community Monitoring, Reporting and Verification (CMRV) involves the training of local people in the Monitoring, Reporting and Verification (MRV) process, for measurement and monitoring of the impacts of the drivers of forest change at the community level. It has the potential to add immensely to the REDD+ movement in Guyana, especially the national MRV system. The GFC continues to build the capacities of local communities to undertake CMRV as part of the process of participating in REDD+. The GFC is seeking to create synergies between CMRV and the national MRVS whereby communities will be part of the reporting framework so as to inform annual national reporting; and will also be able to verify what is reported by the GFC.

For this process to realize its full potential, local capacities would need to be built within the communities that are interested in participating in a REDD+ scheme. This would include the theoretical and practical aspects of REDD+, MRV and CMRV, as well as training in the use of field equipment and devices and the management of financial and human resources.

The data generated from CMRV can be used to prepare baseline information on forest resources available and changes, which can assist local communities and government institutions in the management of their natural resources, adapting land use practices and other socio-economic issues.

3.5 Other interagency applications

The use of the MRVS in the areas outlined above led the GFC to conduct a national review aimed at identifying all the current applications of the MRVS data and outputs; and then identifying potential options for further use for purposes that are broadly aligned or otherwise complementary with national REDD+ objectives and forest resources policy and management in Guyana. The national review sought to:

a) ascertain how MRVS outputs are supporting existing land use policies and regulations, including land use allocation processes in Guyana;
b) establish how well the range of existing national policies and regulations are aligned in relation to the application of MRVS outputs across sectors;
c) identify the potential for broader application of the MRVS for the benefit of other sectors and the management of Guyana's natural resources; and
d) provide findings and recommendations from the review, with a principal focus on opportunities for strengthening the monitoring functions of the MRVS, and the planning and enforcement functions for Government agencies.

In this review, conducted in 2018, a broad range of national stakeholders were consulted, many of whom indicated keen interest in gaining access to the current MRVS data. As a result, the GFC was able to build on or in some cases establish relationships with agencies and academia, involved in environmental management, mangrove protection, national infrastructural development and disaster management among others. The applications for MRVS data applied to date are outlined in the summary below (Table 5).

Application	Impacts and beneficiaries
Planning for, mapping and monitoring of forest cover change within Protected Areas	*Impacts*: Enhanced capabilities within the PAC for planning, monitoring and measuring forest change by driver within Protected Areas *Beneficiaries*: PAC, Department of Environment
Capacity building, academic research and studies on forest change using GIS and Remote Sensing within the University of Guyana (UoG)	*Impacts:* Scientific research, papers and spatial products produced by the UoG that can be used to inform policy, natural resources planning, management and usage, and conservation efforts. *Beneficiaries*: UoG
Managing and protecting the natural environment, specifically regarding activities being undertaken within Guyana's forests	Impacts: Informing a compliance monitoring programme for the activities undertaken within the forested areas of Guyana. *Beneficiaries*: EPA, Department of Environment, MNR
Informing planning and decision making for infrastructure development and climate resilience	*Impacts:* National infrastructural development informed by data that helps to plan for and reduce the impact on forests. *Beneficiaries*: Ministry of Communities, Ministry of Public Infrastructure, Guyana Energy Agency
Development of flood early warning systems	*Impacts:* Improved disaster preparedness and response *Beneficiaries*: Hydrometeorological Department, Ministry of Agriculture

Application	Impacts and beneficiaries
Mangrove mapping for Guyana	*Impacts:* Baseline map will allow for monitoring of changes in mangroves along Guyana's coast; to inform policy regarding mangrove protection and replanting. *Beneficiaries*: Ministry of Agriculture, National Agricultural Research and Extension Institute
Informing national policy and reporting on forest at the level of Sustainable Development Goals, International Commitments (Paris Agreement), and Bilateral Agreements	*Impacts:* A data agnostic system built by Guyana with strong local capacities to operate, that serves national and international monitoring requirements, and which inform decision making and policy setting for the natural resources sector in Guyana. *Beneficiaries*: Government of Guyana

Table 5 Summary of current MRVS applications

4. POTENTIAL BROADER APPLICATIONS OF MRVS

The main finding of the national review was the considerable interest across multiple Government agencies and non-government organisations in the use of MRVS data, for existing uses and a broader range of potential applications linked to their areas of work and broader than the current uses.

The review also noted the scope for these additional applications to be considered in the context of a range of key considerations that include (but were not limited to):

- The level of effort in terms of technical staff, time and costs involved to incorporate the additional functionality into the MRVS, where applicable;
- The extent to which this development would enhance the MRVS overall;
- The extent to which these applications will provide benefit for Guyana more broadly i.e. through improving land management across land tenures by providing increased access for Government agencies, NGOs and other stakeholders to current spatial data information and analysis.

In this context, the additional potential applications were compared and categorised into three distinct groupings that relate specifically to the level of effort required for applying or otherwise using the MRVS outputs for additional value-added purposes. These groupings comprise:

a. Options requiring a relatively low level of effort for implementation:
   - **Development of Land Use Plans:** MRVS data could contribute to the refinement of the National Land Use Plan (NLUP) and preparation of land use plans. Low enhancement to the MRVS (data sharing). Benefit to Guyana: Medium to High.
   - **Monitoring of agricultural land use changes (LUC) in State lands:** MRVS data can be used in the assessments of land under agricultural leased and permits, and to monitor trends in LUC and area availability. MRVS data is updated annually and this may be limiting for ongoing monitoring. Low enhancement to the MRVS (data sharing).
   - **Supporting management and monitoring of Protected Areas:** MRVS data can support the monitoring of protected areas, identification of hotspots, and refinement of the strategy to add 2 million hectares of protected areas as per NDC commitments. Low enhancement to the MRVS (data sharing). Benefit to Guyana: High
   - **Monitoring forest biodiversity:** MRVS data may contribute to monitoring biodiversity loss associated with forest loss and changes and impact on canopy cover. Low enhancement to the MRVS (data sharing). Benefit to Guyana: Medium to high
   - **Supporting EPA with compliance assessments:** MRVS data can assist EPA to target auditing activities related to compliance on environmental management within concession areas (e.g. for mining or timber harvesting). As MRVS data is updated annually and this may be a limitation. Low enhancement to the MRVS (data sharing). Benefit to Guyana: Medium to high

b. Options requiring a medium level of effort for implementation:
   - **Mangroves mapping and monitoring:** MRVS could be extended to include mangroves and provide information to assist in planning restoration and interventions. Medium to high scale MRVS enhancement with new methodologies required. Benefit to Guyana: Medium to high.
   - **Assistance in reporting obligations under the United Nations Convention to Combat Desertification (UNCCD):** MRVS data can be shared to contribute to Guyana's reporting requirement commitments under this convention. The enhancement to the MRVS will be medium. Benefit to Guyana: High
   - **Deployment of GFC-MRVS personnel to assist other agencies:** Through the MRVS development, GFC has built a relatively high level of personnel expertise in the development of systematic monitoring systems, and the analysis and application of satellite imagery. GFC would be assisting other agencies primarily through training and extension, and technical advisory support. GFC effort required could be at a low to medium level, and enhancements to the MRVS would be low to medium. Benefit to Guyana: High.
   - **Integrating Community MRV data:** GFC has worked with communities to build capacities and test the efficacy and value of community-based monitoring to the national MRVS. These data would provide at a minimum ground truthing or verification data to the national assessment. The potential enhancement to the

MRVS is considered at a medium to high level. Benefit to Guyana: High.

c. Options requiring a relatively high level of effort for implementation:
   - **National Forest Inventory (NFI)**: GFC is set to undertake a second NFI. This would complement the second phase of MRVS implementation and provide updated quantification of forest biomass across all forest types, ground truthing information to quantify forest area, and forest biomass data to improve emission factors according to the stratification used. High level enhancement of the MRVS, and information across government. Benefit to Guyana: High.
   - **Contribution to forest monitoring systems to support FLEGT VPA with the EU:** The MRVS would support the monitoring functions to help Guyana on its monitoring obligations under the FLEGT VPA[6] signed in 2018. This will require a medium level of enhancement of the MRVS, with tailored reports to meet the requirements. Benefit to Guyana: Medium.
   - **Identification of forest ecosystem services and valuation of natural capital:** Identified as a priority for Guyana, as forests are a dominant component of its natural capital. Substantial enhancement of the MRVS as new attributes for forest cover and non-forest areas would be required. Benefit to Guyana: High.

4.1 Guyana's Nationally Determined Contributions and the Sustainable Development Goals

Guyana's Nationally Determined Contributions outlined domestic conditional and unconditional contributions to the global effort to combat climate change. In its NDC, it is noted that "*...Guyana is offering to defer the pursuit of our historic natural resource exploitation-based economy, and offer our huge carbon stocks to the world if, in return, our sustainable human development needs and the cost of deferring business as usual are met by financial support from the international community in a predictable, just and equitable manner.*" The NDC commits Guyana to the utilisation of a combination of forest conservation and sustainable forest management and utilisation.

The findings from the MRVS were used to inform the design of the commitments of the NDC, by providing information on the effects of the direct drivers of forest change. The MRVS was further used to inform the contribution of the additional 2 million hectares for conservation through Guyana's National Protected Area System (NPAS) and other effective area-based conservation measures. This led to the NDC commitments focusing on the forest and energy sectors, which account for the majority of current and historic emissions.

The MRVS has been given prominence within the NDC as a conditional contribution, whereby Guyana has committed to the completion and maintenance of building the national MRVS, provided that adequate financial resources are available to do so. More importantly, the MRVS will play a major role as the tool by which the progress in implementation of the NDC activities, in terms of emissions reduction, will be measured and reported to the UNFCCC.

Guyana has committed to the implementation of the Sustainable Development Goals (SDGs) as part of the 2030 Agenda for Sustainable Development. Guyana is using a multi-stakeholder approach to implement programmes and projects to achieve the SDG targets and goals. The First Voluntary National Review was conducted in 2019, with the GFC being tasked with reporting on progress across multiple SDG indicators.

With the data already collected by the MRVS, the GFC was able to report on the progress of the indicators under SDG 15 – Protect, Restore and Promote Sustainable Use of Terrestrial Ecosystems, Sustainably Manage Forests, Combat Desertification, and Halt and Reverse Land Degradation and Halt Biodiversity Loss. Through the MRVS, Guyana was able to provide consistent information on the forest indicators for the review period.

The GFC will continue to use the MRVS data for future reporting on the progress of SDGs that relate to forest and climate change in Guyana.

4.2 Potential applications in the Private Sector

While the MRVS lends itself to applications in the public sector, specifically in relation to land use governance across predominantly public land in Guyana, the review also identified several opportunities for increased use of MRVS data by the private sector.

Most notably, the Forest Products Association (FPA) of Guyana, which represents over 200 forest industry members including small scale concession holders, contributed to the national review by observing its members have used the MRVS data on an ad hoc basis to look at activity within and around their concessions designed for timber harvesting activities, and to identify planned or unplanned activities and the impact on forest condition. Concessionaires have also used the imagery acquired for the MRV data to assist on occasion in planning the management of specific concessions; including access arrangements and managing proximity to other land uses.

The change in the MRV program to use of freely acquired Sentinel data will enable more ready access for this purpose, in contrast to the RapidEye imagery, which required licensing. The FPA has indicated that its members would like to make more use of this data than in the past; and is looking for the means and resources to access and analyse this data on a more regular basis.

5. A BROADER VISION

The recent national review highlighted the scope for the GFC and the MRVS to provide a broader range of applications that support land use policy and allocation and foster inter-agency collaboration. To do so, continued

---

[6] Guyana's Revised Intended Nationally Determined Contribution, 2015:https://www4.unfccc.int/sites/ndcstaging/PublishedDocuments/Guyana%20First/Guyana%27s%20revised%20NDC%20-%20Final.pdf

improvements in the System are required. The GFC and other program stakeholders have identified a set of key improvements that generally relate to making the System more efficient as well as enabling broader applications of the MRVS data.

One key improvement is the move to the use of freely available imagery. From the outset, the GFC has made incremental gains through the inclusion of new sources of satellite data and refinement of mapping and reporting processes. The GFC has tried and tested a range of satellite imagery data sources, both freely available and those paid for by the GFC. The data that has been paid for are governed by contract and licensing restrictions that prevent the GFC from sharing these images.

From 2016 to present, the GFC has been using freely available imagery (i.e. Sentinel and Landsat) to report on forest change. These images are not bound by licensing restrictions and can be shared with partners.

Furthermore, as the GFC continues to implement the MRVS, it will aim towards developing operational procedures for implementation of forest monitoring that can be sustained as a routine function of the GFC, with or without REDD+ financing. This will be done though the continued exploration of no cost and low-cost alternatives to allow for the MRVS to be more dynamic in terms of implementation for other functions and in the absence of a full cost budget.

Another means by which the GFC is working to make the MRVS more accessible and user–friendly is through the development of a centralised information platform. Using the MNR as the focal point for this information platform, the data and results generated under the MRVS can be applied to wider land use management and policy for planning as well as addressing the key drivers of deforestation and forest degradation in the country. The GFC has commenced the exploration of options to establish linkages within the MNR, whereby the data and results obtained through annual assessments under the MRVS can be made available through an online portal to inform land use planning and management. This will effectively strengthen forest management by assisting forest owners and managers to proactively make better decisions for the management of Guyana's forests, and consequently strengthen the national REDD+ programme.

The GFC continues to move the MRVS towards more near real-time monitoring. This has been identified as one of the major improvements to the System that would enable wider applications of the MRVS data. The GFC is exploring the use of Sentinel imagery as an option for continuous resources monitoring. The Sentinel mission provides global coverage of the Earth's land surface every 10 days with one satellite and every five days with two satellites, making the data a valuable resource for monitoring forest change. Increased use of satellite data will allow for an increase in temporal resolution, thus having access to more data to confirm an observation.

With the responsibilities for reporting on the progress of implementation of the NDC commitments and the SDG indicators specific to forests, the scope of the MRVS has thus been widened. Additional capacities would have to be built to allow for mainstreaming of reporting in some of these areas. This would further necessitate a more real time and continuous monitoring system.

Along with the MRVS data and results, the GFC continues to prepare non-technical communication materials such as videos, booklets and other printed materials on the on the MRVS design, results and impact. These materials are developed using non-technical and user-friendly language and are used to promote stakeholder awareness and understanding of the MRVS. These materials are made available online as well as are distributed at relevant national, regional and international fora.

ACKNOWLEDGEMENTS

The Guyana Forestry Commission (GFC) would like to thank the Government of the Kingdom of Norway for supporting Guyana's work on the MRVS development from 2010 to present. The GFC also recognises its close partners in the development of the MRV System: Indufor Asia Pacific and Winrock International.

REFERENCES

Government of the Co-operative Republic of Guyana, 2019. Guyana First Voluntary National Review of the Sustainable Development Goals, July 2019.

Government of the Co-operative Republic of Guyana, 2015. Guyana's Revised Intended Nationally Determined Contribution.

Government of Guyana and United Nations Environment Programme (UNEP), 2017. *Framework of the Guyana Green State Development Strategy and Financing Mechanisms*, March 2017.

Government of Guyana and the Food & Agriculture Organization (FAO), 2017. Mainstreaming Sustainable Land Development and Management in the Co-operative Republic of Guyana, August 2017.

Government of Guyana, 2014. The Reference Level for Guyana's REDD+ Program, December 2014.

Guyana Forestry Commission, 2018. Guyana REDD+ Monitoring Reporting & Verification System (MRVS) Year 7 Interim Measures Report, November 2018.

Guyana Forestry Commission, 2014. Terms of Reference for continuing to Develop Capacities for a National Forest Monitoring and Measurement, Reporting and Verification System to Support REDD+ Participation of Guyana, Roadmap Phase 1 Achievements, Evolving Requirements and Roadmap for Phase 2 Activities, September 19, 2014.

Guyana Forestry Commission, 2009. Terms of Reference for Developing Capacities for a national Monitoring, Reporting and Verification System to support REDD+ participation of Guyana Background, Capacity Assessment and Roadmap, November 13, 2009.

Government of Guyana, 2012. Joint Concept Note, March 31, 2012.

DATA IMBALANCE IN LANDSLIDE SUSCEPTIBILITY ZONATION: UNDER-SAMPLING FOR CLASS-IMBALANCE LEARNING

S. K. Gupta[1,*], Muskan Jhunjhunwalla[2], A. Bhardwaj[1], D. P. Shukla[1]

[1] School of Engineering, Indian Institute of Technology Mandi, India (*sharadgupta27@gmail.com)
[2] Department of Computer Science, National Institute of Technology Hamirpur, India

KEY WORDS: Landslide susceptibility zonation (LSZ), Imbalanced learning, Under-sampling, Artificial neural network (ANN), Fisher Discriminant Analysis (FDA), Logistic Regression(LR)

ABSTRACT:

Machine learning methods such as artificial n eural n etwork, s upport v ector m achine e tc. r equire a l arge a mount of training data, however, the number of landslide occurrences are limited in a study area. The limited number of landslides leads to a small number of positive class pixels in the training data. On contrary, the number of non-landslide pixels (negative class pixels) are enormous in numbers. This under-represented data and severe class distribution skew create a data imbalance for learning algorithms and suboptimal models, which are biased towards the majority class (non-landslide pixels) and have low performance on the minority class (landslide pixels).

In this work, we have used two algorithms namely EasyEnsemble and BalanceCascade for balancing the data. This balanced data is used with feature selection methods such as fisher d iscriminant a nalysis (FDA), l ogistic regression (LR) and artificial n eural n etwork(ANN) t o g enerate L SZ m aps T he r esults o f t he s tudy s how t hat A NN w ith balanced data has major improvements in preparation of susceptibility maps over imbalanced data, where as the LR method is ill-effected b y d ata b alacing a lgoritms. T he F DA d oes n ot s how s ignificant ch anges be tween ba lanced an d imbalanced data.

1. INTRODUCTION

Landslides are amongst the most devastating natural disasters, which cause billions of dollars in property damage and thousands of deaths every year worldwide. India has more than 15% of its land area prone to landslides, hence mapping of these areas for the presence/absence of landslides is of utmost importance. Numerous studies have contributed to reduce the damage from landslides through modelling and production of susceptibility maps (Roodposhti et al., 2019, Jhunjhunwalla et al., 2019, Gupta et al., 2018, Shukla et al., 2016). The susceptibility mapping can be a crucial tool for a wide range of end-users, from both private and public sectors, aimed at hazard mitigation purposes at both local and international levels. Landslide susceptibility zonation (a.k.a. LSZ) maps give approximate information about the occurrence of landslides. The susceptibility mapping requires data of various factors responsible for slope instability. In this work we have considered seven causative factors such as aspect, elevation, plan curvature, profile c urvature, slope, tangential curvature, topographic wetness index.

In recent years, there is an increasing application of machine learning techniques to complex real-world problems. The application ranges from daily life problems to nation's security, processing of the information to decision making support system and from micro-scale analysis of data to macro-scale discovery of knowledge (Stumpf et al., 2012). Most standard machine learning algorithms presume or expect balanced class distributions or equal misclassification c osts (He, G arcia, 2 009) a nd suffers data imbalance (Stumpf et al., 2014). Data imbalance refers to a scenario where majority classes dominate or overpower minority classes. In simple words, there is disproportionate distribution of observations in each class. It leads to the classifier b eing m ore b iased t owards the dominating class (Pradhan et al., 2014). Generally, the data is imbalanced when the class ratio is of the order of 1:100, 1:1000 or 1:10000 etc. i.e.number of points in one-class are 100 times or 1000 times or 10000 times less than that in another class (Liu et al., 2009). The imbalance level can be as high as 10^6. In this research, the class ratio is 1:300, i.e. for each landslide pixel we have more than 300 non-landslide pixels. When we use various machine-learning approaches for the generation of LSZ maps then the algorithms do not classify the landslide pixels correctly. Therefore, it is necessary to reduce the imbalance in the susceptibility mapping data. There are two major data balancing techniques, which are oversampling of a minority class and under-sampling of majority class (He, Garcia, 2009). The minority oversampling cannot be applied, as it will create false landslide pixels. We under-sample the majority class (i.e., non-landslide pixels) using Balance Cascade and Easy Ensemble methods. Some of the techniques used by various authors to overcome data imbalance are random oversampling and under sampling, informed under sampling, Synthetic sampling with data generation etc. (Haixiang et al., 2017, Stumpf et al., 2014, Stumpf et al., 2012, Galar et al., 2012, Chawla, 2010, Liu et al., 2009, He, Garcia, 2009).

This work aims at first, b alancing d ata u sing two different d ata b alancing t echniques i .e. EasyEnsemble and BalanceCascade. This balanced data is used for computing the weights using various methods methods such as fisher d iscriminant a nalysis (FDA), l ogistic regression (LR) and artificial n eural n etwork(ANN) t o g enerate LSZ maps. Furthermore visual analysis, statistical quantities, Heidke Skill Score (HSS) and Recall is used to assess the quality of susceptibility maps. Based on these observations, We can make assertions as how accuracy is affected when balancing techniques are applied to our data w.r.t. imbalanced data (Jhunjhunwalla et al., 2019).

2. STUDY AREA & DATA RESOURCES

A small part of Mandakini river basin of Garhwal Himalaya in Uttarakhand has been considered for the study as shown in Figure-1. Mandakini river originates from the Chorabari Glacier near Kedarnath in Uttarakhand, India. The study area covers about 275.60 sq. km area and lies between 30°19'00"N to 30°49'00"N latitude and 78°49'00"E to 79°21'13"E longitude. The study area falls in the Survey of India toposheet no. 53J and 53N. This region is highly prone to landslides during the monsoon season. This region has highly rugged topography, deep

*Corresponding author

gorges, high peaks where higher areas are mostly snow-covered forming the U-shaped wide valleys of glacial landscape. The study area is highly prone to landslide as every year many landslides occur in this area. In this research we have used 30m shuttle radar topography mission (SRTM) DEM, which was downloaded from EarthExplorer (www.earthexplorer.usgs.gov).

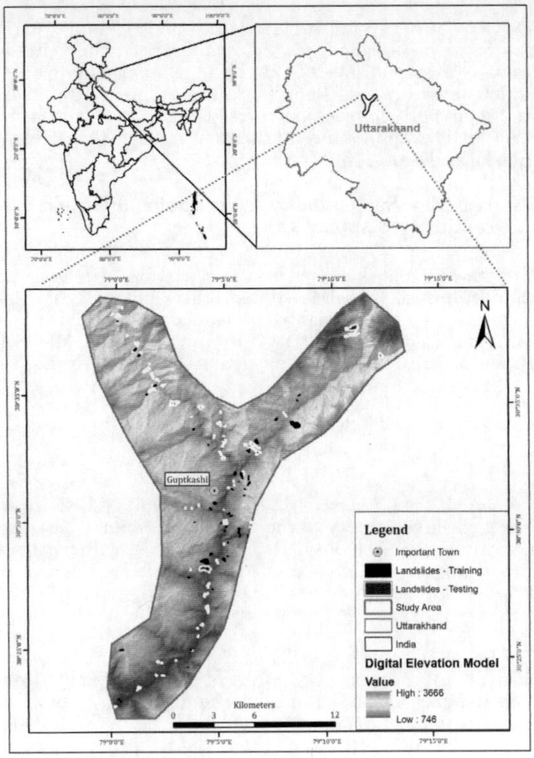

Figure 1. Map of study area (Source: Boundary of India and Uttarakhand provided by Survey of India, DEM provided by USGS)

3. DATA PREPARATION & METHODS

There are total 122 landslides, which have occurred between 2004 and 2017. The landslides occurred within 2004 - 2012 (46 landslides with 1203 pixels) are used for training of the models and 2013 - 2017 (76 landslides with 2743 pixels) are used for testing of the models. A landslide inventory is prepared to show landslide occurrences. The landslide inventory is binary in nature where one ('1') shows the occurrences of landslide and zero ('0') shows non-occurrences of landslide. The inventory was prepared manually from the past satellite images in GIS environment. This task is limited by the resolution of satellite images.

The landslide causative factors can be either categorical, which can be classified i nto fi nite nu mber of groups/classes (e.g. soil types in an area), or continuous (e.g. slope or elevation of the mountain). In this study, we have used only continuous data. Seven causative factors/layers i.e. aspect, slope, topographic wetness index, elevation, profile c urvature a nd p lan c urvature are considered for preparation of LSZ maps. These layers have been prepared from 30 m spatial resolution SRTM elevation model using ArcGIS and QGIS software. The size of all the layers is 1028 × 801 pixels. The layers are converted into ASCII format for training of the models. These layers have been transformed to column vector of size 823428× 1 and are stacked to generate a matrix of size 823428×7. In this work, we have used three algorithms i.e. FDA, LR and ANN for finding t he weights of various factors, which are further used for finding susceptibility index using weighted linear combinations.

3.1 Data Balancing Algorithms

The following data balancing algorithms have been applied to the initial experimental data set to obtain a balanced data set.

3.1.1 EasyEnsemble This method works on samples of majority class. It reduces the number of observations from majority class to make the data set balanced. This method is best to use when the data set is huge and reducing the number of training samples helps to improve run time and storage troubles. EasyEnsemble is an example of informed under sampling as it explore subsets of majority class by independent replacement of subsets (He, Garcia, 2009, Liu et al., 2009). This method is an unsupervised learning algorithm as it explores subsets of majority class by independent random sampling with replacement.

Algorithm - EasyEnsemble (Liu et al., 2009)

i Suppose, we have a majority class training data set M and minority class data set as N, where $|N| < |M|$. We can have T different randomly sampled subsets of majority class as $M_1, M_2, ..., M_T$ which can be represented as M_i, where $|M_i| << |M|$ & $|M_i| = |N|$.

ii Easy Ensemble uses different weak classifiers C_i, the number of classifiers being same as the number of subsets of majority class. The classifier C_i is trained by N minority class samples and M_i majority class samples. The decisions generated by weak classifiers C_i are combined using AdaBoost ensemble (H_i) using the Eq-1 (Liu et al., 2009)

$$H_i(x) = sgn(\sum_{j=1}^{I_i} \alpha_{i,j} C_{i,j}(x) - \beta_i) \quad (1)$$

where, I_i is the number of iterations, $C_{i,j}$ is the weak classifier/learner, $\alpha_{i,j}$ is the weights of $c_{i,j}$, β_i is the ensemble's threshold.

iii This uses boosting algorithm which is an repetitive technique to tune weights of an observation based on previous classification. This is useful in reducing the bias error in machine learning models. Then instead of voting the decisions of all the classifiers to select one, boosting is used to combine the results of all T classifiers and make an improved final decision (Liu et al., 2009). The output ensemble is obtained as Eq-2.

$$H(x) = sgn(\sum_{i=1}^{T} \sum_{j=1}^{I_i} \alpha_{i,j} C_{i,j}(x) - \sum_{i=1}^{T} \beta_i) \quad (2)$$

3.1.2 BalanceCascade In this method we create several subsets of data which are balanced, and a weak classifier is trained for each subset. This method reduces the majority class training sets at every step by removing all the examples that are correctly classified. I t i s d ifferent from Easy Ensemble in two steps. Primarily the weights are adjusted based on false positive rates that a classifier have to achieve. Second, the samples are removed which are correctly classified. This sequential dependence mainly focuses on reducing the redundant information in majority class.
In Balance Cascade, the training can finally b e stopped

when size of majority class($|M|$) is less than size of minority class($|N|$), as size of majority class is getting shrunk at every iteration. The main advantage of Balance Cascade is that it generate the restricted sample space to extract as much useful information possible (Liu et al., 2009).

Algorithm - BalanceCascade (Liu et al., 2009)

i Suppose, we have a majority class training data set M and minority class data set as N, where $|N| < |M|$. We can have T different randomly sampled subsets of majority class as $M_1, M_2, ..., M_T$ which can be represented as M_i, where $|M_i| << |M|$ & $|M_i| = |N|$.

ii Let the false positive rate f (the error rate of misclassifying majority class example to the minority class that H_i should achieve), be $\sqrt[T-1]{\frac{|N|}{|M|}}$.

iii BalanceCascade uses different weak classifiers C_i, the number of classifiers being same as the number of subsets of majority class. The classifier C_i is trained by N minority class samples and M_i majority class samples. The decisions generated by weak classifiers C_i are combined using AdaBoost ensemble (H_i) using the Eq-3 (Liu et al., 2009).

$$H_i(x) = sgn(\sum_{j=1}^{I_i} \alpha_{i,j} C_{i,j}(x) - \beta_i) \quad (3)$$

where, I_i is the number of iterations, $C_{i,j}$ is the weak classifier/learner, $\alpha_{i,j}$ is the weights of $c_{i,j}$, β_i is the ensemble's threshold.

iv We adjust β_i such that H_i's false positive rate is f.

v Similar to EasyEnsemble method, the output ensemble is obtained as Eq-4

$$H(x) = sgn(\sum_{i=1}^{T} \sum_{j=1}^{I_i} \alpha_{i,j} C_{i,j}(x) - \sum_{i=1}^{T} \beta_i) \quad (4)$$

3.2 Methods of Weights Assignment

The following filter and wrapper methods have been used for finding the weights of factors.

3.2.1 Fisher Discriminant Analysis (FDA): This method is used in pattern recognition, statistics and machine learning for finding the linear combination of features to distinguish two or more classes of events or objects (Jhunjhunwalla et al., 2019). In this method, all the factors/layers are projected in one dimension corresponding to landslide occurrences. A multiplicative factor is required for projection of the data. This multiplicative factor is used for giving weights to all the thematic layers (Gupta et al., 2018).

3.2.2 Logistic Regression (LR): It is a special case of linear regression and predicts the probability of the occur-rence of an event by using logit function (Gupta et al., 2018). In this method, the probability of presence or ab-sence of a binary outcome (1 = landslide and 0 = no land-slide) is modelled based on the values of predictor variables (Shukla et al., 2016). The independent variable can be in-terval or categorical while the dependent variable can be multinomial or binary. The LR coefficients are used for giving weights to all the factors.

3.2.3 Artificial Neural Network (ANN): ANN is a computational system, which is inspired by the human brain. The network consists of set of neurons, an input layer, an output layer and few hidden layers. The layers are interconnected and input data is passed to output layer by means of the hidden layers. There can be one or more hidden layers depending on the complexity of data. It is useful when a complex relationship exists between the data and the responses such as between landslide factors and landslide occurrences (Jhunjhunwalla et al., 2019).

ANN require several architectural and training parameters to be selected prior to analysis. The optimal number of hidden layers and the number of neurons per hidden layer are not known apriori. These parameters are empirically determined through rigorous experimentation and examination of different parameter settings (Blackard, Dean, 1999).

3.3 Landslide Susceptibility Index Computation and Susceptibility Mapping

The weights obtained in previous section are used for computation of landslide susceptibility index (LSI). LSI is calculated using Weighted linear combination method (Jhunjhunwalla et al., 2019, Gupta et al., 2018, Michael, Samanta, 2016). LSI can be calculated as given in Eq-5

$$LSI = \sum attributes * weights \quad (5)$$

LSI can be classified into five different zones i.e LSZ (from very high susceptibility to very low susceptibility) based on natural break in the data. Results obtained with/without data balancing are compared using statistical quantities and visual quality analysis. The methodology used in the current study has been shown in Figure-2.

The EasyEnsemble and BalanceCascade methods was applied on total 30 subsets, whihch were randomly taken from the non-landslide pixels by the algorithm in python. After the data balancing, the FDA, LR and ANN methods were applied on these 30 subsets of data, which resulted in 30 LSI images for each of the methods. The mean and median of these 30 LSI images for each method was taken to generate mean image and median image for all the methods.

3.4 Accuracy Assessment

3.4.1 Heidke Skill Score: The accuracy of the LSZ maps is measured using Heidke Skill Score (HSS). It is the measure of skill of prediction and lies between 0 and 1. HSS can be defined as given in the following Eq-6 (NDFD Verification Score Definitions, 2017, Hyvärinen, 2014).

$$HSS = \frac{NC - E}{T - E} \quad (6)$$

where NC is the number of correct predictions, i.e. number of times the prediction and observation match, E is the number of predictions expected to verify based on chance, i.e. incorrect predictions and T is the total number of observations.

3.4.2 Recall: Recall can be defined as the ratio of relevant instances (landslides) predicted correctly by the model to actual number of relevant instances. It is also known as "sensitivity". This gives us a measure of how accurately the model predicts with respect to the actual instances of that class. We can compute recall using the expression in Eq-7

$$Recall = \frac{TruePositive}{TruePositive + FalseNegative} \quad (7)$$

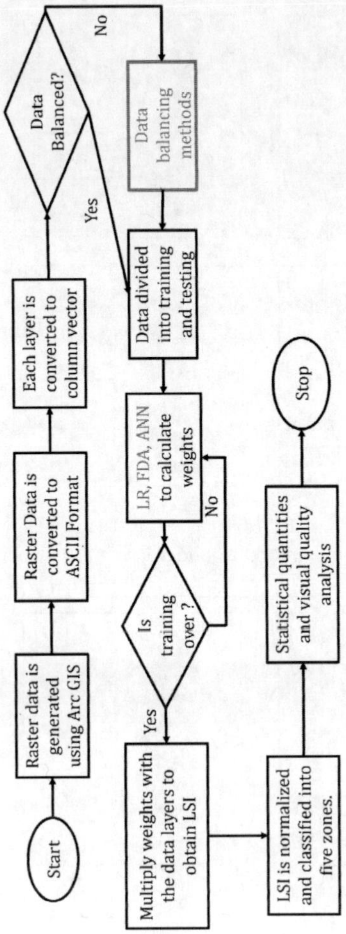

Figure 2. Flowchart of the study

4. RESULTS AND DISCUSSIONS

4.1 With Imbalanced Data

The weights and ranking of factors is important for preparation of landslide susceptibility zonation maps. Weights and ranking obtained by FDA, LR and ANN for all seven causative factors without using the data balancing algorithms are given in Table-1 (Jhunjhunwalla et al., 2019).

Table 1. Weights calculated from different models using imbalanced data

Causative Factors	LR	FDA	ANN
Aspect	-1.11	0.31	0.66
DEM	-3.72	-3.69	0.38
Plan Curvature	1.00	4.59	0.17
Profile Curvature	-2.35	0.80	0.05
Slope	5.04	3.36	0.36
Tangential Curvature	0.51	-3.49	-0.43
TWI	2.42	0.72	0.72

The weights obtained in Table-1 are multiplied with the corresponding causative factor layer to obtain the LSI values. The LSI values are normalized between 0 and 1 and classified into five different zones i.e. from very high susceptibility to very low susceptibility (as shown in Figure-3, 4(a) and 5(a)) based on natural break in the data. The mean and median of LSI values for all the methods (except ANN) is observed to be near 0.55 (refer Table-2) so 0.55 is set as a threshold for classification.

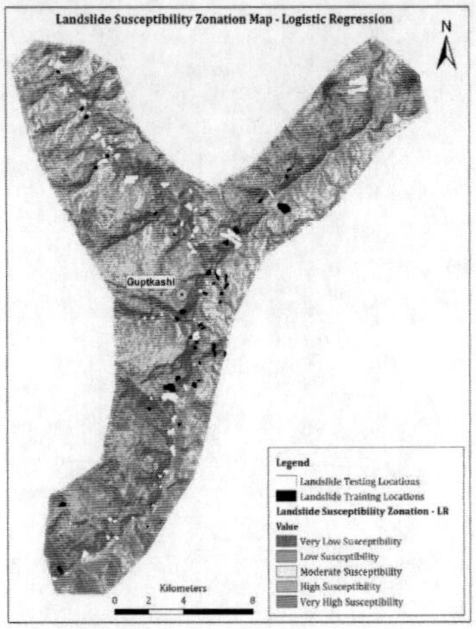

Figure 3. LSZ map obtained using weights from LR, without data balancing

The landslides with LSI greater than 0.55 are considered to be correctly classified and below 0.55 are considered to be falsely classified. As we can see from Table-2 that ANN has mean/median values significantly less than 0.55 and hence the susceptibility maps generated by ANN have maximun area lying in low or very low susceptibility zones. Hence, after data balancing, the mean/median value of LSI for ANN should increase significantly.

Table 2. Statistics (mean, median & standard deviation) for all the three methods using imbalanced data

Method	Mean	Median	Standard Derivation
LR	0.58	0.58	0.11
FDA	0.55	0.56	0.12
ANN	0.43	0.42	0.17

4.2 With Balanced Data

The imbalanced data is provided to EasyEnsemble and BalanceCascade algorithms and the data is balanced to match the size of minority class pixels. The feature selection algorithms are applied to balanced data and the weights are obtained. These weights are multiplied with the different layers and susceptibility index map are generated. The statistical quantities for LSI using all three methods are given in Table-3. The mean and median for ANN has increased significantly using balanced data, however, the values using LR has been reduced and mean/median is very small compared to that without balancing the data. The mean/median obtained for LSI values using FDA does not show significant changes. The susceptibility maps obtained using balanced data are shown in Figure-4(b) and 5(b).

The results of accuracy assessment using HSS and Recall are given in Table-4 & 5. The threshold value of 0.55 was considered for calculating the number of correctly classified landslides (for computation of HSS) and number of correctly classified pixels of those landslides (for computation of recall). It is evident from Table-4 & 5 that LSZ maps prepared using mean of weights in ANN with the data balanced using EasyEnsemble method gives the highest Score and Recall value and hence the highest accuracy of landslide zonation. Results obtained using FDA

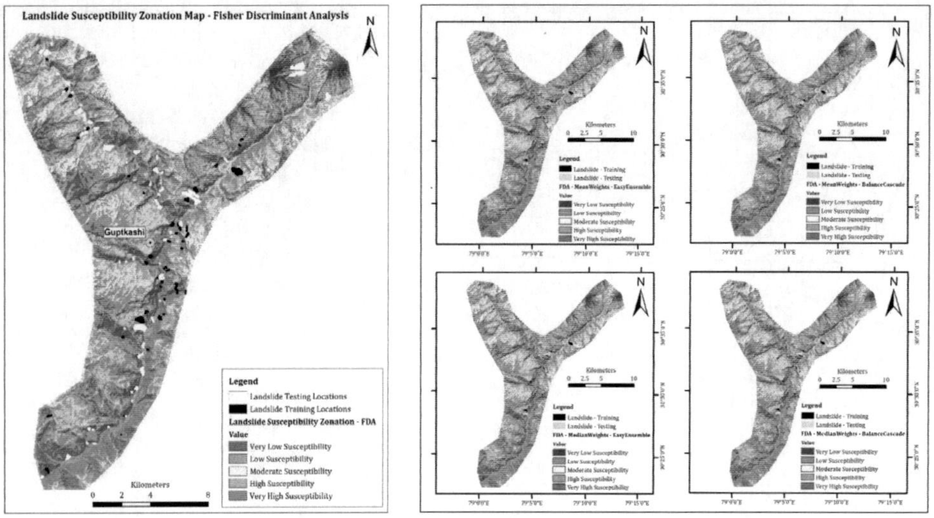

Figure 4. LSZ map obtained using weights from FDA, without data balancing (a) and with data balancing (b)

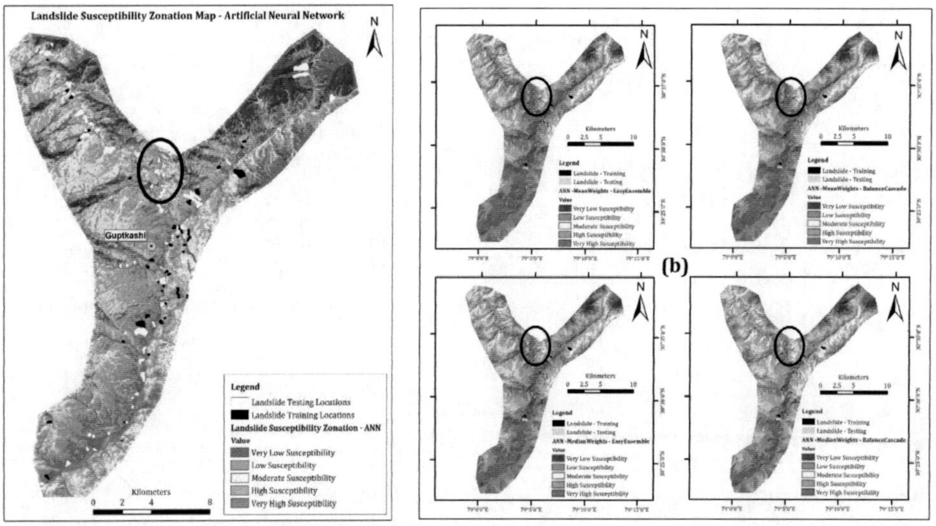

Figure 5. LSZ map obtained using weights from ANN, without data balancing (a) and with data balancing (b). The sample location with significant changes are marked with black circle.

Table 3. Statistics (mean, median and standard deviation) for all the three methods using balanced data

Balancing Method	Statistical Quantities	LR		FDA		ANN	
		Mean Image	Median Image	Mean Image	Median Image	Mean Image	Median Image
Easy Ensemble	Mean	0.2125	0.3834	0.5368	0.5558	0.6147	0.5822
	Median	0.2141	0.3870	0.5684	0.5604	0.6289	0.5948
	Standard Deviation	0.0348	0.0775	0.1140	0.1163	0.1442	0.1364
Balance Cascade	Mean	0.2337	0.2934	0.5568	0.5518	0.5926	0.5455
	Median	0.2358	0.2960	0.5614	0.5562	0.6064	0.5582
	Standard Deviation	0.0438	0.0565	0.1149	0.1151	0.1475	0.1268

also shows comparable HSS and recall values. The zonation of LSI obtained using LR can not prepared due to the smaller mean values. The natural break limits for different zones can not be applied on it. The results obtained in this study validates the fact that LR does not require the data balancing, given that we have sufficient samples in positive class (Crone, Finlay, 2012, King, Zeng, 2001). The FDA method may or may not not show major changes in the results with/without data balancing, which also agree with the results of few studies (Xue, Hall, 2015, Xue, Titterington, 2008).

5. CONCLUSIONS

Data balancing methods improves accuracy for machine learning based methods such as ANN, Support vector machine etc. The EasyEnsemble method coupled with mean of weights seems to overpredict the high susceptibility zones whereas BalanceCascade method with median of

Table 4. Accuracy assessment of FDA and ANN on balanced data using Heidke Skill Score for total 76 landslides from 2013 to 2017

Balancing Method	Methods	Correctly Classified Landslides (LSI >= 0.55)	Wrongly Classified Landslides (LSI <0.55)	HSS
EasyEnsemble	FDA Mean	73	3	0.9589
	FDA Median	70	6	0.9143
	ANN Mean	73	3	0.9589
	ANN Median	73	3	0.9589
BalanceCascade	FDA Mean	69	7	0.8986
	FDA Median	69	7	0.8986
	ANN Mean	72	4	0.9444
	ANN Median	68	8	0.8824

Table 5. Accuracy assessment of FDA and ANN on balanced data using Recall for total 2743 pixels in 76 landslides from 2013 to 2017

Balancing Method	Methods	Correctly Classified Pixels (LSI >= 0.55)	Wrongly Classified Pixels (LSI <0.55)	Recall
EasyEnsemble	FDA Mean	2192	551	0.7991
	FDA Median	2114	629	0.7707
	ANN Mean	2254	489	0.8219
	ANN Median	2198	545	0.8013
BalanceCascade	FDA Mean	2125	618	0.7747
	FDA Median	2074	669	0.7561
	ANN Mean	2205	538	0.8039
	ANN Median	2011	732	0.7331

weights generated by ANN gives most appropriate LSZ map based on visual analysis. Using statistical quantities, the LSI generated from both the data balancing methods show mean and median value greater or very near to 0.55, which also justifies the importance of data balancing before using ANN. The HSS and recall value shows the superiority of LSZ maps prepared using mean of weights in ANN with the data balanced using EasyEnsemble method. The Balanced data do not show good results with Logistic regression as the LR method is not able to model the underlying probability distribution. Balanced data does not effect FDA method and have approximately same accuracy as before. The landslide data is highly imbalanced in nature, hence balancing algorithms must be applied before preparation of LSZ maps using machine learning methods. However the data driven methods do not need balancing as seen from the results.

REFERENCES

Blackard, J. A., Dean, D. J., 1999. Comparative accuracies of artificial neural networks and discriminant analysis in predicting forest cover types from cartographic variables. Computers and Electronics in Agriculture, 24(3), 131–151.

Chawla, N. V., 2010. Data Mining for Imbalanced Datasets: An Overview. O. Maimon, L. Rokach (eds), Data Mining and Knowledge Discovery Handbook, 2nd edn, Springer US.

Crone, S. F., Finlay, S., 2012. Instance sampling in credit scoring: An empirical study of sample size and balancing. International Journal of Forecasting, 28(1), 224–238.

Galar, M., Fernandez, A., Barrenechea, E., Bustince, H., Herrera, F., 2012. A Review on Ensembles for the Class Imbalance Problem: Bagging-, Boosting-, and Hybrid-Based Approaches. IEEE Transactions on Systems, Man, and Cybernetics, Part C (Applications and Reviews), 42(4), 463–484.

Gupta, S. K., Shukla, D. P., Thakur, M., 2018. Selection of weightages for causative factors used in preparation of landslide susceptibility zonation (LSZ). Geomatics, Nat. Hazards Risk, 9(1), 471–487.

Haixiang, G., Yijing, L., Shang, J., Mingyun, G., Yuanyue, H., Bing, G., 2017. Learning from class-imbalanced data: Review of methods and applications. Expert Systems with Applications, 73, 220–239.

He, H., Garcia, E. A., 2009. Learning from Imbalanced Data. IEEE Transactions on Knowledge and Data Engineering, 21(9), 1263-1284.

Hyvärinen, O., 2014. A Probabilistic Derivation of Heidke Skill Score. Weather and Forecasting, 29(1), 177-181.

Jhunjhunwalla, M., Gupta, S. K., Shukla, D. P., 2019. Landslide Susceptibility Zonation (LSZ) Using Machine Learning Approach for DEM Derived Continuous Dataset. K. Santosh, R. S. Hegadi (eds), Recent Trends on Image Processing and Pattern Recognition, Springer Singapore, 505–519.

King, G., Zeng, L., 2001. Logistic Regression in Rare Events Data. Political Analysis, 9(2), 137–163.

Liu, X.-Y., Wu, J., Zhou, Z.-h., 2009. Exploratory Undersampling for Class-Imbalance Learning. IEEE Transactions on Systems, Man and Cybernetics, 39(2), 539–550.

Michael, E. A., Samanta, S., 2016. Landslide vulnerability mapping (LVM) using weighted linear combination (WLC) model through remote sensing and GIS techniques. Model. Earth Syst. Environ., 2(2), 1–15.

NDFD Verification Score Definitions, 2017.

Pradhan, B., Abokharima, M. H., Jebur, M. N., Tehrany, M. S., 2014. Land subsidence susceptibility mapping at Kinta Valley (Malaysia) using the evidential belief function model in GIS. Nat. Hazards, 73(2), 1019–1042.

Roodposhti, M. S., Aryal, J., Pradhan, B., 2019. A novel rule-based approach in mapping landslide susceptibility. Sensors (Switzerland), 19(10), 2274.

Shukla, D. P., Gupta, S., Dubey, C. S., Thakur, M., 2016. Geo-spatial Technology for Landslide Hazard Zonation and Prediction. M. Marghany (ed.), Environmental Applications of Remote Sensing, InTech, Rijeka, Coratia, 281–308.

Stumpf, A., Lachiche, N., Kerle, N., Malet, J.-P., Puissant, A., 2012. Adaptive spatial sampling with active random forest for object-oriented landslide mapping. 2012 IEEE International Geoscience and Remote Sensing Symposium, IEEE, 87–90.

Stumpf, A., Lachiche, N., Malet, J.-P., Kerle, N., Puissant, A., 2014. Active Learning in the Spatial Domain for Remote Sensing Image Classification. IEEE Transactions on Geoscience and Remote Sensing, 52(5), 2492–2507.

Xue, J. H., Hall, P., 2015. Why does rebalancing class-unbalanced data improve AUC for linear discriminant analysis? IEEE Transactions on Pattern Analysis and Machine Intelligence, 37(5), 1109–1112. http://ieeexplore.ieee.org/document/6906278/.

Xue, J.-H., Titterington, D. M., 2008. Do unbalanced data have a negative effect on LDA? Pattern Recognition, 41(5), 1558–1571.

MANAGING URBAN SPRAWL USING REMOTE SENSING AND GIS

Krishnaveni K.S. [1], Dr.Anilkumar P.P. [2]
[1] Research Scholar, Department of Architecture & Planning, NIT Calicut, Kerala, India- kveni07@gmail.com
[2] Professor & Head, Department of Architecture & Planning, NIT Calicut, Kerala, India- ppa@nitc.ac.in

KEYWORDS: Urban sprawl, remote sensing, GIS, built-up index, Shannon's entropy, sustainability

ABSTRACT:

Indian cities, like several other developing cities around the world, are urbanizing at an alarming rate. This unprecedented and uncontrolled urbanization may result in urban sprawl, which is characterized by low-density impervious surfaces, often clumsy, extends along the fringes of metropolitan areas with unbelievable pace, disperse, auto-dependent with environmentally and socially impacting characteristics. The ill-effects of urban sprawl in developing countries scenario is a bit complicated compared to that of developed countries because of uncontrolled population growth and haphazard urbanization. This paper attempts to investigate the capabilities of remote sensing and GIS techniques in understanding the urban sprawl phenomenon in a better way compared to time-consuming conventional methods. An overview of the enormous potential of remote sensing and GIS techniques in mapping and monitoring the Spatio-temporal patterns urban sprawl is dealt with here. The spatial pattern and dynamics of the urban sprawl of Kozhikode Metropolitan Area (KMA, Kerala, India) during the period from 1991 to 2018 using the integrated approach of remote sensing and GIS are attempted here. Index derived Built-up Index (IDBI) which is a thematic index-based index (combination of Normalized Difference Built-up Index (NDBI), Modified Normalized Difference Water Index (MNDWI) and Soil Adjusted Vegetation Index (SAVI)) is used for the rapid and automated extraction of built-up features from the time series satellite imageries. The extracted built-up areas of each year are then used for Shannon's entropy calculations, which is a method for the quantification of urban sprawl. The results of IDBI and Shannon's entropy analysis highlight the fact that there occurs an alarming increase in the built-up areal extent from 1991 to 2018. The urban planning authorities can make use of these techniques of built-up area extraction and urban sprawl analysis for effective city planning and sprawl control.

1. INTRODUCTION

The 2019 review of the World Population Prospects by the United Nations Population Division sets out that by the year 2030, the World's urban population will increase to 4.9 billion, whereas the World's rural population is projected to reduce by some 28 million between 2005 and 2030. Indeed, by 2050, very few nations are anticipated to have higher rural shares than urban ones. More than half of the world's population currently lives in metropolitan regions and is projected to reach two-thirds by 2050. (Web-1). Therefore, at a global scale, all potential population rises will be focused more in towns and cities, majority of them will be in developing countries. This massive increase in the urban population, together with an unprecedented rate of urbanization, may cause unplanned, uncontrolled or unrestricted growth/spreading of urban areas into the peripheries may be recognized as urban sprawl.

The notion of urban sprawl has been interpreted and defined in countless ways, and as a consequence, there has been no agreement on the definition of urban sprawl. However, a widely accepted definition is uncontrolled, unlawful, unplanned, and uncoordinated low-density single-use development facilitates that does not provide for a functional and appealing combination of land use mixes and/or is not functionally connected to neighboring land-uses and which variously appears as leapfrog, or isolated development, ribbon or strip with scattered residential neighbourhoods and commercial strip development, with automobile dominance which are aesthetically displeasing in nature (Nelson, 1995,Hiller, 2013). There is an enormous number of journals and publications that explore the adverse effects of urban sprawl on metropolises. Freilich (1999) identified seven adverse sprawl effects, including impacts on society, impacts on housing, impacts on jobs, fiscal impacts, political impacts, impacts on transport, impacts on agriculture, and open space.

It is imperative to discover the appropriate methods to analyze the spatial features, and unique mechanism of urban sprawl within the Indian context as the consequences of urban sprawl in India are on the rise. The measurement, mapping, and tracking of urban sprawl are essential for public authorities and planners in any region to accomplish adequate management of urban sprawling. The process of urban sprawl measurement has been a hot topic of research among the academic fraternity. In this context, remote sensing and geographical information system (GIS) methods are outlined in this study to monitor and evaluate urban sprawl. To monitor dynamic changes in urban sprawl, the built-up regions were extracted from Landsat classified images of four distinct periods. Across the world, researchers are focusing on various multi-dimensional indicators of urban sprawl by employing remote sensing and GIS-based analysis or descriptive statistical analysis. (Ewing et al., 2002) (Frenkel, M. A. 2005) (Galster, G. et al. 2001) (Barnes et al. 2001) (Song et al. 2004). This study attempts to find an easy way to extract the built-up features using an index derived built-up index through the techniques of remote sensing and GIS.

Among the various indicators of urban sprawl, this study considers the built-up features as an indicator of urban sprawl and attempted to map all the built-up features within the study area using a completely automated method of built-up extraction known as Index-derived built-up index (IDBI) which was proposed by Xu in 2008. The three thematic indices used for the generation of IDBI are the Soil Adjusted Vegetation Index (SAVI, the Normalized Difference Water Index (NDBI),), and the Modified Normalized Difference Water Index (MNDWI). IDBI helps to extract built-up features very quickly by effectively eliminating background noise caused by water, soil, etc.

2. URBAN SPRAWL: CAUSES AND CONSEQUENCES

Although there are various measures and dimensions for the phenomenon of urban sprawl, there is no common universally accepted definition for the process of urban sprawl. It should always be better to describe rather than to define the concept of urban sprawl. (Galster et al. 2000, Wilson et al.). Generally, urban sprawl is characterized by an example of a sprawling city like Los Angeles which symbolizes sprawl characteristics or as an aesthetic condition or as the cause of an unwanted externality or as selected patterns of land use/ land development or even defined as a process of development over time and space (Galster, 2010). A widely accepted definition of urban sprawl is low-density, dispersed, decentralized, polycentric sub-urban development, which is mainly catalyzed by low-cost auto travel and technological innovations (Ewing, 1997). The notable characteristics of urban sprawl are discontinuous development with low-density residential patterns, fragmentation of land uses, transportation dominated by private vehicles, excessive land consumption, and social segregation with aesthetically displeasing type of development. Based on its spatial form, sprawl can be classified as linear strip\ribbon development (occur as a result of rail/road development), leapfrog development, single-use development, and continuous low-density sprawl and single-use development. (Batty et al., 2003).

There are a large number of factors that add fuel to the process of urban sprawl. They are urban expansion coupled with population growth, rural-urban migration, economic growth, industrialization, lower land value in the peripheries, lower tax rates in the fringes, lack of affordable housing at the city core, physical geography, emergence of Information and Communication Systems (ICT) and other technological innovations, desire for living in greener surroundings, lack of spatial planning and failure to enforce planning etc. These causative factors vary from place to place. In a developed country, people see sprawl as a choice, whereas in a developing country, sprawl is mainly seen as a result of necessity (Kew, 2013).

The damaging dimensions of urban sprawl are very much critical, and it is inevitable to tame it for the sustainability of our environment. Urban sprawl consequences can be broadly categorized into three groups; social/community impacts, economic impacts, and environmental/ecological impacts. Table 1 summarizes the detrimental impacts arising from uncontrolled urban sprawl. The measurement, mapping, and surveillance of the extent of urban sprawl are exceptionally crucial for government officials and planners in any region to manage the undesired effects resulting from urban expansions.

Table 1: Impacts of uncontrolled Urban Sprawl

Community Impacts	Economic Impacts	Ecological Impacts
1.Segregated residential development 2.An overall reduction in quality of life 3.Less social interactions & more prolonged time to commute 4.A large proportion of single households with intensive resources utilization 5.Increased health issues (Eg. Asthma because of air pollution, Insomnia and other health effects because of noise pollution and urban heat island effect) 6.Increased obesity, stress and physical activity 7. spatial and functional demarcation of areas where people live, work, and play/leisure.	1.Economic discrimination 2. Increase in transportation cost for commuting to households 3. Higher cost because of traffic congestion 4.Higher energy and land consumption per person 5.A rise in public service costs and taxes 6.Higher costs for the substitution of various ecosystem related services with technology 7. Higher cost for extending urban infrastructure into newly developed areas	1. Heavy dependence on automobiles resulted in higher air pollution and depletion of fossil fuels. 2.Water pollution and hydrological alterations of watersheds, surface watercourses etc. because of an increase in impervious surfaces 3.Higher noise pollution 4. Loss of arable soil 4.Disruption of critical natural habitats and environmentally sensitive areas like wetlands, sacred groves, wildlife corridors, etc. 5.Fragmentation of landscape and loss of existing green infrastructure 6. Reduced resilience of ecosystem 7.Loss of agricultural land, built heritage, natural open spaces, and loss of recreational areas. 8.Climate alteration 9. Reduces landscape attractiveness

3. URBAN SPRAWL MANAGEMENT USING REMOTE SENSING NAD GIS

In order to understand the rapid changes that are happening to our cities, mapping and monitoring of the urban evolution at spatial and temporal scales are inevitable. Uncontrolled urbanization, together with unprecedented population growth, results in detrimental land-use land-cover changes, which can be considered as a threat to urban sustainability. Remote sensing and GIS techniques are innovative tools widely used to track and monitor urban growth, changes in land use land cover features and urban transition. (Bhatta, 2012).

Remote sensing is the art and science of obtaining information about an object, an area or phenomenon, without direct contact with the object or target under consideration (Lillesand & Keifer). The electromagnetic radiation is used as a critical information source in remote sensing. To study the properties of a target, the technique of remote sensing makes use of the energy that is emitted/ reflected/ absorbed by the destination under consideration. A Geographical Information System (GIS)is a "computer-based system that provides the following four sets of capabilities to handle georeferenced data: input, data management (storage and retrieval), manipulation and analysis, output" (Aronoff, 1989). GIS is composed of a set of tools for capturing, storing, retrieving, analyzing and displaying information which are spatially referenced to the earth. GIS offers forward mapping features like a database management system for displaying geographical information, as well as backward mapping functions for carrying out "querying." In addition to the quick data production capabilities, GIS technologies are an immense source of help for the policy implementers.

With the integrated use of remote sensing and GIS techniques, the urban sprawl phenomenon can be mapped effectively and monitored regularly. Recent advances in satellite remote sensing sensor capabilities help to create urban maps with even tiny details and help planners and policymakers to get a more in-depth understanding of urban expansion, urban sprawl, and urban land management issues (NASA 2001). In the beginning, remote sensing was dominated by aerial photography. The first launch of the Landsat satellite in 1972 paved the way for satellite based remote sensing techniques. Later, the emergence of very high-resolution satellites paved the way for a vast array of urban applications with better results. Urban remote sensing with space-borne sensors is, in fact, an emerging area of research for the remote sensing community, urban planners, etc. Integrated remote sensing and GIS technologies can contribute much to the generation of digital maps with detailed data on urban growth and the evolution of urban sprawl of a city or region. The magnitude and level of urban sprawl, the various forms of sprawl and its existence, etc can be easily identified using advanced remote sensing and GIS techniques.

Remote sensing and GIS applications will be immensely beneficial to developing countries where there are restricted funding and little data on urban land characteristics and their modifications over time. With these techniques, developing countries can establish an urban spatial information system by integrating conventional data (hard copy maps and reports), ground measurements, and digital data (satellite images, digital maps, attribute data, etc.). The valuable information gained from remote sensing can contribute much in the domains of monitoring, quantifying, modeling and ultimately predicting the phenomenon of urban sprawl. The implications of combined remote sensing and GIS technologies are effective and timely tracking of urban sprawl along with a better understanding of various benefits applied in the areas of urban planning and management.

The extensive availability of freely downloadable satellite imageries and other spatial data, together with numerous user-friendly geospatial tools, resulted in a large number of applications relevant urban sprawl studies. Typical research of urban sprawl includes an effort to quantify the built-up features and quantity of paved surface in a region using remotely sensed information from satellite imageries. In general, the built-up area and the amount of paved surface are considered as indicators of urban sprawl. The changes that are happening to the built-up and paved features will be used as a proxy to quantify the nature and extent of urban sprawl. In other terms, the mapping and monitoring of urban sprawl using geospatial tools will enable us to visualize how and where the city is growing in an uncontrolled and unplanned manner. This will help the concerned local planning authorities to get an easy understanding of the development that can be sanctioned or need to be avoided in a particular region. Additionally, the identification and preservation of pristine natural environmental resources, natural open space, environmentally sensitive areas like wetlands, sacred grooves, etc. will help in the reduction of urban sprawl and thereby enhance urban sustainability.

The advantages of remote sensing applications in urban sprawl mapping and monitoring are listed below.

- Possibility of incorporating multiple sensor data with the same or different spatial, spectral or temporal resolutions.
- Extensive areal coverage and quick data acquisition
- Availability of temporal datasets/repetitive data acquisition
- Can process the data digitally and analyze it efficiently compared to conventional methods
- It can integrate with GIS/GNSS, database management systems, etc.
- Cost-effective
- Technologically sound
- It provides an accurate and permanent recording of spatial data and processes.

Despite the countless benefits of remote sensing and GIS applications in urban sprawl studies, there are some challenges/ limitations associated with it. Poor atmospheric conditions become a real hindrance to the data availability of cloud-free images of specific scenes and time period. High spatial resolution images are often costly, and there are chances for the occurrence of spectral variability within one class. In many cases, the shadows of trees and buildings pose a greater threat to classification accuracies by reducing the class separability. There will be spectral ambiguity arising from mixed pixels between urban land use and land covers such as bare soil and dry mud with impermeable surfaces, vegetation with forest, etc. The complexity and dynamics of the urban landscape is also a challenge for urban geospatial applications. It is a cumbersome task to integrate multi-source data especially in the case of temporal studies on urban dynamics. There are chances for errors while integrating satellite data with other supplementary data like census data. Besides, the presence of various built-up features in a small region results in the generalization of pixels, ultimately contributing to errors in classification. It is, therefore, necessary to create new strategies and further refine prevailing tools and techniques to take complete advantage of the enormous potential of remote sensing data in the urban setting.

4. CASE STUDY: URBAN SPRAWL QUANTIFICATION USING REMOTE SENSING AND GIS

The unprecedented population growth, together with uncontrolled urbanization, force the cities to expand further to accommodate the urban growth. The urban expansion towards the periphery is regarded as urban sprawl. An rise in the built-up features especially in the urban fringes can be considered as common characteristic feature of urban sprawl. The built-up area is a prospective and comparatively precise parameter for urban sprawl measurement. The dynamics of built-area change over the years give vital information regarding the nature, extent, and development of urban sprawl. In other words, urban sprawl is linked to the growth of built-up features within a region. In this study, a thematic index-based built-up index known as the index-derived built-up index (IDBI) is used to extract the built-up feature class quickly and efficiently. This index is composed of three indices viz., the Normalized Difference Built-Up Index (NDBI), the Soil Adjusted Vegetation Index (SAVI) and the Modified Normalized Difference Water Index (MNDWI). The objectives of the study are to extract the built-up areas and quantify its areal extend from 1991 to 2000 and 2018 using IDBI (Index derived built-up index) and to measure the degree of dispersion using Shannon's entropy. The detailed study of urban sprawl using IDBI is discussed with the built-up area changes from 1991 to 2018 in the subsequent sections.

4.1 Study Area

Kozhikode Urban Agglomeration (KUA) in Calicut district of Kerala state, India is chosen as the study area for smart growth suitability analysis using GIS. As it is composed of contiguous counties, it is better to take urban agglomeration as an appropriate unit of analysis. It includes Kozhikode City Corporation, Municipalities like Feroke, Koduvally, Payyoli, Koyilandy, Ramanattukara, and Vadakara, together with some census towns and outgrowths as per Census 2011. Figure 1 shows the map of the study area.

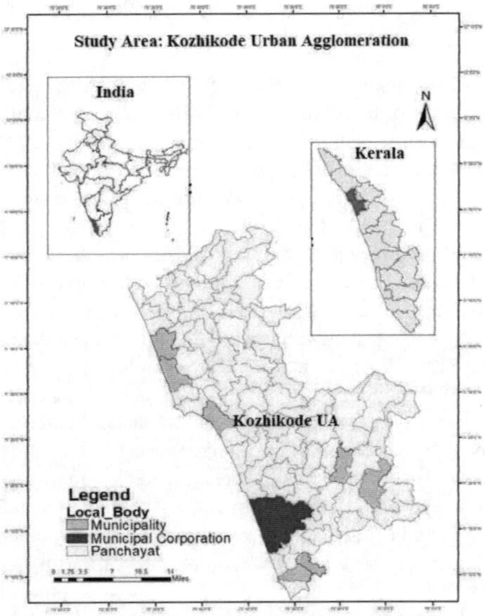

Figure 1: Map of Study Area

4.2 Methodology

4.2.1. Data and pre-processing

The satellite data used for this study were downloaded freely from the USGS Earth Explorer website(Web-2). Landsat temporal images of the years 1991, 2000 and 2018 were used for this study. Landsat Thematic Mapper (TM) of 1991, Landsat Enhanced Thematic Mapper (ETM+) of 2000, and Landsat 8 of 2018 were made use of. (Table 1). To minimize the errors from temporal variation, only the imageries of the same time period (February-March) were considered in this study. The downloaded imageries were in GeoTIFF format, and data were initially pre-processed and finally cropped to the extent of the study area (Figure 2). Erdas imagine software is used for all tasks related to image processing and index derivation, whereas mapping and display related works are carried out using ArcGIS. Besides, the Google earth application is also made use in this study for various applications related to accuracy assessment.

Figure 2: Landsat images of the study area for various years

Table 1. Details of Satellite imageries used

Date	Sensor	Path/row	Bands with resolution (in meters)
1991 March 3	Landsat 5-TM	145/052	Blue (30), Green(30m), Red (30), NIR (30), SWIR1(30), TIR (120(30)), SWIR2(30)
2000 February 22	Landsat 7 ETM+	145/052	Blue (30), Green (30), Red (30), NIR (30), SWIR1(30), TIR (60(30)), SWIR2(30), PAN (15)
2018 March 10	Landsat 8	145/052	Coastal aerosol (30), Blue (30), Green (30), Red (30), NIR (30), SWIR1(30), SWIR2(30), PAN (15), Cirrus (30), TIRS1(100), TIRS-II (100)

4.2.2. Automated built-up area extraction using Index derived built-up index

The methodology adopted for the rapid extraction of built-areas areas from time series of Landsat satellite imageries is given in Figure 3. From the website of USGS Earth explorer satellite imageries of interest were freely downloaded. It includes Landsat 5-TM of 1991, Landsat 7 ETM+ OF 2000, and Landsat 8 of 2018. Then all these images were clipped/ subsetted by using the shapefile of Kozhikode Urban Agglomeration as input for the extent of the study area to be clipped. The data were initially pre-processed and rectified after subsetting it to the required areal extent of Kozhikode Urban Agglomeration.

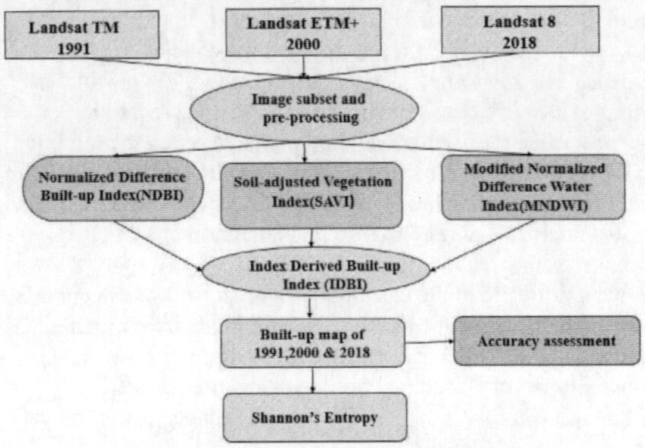

Figure 3: Flow chart of the methodology

a. Normalized Difference Built-up Index (NDBI)

$$NDBI = \frac{MIR - NIR}{MIR + NIR}$$

Proposed by Zha et al. (2003). In the Mid Infrared (MIR) region, an increased spectral reflectance is shown by the built-up features compared to the Near Infrared (NIR) region of the electromagnetic spectrum. There are chances for the mixing of vegetation noise with the built-up land information. In Landsat TM and ETM+, band 4 denotes NIR, and band 5 denotes MIR, whereas, in Landsat 8, band5 and 6 denote NIR and MIR, respectively.

b. Soil Adjusted Vegetation Index (SAVI)

$$SAVI = \frac{(NIR - Red)(1 + L)}{(NIR + Red + L)}$$

Where L is the soil adjustment factor (Normally value=0.5). In Landsat TM and ETM+, band 3 denotes Red, and band 4 denotes NIR, whereas, in Landsat 8, band4 and 5 denote Red and NIR, respectively. SAVI is preferred over NDVI because it is more sensitive in vegetation detection even in urban areas with less vegetation coverage. A study conducted by Ray (2006) on vegetation indices concluded that NDVI shows better performance of vegetation mapping in areas with plant cover 30% or more, whereas SAVI performed with better accuracies in areas with plant cover as low as 15%. SAVI is adopted here because it is more suitable for an urban area.

c. Modified Normalized Difference Water Index (MNDWI)

$$MNDWI = \frac{(Green - MIR)}{(Green + MIR)}$$

Bands 2 and 5 denote Green and MIR respectively in Landsat TM and ETM+ imageries, whereas bands 3 and 6 denote Green and MIR in Landsat 8 Imagery. MNDWI is the modified version of the Normalized Difference Water Index (NDWI) developed by McFeeters in the year 1996. NDWI is obtained by substituting middle infrared (MIR) with near infra-red(NIR) band. MIR is more sensitive to water information compared to NIR, and hence MNDWI offers better performance in extracting water features compared to NDWI. It helps in the enhancement of open water features while effectively suppressing noises from built-up land, vegetation, and soil features.

d. Index derived Built-up Index (IDBI)

$$IDBI = \frac{[NDBI - \frac{SAVI + MNDWI}{2}]}{[NDBI + \frac{SAVI + MNDWI}{2}]}$$

IDBI is an index derived from three indices, namely, NDBI, MNDWI, and SAVI. In this index, the spectral clusters of the built-up, water body, and vegetation are well isolated, and there is a significant reduction in redundancy between original imagery and thematic band imageries. Its value ranges from -1 to +1 wherein, built-up features exhibit positive values, and suppressed vegetation or water noise should have zero to negative values. A notable feature of IDBI is that the image can be used as itself without geometric corrections for further computations.

4.3 Results and Discussions

4.3.1. Built-up Area Extraction Using Index –based Built-up Index (IDBI)

Here, an increase in built-up areas is considered as a proxy for the rise in urban sprawl. A thematic index-based index called Index Derived Built-up Index (IDBI) is used to perform the automatic extraction of the built-up classes from the landsat imageries of 1991, 2000 and 2018. The built-up features were displayed in a light grey to white colour in the IDBI image whereas the vegetation and water features were substantially obscured and outlined in dark grey to black shade. To extract the urban built-up classes of 1991, 2000, and 2018, threshold values that best discriminate built-up features were manually determined. In general, water and vegetation classes showed negative IDBI values whereas built-up features exhibited positive values. Pixels with values above the threshold value have been categorized as built-up land and given a value of 1, whereas pixels with values less than or equal to the threshold are labelled as non-built up features and assigned a value equal to zero. The resulting image is, therefore, a binary image with value one as extracted built-up information and zero as other classes. Figure 4 shows the final map of built-up areas of 1991,2000 and 2018.

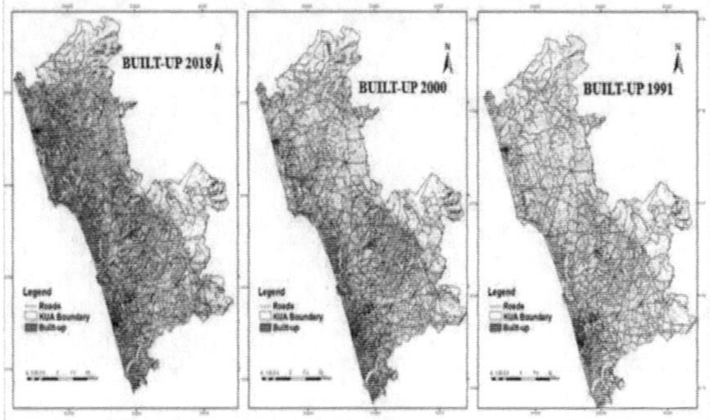

Figure 4: Built-up area map of 2018, 2000 and 1991 derived from IDBI

4.3.2. Accuracy Assessment

For accuracy assessment, field data points were selected, and the spatial locations of these points were identified with the help of a handheld GPS. About 610 ground truth points were selected. About 310 points are selected for built-up features, and 300 points for non-built-up features were collected using handheld GPS. The ground sample points are overlaid on each of the maps to determine the classification accuracies. For this, all the built-up features raster maps were converted into a vector format. Then intersection is carried out between sample data points and all built-up maps. The accuracy assessment of built-up features derived from IDBI is shown in Table 2. The analysis shows that with a kappa coefficient of 0.812, the overall accuracy is 90.65%.

Table 2: Accuracy assessment of IDBI

	Built-up land	Non-built-up land	Total	User's accuracy
Built-up land	285	32	317	89.90%
Non-built up land	25	268	293	91.46%
Total	310	300	610	
Producer's accuracy	91.93%	89.33%		
Overall accuracy	90.65%		kappa	0.812

The total areal extent of Kozhikode Urban Agglomeration is 1803km^2. After converting the raster to polygon, the areal extent of built-up classes for different years is obtained. For the year 1991, the built-up areal extent of KUA was about 324.54 km^2, which was 18% of its total land area. In the year 2000, the built-up areal extent increased to 522.87km^2 (29 % of the total geographical area of KMA). When it reaches 2018, the built-up areas increased to an areal extent of 829.38 km^2 (46% of KMA's total area). Figure 5 shows the built-up area change for the years 1991, 2000 and 2018.

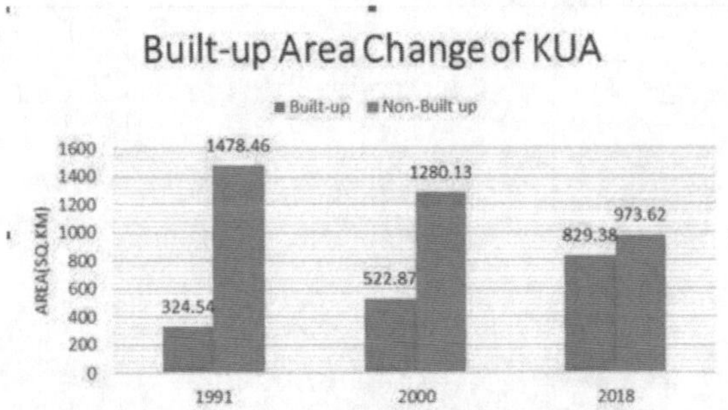

Figure 5: Change in the areal extent of built-up and non-built-up classes for 1991, 2000 and 2018.

From the above graph, it is clear that there is an alarming increase in the built-up area from 1991 to 2018, which indirectly highlights that there is severe urban sprawl/urban growth in the study area. The method of extracting built-up features from various indices is found to be very effective and efficient compared to traditional methods of built-up land extraction like supervised image classification which is tedious and time-consuming.

4.3.2. Urban Sprawl Analysis Using Shannon's Entropy

Rudolph Clausius, in 1867, proposed the concept of entropy from the second law of Thermodynamics. (Claudius,1867). In the year 1948, Shannon has developed the concept of entropy as a measure of randomness and disorganization. In view of the complexities of urban growth, sprawl and urban evolution, entropy value must remain within the range set by the minimum value below which the system becomes vulnerable and unstable, and the maximum value over which the system becomes unsustainable. (Cabral, 2013). The integration of GIS and remote sensing technologies allows to effectively apply the entropy principle in the calculation of urban sprawl. Shannon's entropy measure is one of the most commonly used urban sprawl indexes to describe the degree of dispersion or spatial concentration of a specific variable in a particular area. (Bhatta(2009), Li (2004)).The formula of Shannon's entropy (H_n) is

$$H_n = -\sum_{i=1}^{n} P_i log_e(P_i)$$

Where P_i denotes probability of a phenomenon (variable) occurring in the ith zone ($P_i = \frac{x_i}{\Sigma x_i}$, where x_i is the observed value of the variable in the ith zone), and n denotes the total number of zones/ grids. The range of Shannon's entropy value varies from 0 to loge(n). Values closer to zero means that built-up features are extremely compact in distribution, while values closer to loge(n), indicates the scattered distribution of built-up areas. Higher entropy values highlight sprawl occurrence. The threshold value to categorize sprawl and non-sprawl is taken as the mid-way mark of loge(n). If a city crosses its threshold entropy value, it is said to be a sprawling city. The difference in entropy between two time intervals t1 and t2 can be used to show the magnitude of urban sprawl transition.

Here, the study area is divided into 17 grid zones, each of size 15km x 15km. Considering each grid as a distinct spatial entity, Shannon's entropy has been calculated for all the 17 grid zones. The total number of zones in KMA is thus, n=17 and the upper limit of entropy value is loge(n)= 2.833. Figure 6 shows Shannon's entropy grid map of 1991.

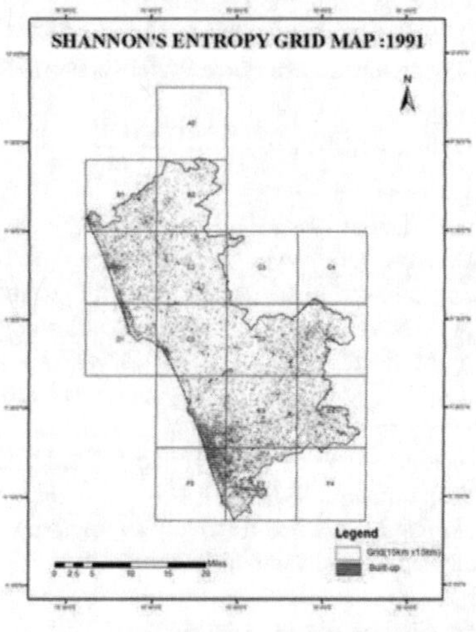

Figure 6: Shannon's Entropy Grid Map of Study Area

For the Kozhikode Urban Agglomeration, the value of Shannon's entropy has been calculated from the built-up areas for all the 17 zones over the years 1991, 2000, and 2018. The results are summarized in Table 3 and Table 4.

Table 3: Shannon's entropy calculation for 17 grids over the years 1991, 2000, and 2018.

	Year	1991	2000	2018	1991	2000	2018
		Built-up Area (sq.km)			$-P_i log_e(P_i)$		
	A2	0	0.15	0.95	0	0.000782	0.003977302
	B1	12.7	28.64	44.78	0.034906371	0.065801	0.091781506
	B2	22.47	38.33	57.32	0.054648662	0.081868	0.109634677
	C1	21.99	43.87	53.76	0.053744626	0.090416	0.104737406
	C2	43.54	55.98	61.98	0.089918121	0.107806	0.115860849
Grid Number	C3	0	0.09	8.55	0	0.000494	0.025376267
	C4	0	0.65	16.37	0	0.002858	0.042688718
	D1	10.579	29.76	59.17	0.030148896	0.067741	0.11213068
	D2	44.87	58.69	79.93	0.091916003	0.111486	0.138139963
	D3	6.7	20.55	43.82	0.020791552	0.050997	0.090340578
	D4	5.5	12.9	31.96	0.017669731	0.035344	0.071484081
	E2	37.33	51.75	78.9	0.080279375	0.101915	0.136927427
	E3	43.001	51.95	78.88	0.089102075	0.102198	0.136903809
	E4	27.36	38.7	47.89	0.063553579	0.082452	0.09637233
	F2	19.73	46.87	82.6	0.049407804	0.094879	0.141249094
	F3	28.77	43.86	78.13	0.066026975	0.090401	0.136016102
	F4	0	0.13	4.39	0	0.000688	0.014652515
Shannon's Entropy, Hn					0.742113771	1.088126	1.568273306

Table 4: Shannon's entropy with the magnitude of urban sprawl change (ΔH_n) for the years 1991, 2000 and 2018

Sl.No	Year	Shannon's Entropy	ΔH_n
1	1991	0.74	-
2	2000	1.09	0.35
3	2018	1.57	0.48
		$log_e(n)$= 2.833	

Figure 7 shows the graphical variation of Shannon's entropy values for the years 1991, 2000, and 2018. We can see an increasing trend in the value of Shannon's entropy values from 1991 to 2000 and to 2018. The increase in the value of Shannon's entropy indicates an increase in the dispersion of built-up areas, which in turn is an indication of the occurrence of the phenomenon of urban sprawl. As the values get more and more closer to the upper entropy limit (2.833), there is more and more dispersion of built-up areas in the region. The difference in entropy between the years 1991 and 2000 is 0.35, and 2000 and 2018 are 0.48. Here also, we can see a tremendous increase in urban sprawl as the years pass.

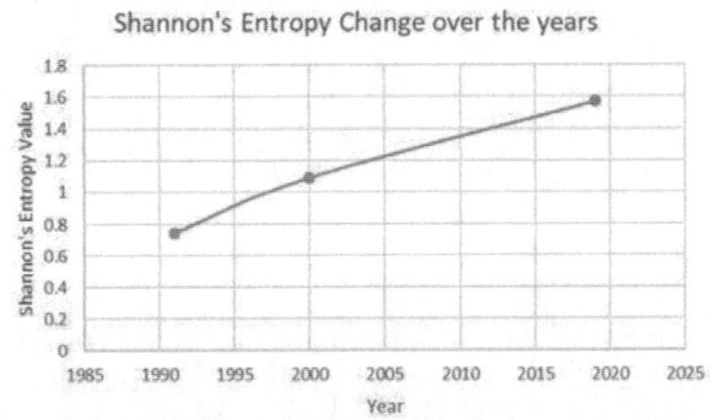

Figure 7 : Graphical Variation of Shannon's Entropy Values Over the Years

The maximum entropy value is 2.833, whereas the threshold value is identified as 1.42 which is exactly the half-way mark of loge(n). The value of entropy is less than the threshold value of 1.42 in the years 1991 and 2000. But in the year 2018, the entropy value exceeds the threshold value which in turn highlights the fact that the city is sprawling. From this figure, we can easily conclude that the city started its tendency to sprawl from the year 2012 onwards.

5. CONCLUSION

The issue of urban sprawl is a matter of concern in many countries across the world, especially in developing countries. If it is left unchecked, the phenomenon of uncontrolled urban sprawl will ultimately result in damaging impacts on urban sustainability. To evaluate and monitor the changes in urban settings, continuous spatial and temporal monitoring of our urban environment is a must. Together with its persistent and synoptic coverage, the multispectral capabilities of satellite remote sensing make it a versatile instrument for mapping and monitoring the changes in the urban core and urban planning applications. This will help to reduce unplanned urban expansion and thus helps in the preservation of sustainability of our urban environments. Adequate spatial inputs which portray the urban sprawl phenomenon can be obtained cost-effectively and efficiently using integrated GIS and remote sensing technologies. The dynamics of built-up area change over the years give noteworthy information about the nature, extent and development of urban sprawl.

In this study, the built-up area is considered as an indicator of urban sprawl. Built-up features of the study area for the years 1991, 2000 and 2018 were obtained from an automated thematic index-based index known as Index derived built-up index (IDBI). IDBI is derived from three thematic indices viz, Normalized Difference Built-up Index (NDBI), Modified Normalized Difference Water Index and Soil Adjusted Vegetation Index (SAVI). Built-up land features can be rapidly extracted, whereas soil, water, and vegetation noises, etc can be effectively suppressed using IDBI. The accuracy assessment showed that the overall accuracy is 90.65%, with a kappa coefficient of 0.812. A commonly used urban sprawl metric called Shannon's entropy is used for the measurement and quantification of the phenomenon of urban sprawl in the study area. The spatial concentration or degree of dispersion in a particular area can be easily tracked using Shannon's entropy coupled with remote sensing and GIS. The results of IDBI and Shannon's entropy analysis highlight the fact that there occurs an alarming increase in the built-up areal extent from 1991 to 2018. The city started its tendency to sprawl from the year 2012 onwards. The urban planning authorities can make use of these techniques of built-up area extraction and urban sprawl analysis for effective city planning and sprawl control. There is a need to create innovative approaches and further refine current methods and techniques in order to take complete benefit of the variety of remote sensing data within the urban setting

REFERENCES

1. Barnes, K.B., Morgan JM, Roberge MC, and Lowe S. (2001). "Sprawl development: its patterns, consequences, and measurement," Towson University, Towson, 1-24
2. Bhatta, B. (2009) Analysis of urban growth pattern using remote sensing and GIS: a case study of Kolkata, India. International Journal of Remote Sensing, 30(18) 4733–4746.

3. Bhatta B. (2010), Analysis of Urban Growth and Sprawl from Remote Sensing Data, Springer, Cham
4. Bhatta, B. (2012). Urban Growth Analysis and Remote Sensing; A Case Study of Kolkata, India 1980-2010, ISSN 2211-4173, Canada: Springer.
5. Pedro Cabral 1, *, Gabriela Augusto 1, Mussie Tewolde 2 and Yikalo Araya 3, Entropy 2013, 15, 5223-5236; DOI:10.3390/e15125223, Review Entropy in Urban Systems
6. Claudius, R. The Mechanical Theory of Heat: With its Applications to the Steam-Engine and the Physical Properties of Bodies; J. Van Voorst: London, UK, 1867; p. 376
7. DiLorenzo, W., 2000. The myth of suburban sprawl. USA Today (May) 128: 54–6.
8. Ewing, R., Pendall R. and D. Chen., (2002): Sprawl Scores for 83 Metropolitan Regions. Washington D. C: Smart Growth America
8. Freilich, R. (1999). From sprawl to smart growth: successful legal, planning, and environmental systems. Chicago: American Bar Association.
9. Frenkel, M. A. (2005). "Measuring Urban Sprawl; How can we deal with it?", ERSA conference papers from the European Regional Science Association
10. Galster, G., Hanson, R., Ratcliffe, M. R., Wolman, H. Coleman, S. & Freihage, J., (2001): Wrestling sprawl to the ground: defining and measuring an elusive concept, Housing Policy Debate, vol. 12, no 4, pp.681–717
11. Gordon, P. and Richardson, H. (1997) Are Compact Cities a Desirable Planning Goal Journal of the American Planning Association, 63, 95-106.
12. Hiller, B.T., Melotte, B. J., Hiller, S. M.2013. Uncontrolled Sprawl or Managed Growth? An Australian Case Study, Leadership and Management in Engineering
13. Li, X., & Yeh, A. G. O. (2004). Analyzing spatial restructuring of land use patterns in a fast growing region remote sensing and GIS. Landscape and Urban Planning, 69, 335–354).
14. McFeeters, S.K. (1996), The use of normalized difference water index in the delineation of open water features, International Journal of Remote Sensing, pp 1425-1432
15. NASA (2001) Satellite maps provide better urban sprawl insight. NASA, USA
16. Nelson, A. C., Duncan, J., Mullen, C., and Bishop, K., 1995, Growth management: Principles and practices, APA Planners Press, Chicago
17. Polidoro, Mauricio, Lollo, Jose A., Barros, Mirian V.F. (2011). "Environmental Impact of Urban Sprawl in Londrina, Parana, Brazil," Journal of Urban and Environmental Engineering, S. 5(1), s. 73-83.
18. Prakash, A., Gupta, R. P. (1998), Land-use mapping and change detection in a coal mining area-a case study in the Jharia coalfield, India, International Journal of Remote Sensing, 19(3), pp. 391-410
19. Ray, T.W. (2006)., Vegetation in Remote Sensing FAQ's. In ER Mapper Applications, pp 85-97, (Perth, Australia, ER Mapper Ltd)
20. Rawashdeh, S. A., Saleh, B. (2006), Satellite monitoring of urban spatial growth in Amman Area, Jordan, Journal of Urban Planning and Development, 132(4), pp. 211-216
21. Song, Y, G.-J. Knaap. (2004). "Measuring Urban Form-Is Portland Winning the War on Sprawl?", Journal of the American Planning Association, 70, 210-225
22. Sudhira, H.S., Ramachandra, T.V., Jagadish, K.S 2004, « Urban Sprawl: metrics, dynamics, and modeling using GIS, » International Journal of Applied Earth Observation and Geoinformation 5: 29-39.7)
23. UN's World Urbanization Prospects (2018), United Nations Department of Economic and Social Affairs: World Urbanization Prospects: 2018, The 2018 Revision, UNDESA.
24.Web-1:
https://www.un.org/en/ecosoc/integration/pdf/fact_sheet.pdf).
25. Web-2: (USGS) portal: https://earthexplorer.usgs.gov/
26. Wilson, Emily H., Hurd, James D., Civco, Daniel, Arnold, Chester L. (2003). "Development of a Geospatial Model to Quantify, Describe and Map Urban Growth," Remote Sensing of Environment, S. 86(3), s. 275-285.
27. Xu, H. (2008), Anew index for delineating built-up land features in satellite imagery, International journal of remote sensing, Vol.29, No.14, pp 4269-4276
28. Zha, Y., Gao, J., Ni, S. (2003)., Use of normalized difference built-up index in automatically mapping urban areas from TM images, International Journal of Remote Sensing, pp 583-594

OPTMIZATION OF BIO-OPTICAL MODEL PARAMETERS FOR TURBID LAKE WATER QUALITY ESTIMATION USING LANDSAT 8 AND WASI-2D

Manuel, A. [1,2], Blanco, A.C.[1,2], Tamondong, A.M.[1,2], Jalbuena, R. [1,2], Cabrera, O. [3], Gege, P.[4]

[1]Department of Geodetic Engineering, University of the Philippines Diliman
[2]Training Center for Applied Geodesy and Photogrammetry, University of the Philippines Diliman, Quezon City, Philippines
e-mail: amimanuel@gmail.com, acblanco@up.edu.ph, amtamondong@up.edu.ph, jalbuena.rj@gmail.com
[3]Institute of Environmental Science and Meteorology, University of the Philippines Diliman
email: olive@iesm.upd.edu.ph
[4]German Aerospace Center (DLR), Remote Sensing Technology Institute, 82234 Weßling Germany
email: peter.gege@dlr.de

KEYWORDS: Remote sensing, inland water, chl-a, bio-optical modeling, inversion

ABSTRACT:

Laguna Lake, the Philippines' largest freshwater lake, has always been historically, economically, and ecologically significant to the people living near it. However, as it lies at the center of urban development in Metro Manila, it suffers from water quality degradation. Water quality sampling by current field methods is not enough to assess the spatial and temporal variations of water quality in the lake. Regular water quality monitoring is advised, and remote sensing addresses the need for a synchronized and frequent observation and provides an efficient way to obtain bio-optical water quality parameters. Optimization of bio-optical models is done as local parameters change regionally and seasonally, thus requiring calibration. Field spectral measurements and in-situ water quality data taken during simultaneous satellite overpass were used to calibrate the bio-optical modelling tool WASI-2D to get estimates of chlorophyll-a concentration from the corresponding Landsat-8 images. The initial output values for chlorophyll-a concentration, which ranges from 10-40µg/L, has an RMSE of up to 10µg/L when compared with in situ data. Further refinements in the initial and constant parameters of the model resulted in an improved chlorophyll-a concentration retrieval from the Landsat-8 images. The outputs provided a chlorophyll-a concentration range from 5-12µg/L, well within the usual range of measured values in the lake, with an RMSE of 2.28µg/L compared to in situ data.

1. INTRODUCTION

1.1 Laguna Lake

Laguna de Bay is the largest freshwater lake in the Philippines and the third in South-East Asia with an area of approximately 900 km^2. The lake has been historically, economically, and ecologically significant to the communities surrounding it. Laguna de Bay lies at the center of urban development in Metro Manila, and it suffers from water quality degradation due to intensified exploitation of the its resources along with decades of industrial, household, and toxic pollution (Delos Reyes and Martens, 1994; Tamayo-Zafaralla et al., 2010).

The Laguna Lake Development Authority is a body mandated to address these problems and aim to improve the lake's ecological state along with its 21 major tributary rivers. With the increasing conflict regarding water use, allocation, and preservation of the lake, LLDA is expected to have a comprehensive resource management and monitoring system (Nauta et al, 2003). Implementation of sustainable management strategies should be supported by frequent monitoring (Gray and Shimshack, 2011). However, sampling and analysis used by field-based monitoring programs are labor, cost, and time intensive (Schaeffer et al, 2013). Traditional sampling methods may also produce unevenly distributed sampling points. Data acquired at various times in a day may result to inaccuracies in water quality assessment (Bresciani et al., 2011). In situ measurements can also hardly capture the temporal and spatial variability of large-scale lake phenomena (Dörnhöfer and Oppelt, 2016).

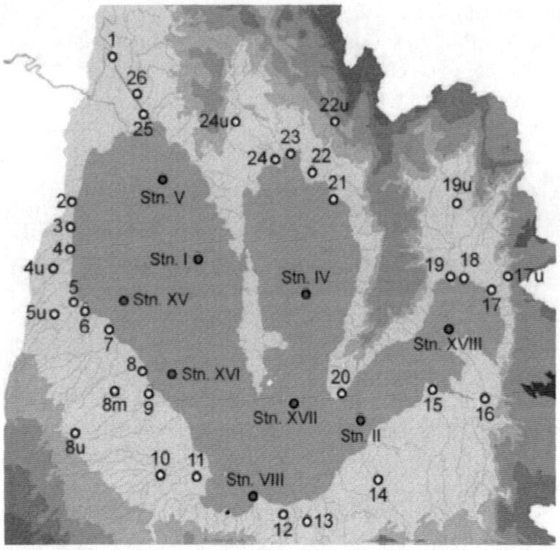

Figure 1. Water quality monitoring stations in Laguna Lake and its tributaries (Source: LLDA)

Water quality monitoring by the Laguna Lake Development Authority are based on 9 sampling stations inside the bay and take measurements once a month (Figure 1). There are instances wherein not all 9 stations are sampled in one day. Even when the sampling is completed within a day, the water quality values are not directly comparable from station to station considering variations within the day. While it is critical to monitor the water quality of an area efficiently and effectively, conventional methods of monitoring produce numerical outputs that are too technical that only few people can understand the implications (Chen, 2009). Also given the extensive area of the lake, the limited number of sampling points is not enough to create a visual output that would be easily comprehensible (Khattab and Merkel, 2014). Adding frequent sampling stations would also require considerable efforts in terms of time, labor, and money. These would still be insufficient to detect the changes of the spatial and temporal coverage compared to a detailed map that could be produced from satellite images.

Remote sensing addresses the need for a synchronized and frequent observation, it provides an efficient way to obtain bio-optical water quality parameters (i.e. chlorophyll-a, CDOM, SPM) over large areas. Remote Sensing is a very useful measurement tool than can greatly help water management authorities (Chen and Feng, 2018).

Furthermore, free and readily downloadable satellite images from recently launched sensors such as Landsat 8 offer spatial and radiometric resolutions that are suitable for inland water applications (Palmer et al., 2015; Manzo et al., 2015).

1.2 Bio-Optical Modeling of Inland Waters

Modelling a relationship between water quality parameters and remote sensing reflectance is a tedious process that requires a deep understanding of hydrologic optics, along with repeated site sampling and simulations that require high computing resources. (IOCCG, 2006) Fortunately, there are already established bio-optical models that have been widely used especially for inland water applications. However, parameters for such models can change regionally and seasonally, thus local calibration and fit tuning of the parameters are needed to get good water quality estimate from the satellite images (Gege, 2014).

Due to the presence of multiple constituents at different compositions, remote sensing for monitoring inland quality has been far less common compared to its application to open oceans. The complex interaction among water constituents is often intensified by anthropogenic actions, which creates uncertainty in the models designed for inland waters. Within the same inland aquatic system, different regions can still be dominated by different constituents (Ogasharawa, 2015).

1.3 Bio-Optical properties and Inherent Optical Properties

According to Morel (2001), remote sensing of inland waters is based on the optical properties of the constituents in the water. These optical properties are divided into two categories, namely, Apparent Optical Properties (AOPs) that depend on the medium and directional structure of the ambient light field, and Inherent Optical Properties (IOPs), which depend only on the medium itself and independent of the light field (Preisendorfer, 1976). Properties such as absorption (a) and scattering (b) coefficients vary on the composition of the medium or constituents itself (Mobley, 2001).

Absorption properties of water constituents (e.g., algal and non-algal particles) are the most commonly used parameters since the components in the water can either absorb the light in a specific part of the electromagnetic spectrum and contribute to the color of the water seen by sensors. There are a lot of studies reporting the absorption coefficients from inland waters and most are for sites in USA, China, and South Africa. (Ogashawara, 2017).

Since IOPs vary not only across geographic regions but also within the same site, the complexities are said to be mainly due to the spatio-temporal variability of the water constituents within the same lake. The dominant constituent at the water column may not only change spatially across short distances, but also across seasons and even daily (Gege, 2014). The variability in composition of constituents and associated IOP's in an aquatic system affect the magnitude and shape of the Remote sensing reflectance (R_{rs}). The variations in the R_{rs} are related to the variations in IOPs locally at each site (Mobley, 2001).

1.4 Inversion of Bio-Optical Models

The forward process in water color optics, is that if the parameters such as particle size, index of refraction of water, distribution of particles, and properties of the dissolved materials are known, the radiance distribution or spectrum of those specific parameters can be simulated by the forward solution of the radiative transfer equation. The inverse modelling, which is used in remote sensing of water bodies, retrieves the Inherent Optical Properties from the radiance and distribution spectrum. From these IOP's, the contribution of constituents such as dissolved matter, particulates from the water spectra can be retrieved (IOCCG, 2006). A simplified diagram of this is shown in Figure 2.

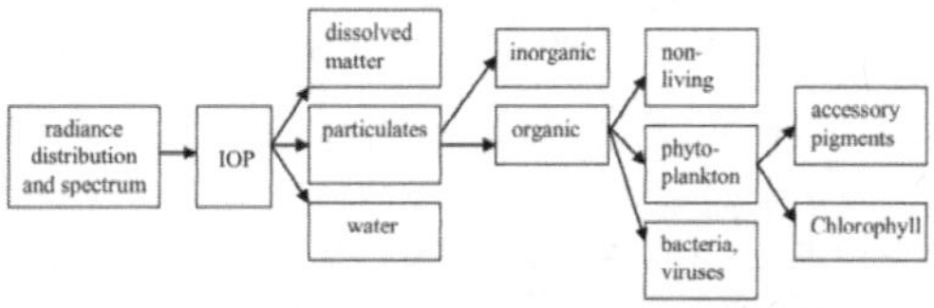

Figure 2. Diagram of inverse radiative transfer elements from Ocean Color Algorithms Working Group Report 5 (IOCCG, 2006)

Inversion of these known relationships determine biogeochemical parameters from the radiance spectrum inversion, however, is inexact and rely on approximations. The inversion is a two-step process: retrieval of IOP's from radiance, then the retrieval of biogeochemical parameters from these IOPs. Both are inexact procedures, which is why spectral inversion must be done with caution and data from field missions (IOCCG, 2006).

Semi-analytical models such as that of Albert and Mobley (2003) and Albert (2004) that are implemented in the Water Color Simulator software rely on the inversion of the Radiative Transfer Equations to establish the relationship among the Remote-sensing reflectance (AOP) and IOPs.

There are existing software that implement these inversion of semi-analytical models, such as the most commonly used for inland waters which is the Water Color Simulator (WASI) software by Peter Gege (2003). The WASI Software tool is developed for analyzing and simulating common types of spectra and allows quantitative analysis of multispectral and hyperspectral images of water bodies. It makes use of the models developed by Albert and Mobley in 2003 and is optimized for inland water applications but can be also applied for oceanic and coastal applications. The two most recent versions WASI 4 and WASI 5 includes the 2D module which applies the spectral inversion models (Figure 3.0) to remote sensing reflectance values in satellite images. This additional component of the software called WASI-2D applies the inverted parameters to atmospherically corrected satellite images and outputs them as estimated water quality concentration maps (Gege, 2014).

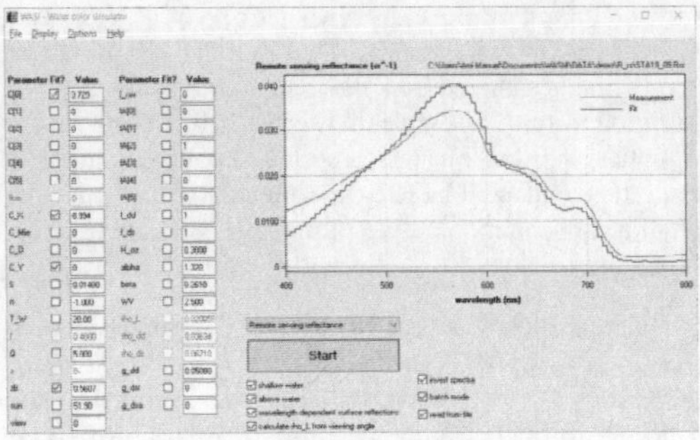

Figure 3. Spectral inversion interface of WASI- Water color simulator

Another example of software implementation of bio-optical models applied to satellite image is the BOMBER software by Giardino, et al. (2012), but it and differs in the implemented bio-optical models. The BOMBER software also lacks the individual spectral matching simulation and the option to select specific parameters to fit.

2. METHODOLOGY

The bio-optical modelling workflow consists of 2 major parts as shown in Figure 4. The first part is the processing of the field reflectance spectra and assessment of the field acquired IOP spectra. The values outputted from the spectral processing were used as initial inputs and constant parameters when running WASI-2D. The second one consists of the application of the bio-optical model inversion on the satellite images incorporating the said values.

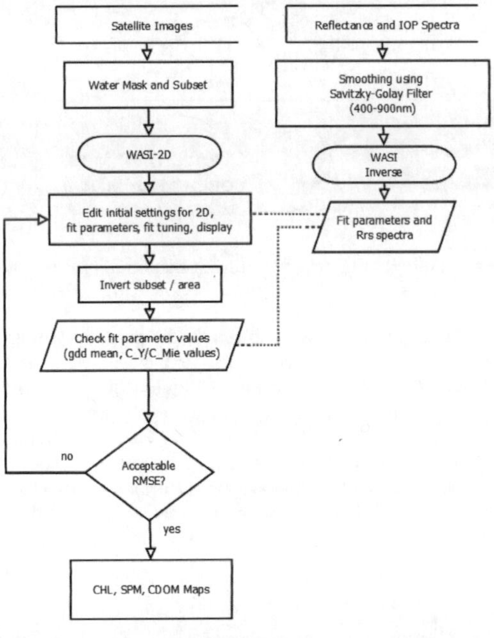

Figure 4. Workflow for the inversion using WASI

2.1 Data and Field Surveys

To acquire ground truth values for verification and aid for the inversion process, a 5-day field mission on the Laguna Lake were done for the Dry Season last November 13-17, 2018. The field team acquired data for forty-eight (48) points across the lake. This was done concurrent with different satellite overpass of Landsat 8, Sentinel 2 and Sentinel 3.

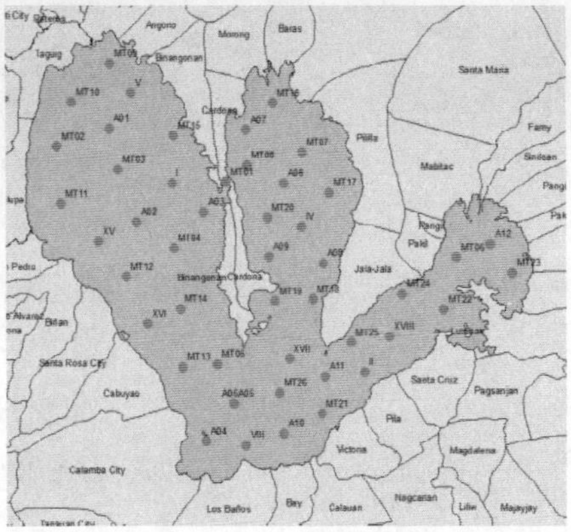

Figure 4. Field Validation Points for Laguna Lake (November 13-17, 2018)

Instruments used in the fieldwork include an Ocean Optics VIS-NIR Spectrometer (for obtaining reflectance spectra), a Fluoroprobe (submersible fluorometer), and a CLW Infinity Optical Sensor, which measures turbidity, and chl-a. The Fluoroprobe also measures microalgal class concentrations in addition to the total chl-a concentration.

The satellite image used for the bio-optical inversion is a cloud, cloud shadow, and land masked Landsat 8 image. Masking non-water pixels before running WASI-2D is crucial as spectral inversion may fail. The initial fit parameter input for a pixel takes on the value of the final fit values of the previously processed pixel.

Figure 5. Cloud and cloud shadow masked Landsat 8 image from December 18, 2018

2.3 Field Spectra and Satellite Image Inversion using WASI

Since spectral inversion can result in an infinite combination of different fit parameters, some of the constant parameter values must be set in accordance to the properties of the water body. This can be done with the spectral inversion of sample spectral signatures of Laguna Lake waters. Before proceeding with the inversion process with WASI, a spectral smoothing using the Savitzky-Golay smoothing filter was applied to the spectral signature data using a python script (Figure 6).

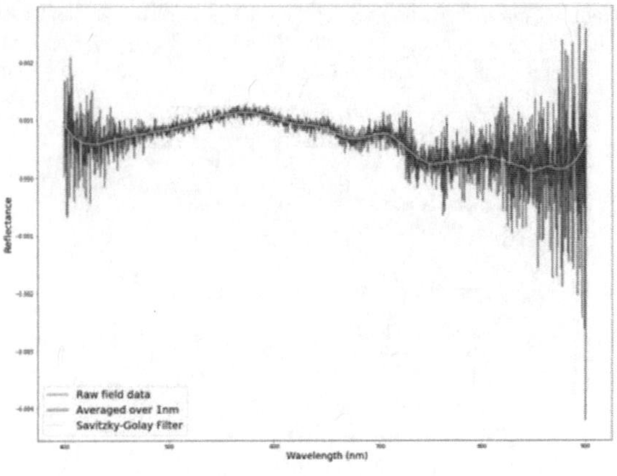

Figure 6. Field sample spectral data smoothing using Savitzky-Golay filter

Initial WASI run gave values and results that do not fall within the field acquired values for chlorophyll. This is due to one spectral shape not necessarily corresponding to a unique set of water quality parameters for the same sun illumination patterns. Therefore, incorporation of a constant value or setting of initial parameters, as well as different fit tuning settings were tested out to improve model performance. In situ Chlorophyll-a measurements was used as a constant to help guide the inversion model that this is the baseline or field acquired value, so it would set the initial simulated spectra closer to the actual field value.

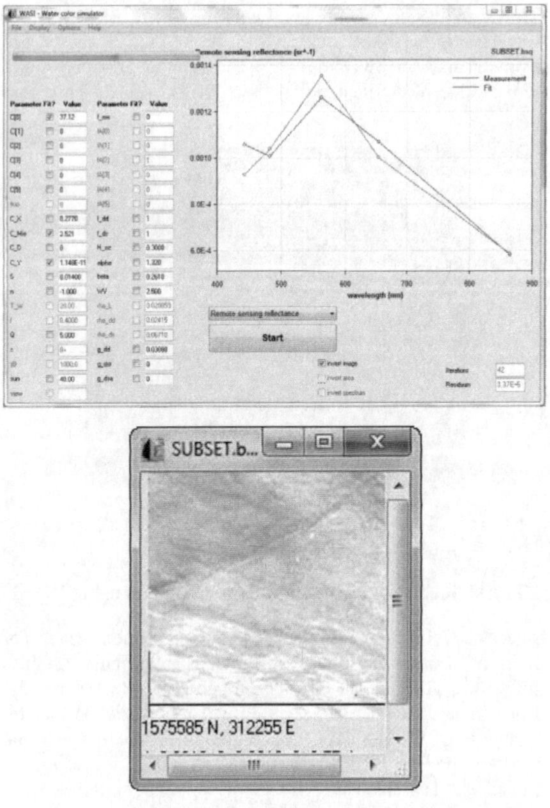

Figure 7. WASI 2-D Interface

Tests for which parameters would be fit and constant as well as setting adequate parameter ranges and initial values were done to get optimal results.

3. RESULTS AND DISCUSSION

Out of the 48 points gathered from the field mission, only 27 were useful for ground truth assessment. An inevitable disadvantage of using optical satellite imagery are cloud covers, as the data underneath the clouds will be irretrievable and masked out. The chlorophyll-a concentration output of the best performing run is shown in Figure 8 below.

The first few runs of the WASI software were overapproximating the values up to 40 µg/L. These yielded a low R^2 value of 0.16 and an RMSE of up to 10 µg/L. Aside from the low correspondence to field data, the resulting fit values for CDOM were at 0. For these initial runs, the contributions for sky radiance specifically the parameter gdd (fraction of sky radiance due to direct solar radiation) were fit and the C_Mie (concentration of non-algal particles of type II) was set at a constant value of 0. Fitting the gdd correction for sky glint might be affecting the scale of the fit spectra thus affecting the estimation of chlorophyll-a.

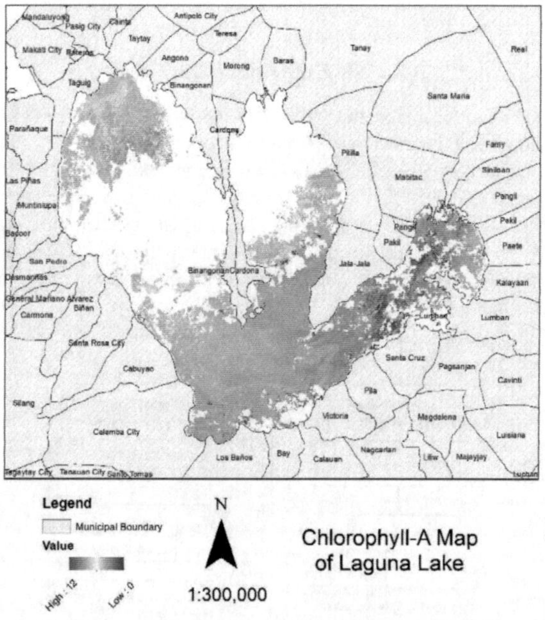

Figure 8. Chlorophyll-a map output from Landsat 8 processing in WASI

Limiting the range of the fit parameter in the program settings and setting the gdd as a constant based on the inverted field spectral values improved the results. Addition of the C_Mie parameter as a fit also resulted in better approximations for chlorophyll-a and gave reasonable values for CDOM. Different runs of the WASI software testing out optimum fit tuning (residuals and maximum iterations) and parameter settings which brought the values within the range of the field monitored values were done.

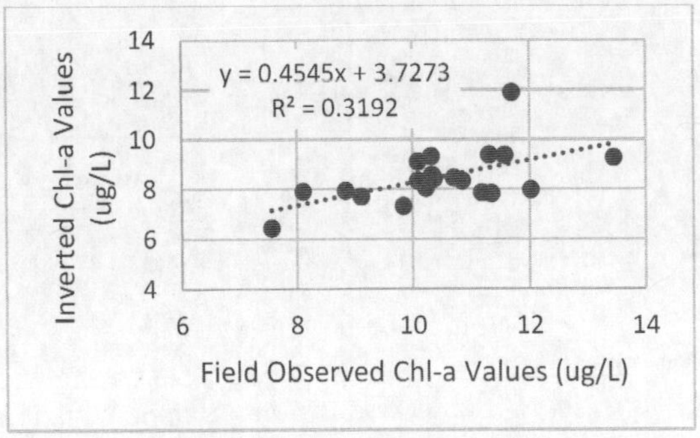

Figure 9. Assessment of Inverted and Field Observed Chlorophyll-a values

The best run yielded an R^2 of 0.32 with an RMSE of 2.28 μg/L, an improvement to the initial runs. The fitted parameters for the best performing run were chlorophyll-a (C[0]), suspended particulate matter type I (C_X), suspended particulate matter type mie (C_Mie), and colored dissolved organic matter (C_Y). SPM Type 1 (C_X) was tested out as a constant value but it did not yield good approximations compared to when it is fit along with C_Mie. The upper left portion of the output image produced inconsistent values for all the runs. This may be due to the presence of haze or thin cloud cover that were not masked out during pre-processing.

Other parameters such as illumination were kept constant based on the satellite overpass situation and values inverted from field spectral data. Water temperature was also set at a constant of 28.5°C. Assessment of SPM and CDOM fit performance were not done because of the limitations in the available ground truth data.

4. CONCLUSION

Setting up the WASI software incorporating field inverted spectra values to the fit tuning and parameter settings improved the fit performance of the bio-optical models. The inversion of the models on Landsat 8 fitting chlorophyll-a gave an R^2 of 0.32 with an RMSE of 2.28 μg/L. The outputs of WASI-2D were able to give a visual of the spatio-temporal variation of the chlorophyll-a concentration from a Landsat 8 image.

SPM and CDOM performance show promising results but inconclusive. Further work relating the turbidity field data to SPM and Fluoroprobe outputs to CDOM should be done to assess the other fit parameter outputs of WASI.

ACKNOWLEDGEMENTS

This study was conducted as a part of the Multi-platform and Cross-sensor Water Quality Monitoring (MCWQM) Project. The Project was implemented by the University of the Philippines Training Center for Applied Geodesy and Photogrammetry (TCAGP), funded by the Department of Science and Technology and the Philippine Council for Industry, Energy, and Emerging Technology Research and Development (PCIEERD). The Project was also done in collaboration with the Laguna Lake Development Authority (LLDA).

Ms. Manuel would also like to acknowledge the Engineering Research and Development for Technology (ERDT) Program and the University of the Philippines Office of International Linkages (UP OIL) for granting her the privilege to present this research in the ISRSE PECORA 21 conference in Baltimore, Maryland, USA.

REFERENCES

Albert, A., Mobley, C.D. (2003). An analytical model for subsurface irradiance and remote sensing reflectance in deep and shallow case-2 waters. Optics Express, 11 (2003), pp. 2873-2890

Albert, A. (2004). Inversion technique for optical remote sensing in shallow water. Ph.D. thesis, University of Hamburg. http://www.sub.uni-hamburg.de/opus/volltexte/2005/2325

Bresciani, M., Stroppiana, D., Odermatt, D., Morabito G., Giardino, C. (2011). Assessing remotely sensed chlorophyll-a for the implementation of the Water Framework Directive in European perialpine lakes. Science of The Total Environment 409(17):3038-91. doi.org/10.1016/j.scitotenv.2011.05.001.

Chen, S., Fang, L., Zhang, L. and Huang, W. (2009). Remote sensing of turbidity in seawater intrusion reaches of Pearl River Estuary – A case study in Modaomen water way, China. Estuarine, Coastal and Shelf Science, 82(1), pp.119-127.

Chen, X. and Feng, L. (2018). Remote Sensing of Lakes' Water Environment. Comprehensive Remote Sensing, pp.249-277.

Delos Reyes, M. and Martens, R. (1994). Geoecology of Laguna de Bay, Philippines I. Techno-commercial impact on the trophic level structure of the Laguna de Bay aquatic ecosystem 1968–1980. Ecological Modelling, 75-76, pp.497-509.

Dörnhöfer, K. and Oppelt, N. (2016). Remote sensing for lake research and monitoring – Recent advances. Ecological Indicators, 64, pp.105-122.

Gege, P. (2004). The water color simulator WASI: an integrating software tool for analysis and simulation of optical in situ spectra. Computers & Geosciences, 30(5). pp. 523-532.

Gege, P. (2014) WASI-2D: A software tool for regionally optimized analysis of imaging spectrometer data from deep and shallow waters. Computers and Geosciences, (62). pp. 208-215.

Gege, P. (2017) Chapter 2: Radiative Transfer Theory for Inland Waters. Bio-optical Modeling and Remote Sensing for Inland Waters. pp. 25-67.

Giardino, C., Candiani, G., Bresciani, M., Lee, Z., Gagliano, S., Pepe, M. (2012). BOMBER: A tool for estimating water quality and bottom properties from remote sensing images. Computers and Geosciences (45). pp 313–318.

Gray, W. and Shimshack, J. (2011). The Effectiveness of Environmental Monitoring and Enforcement: A Review of the Empirical Evidence. Review of Environmental Economics and Policy, 5(1), pp.3-24.

IOCCG (2006). Remote Sensing of Inherent Optical Properties: Fundamentals, Tests of Algorithms, and Applications. Lee, Z.-P. (ed.), Reports of the International Ocean-Colour Coordinating Group, No. 5, IOCCG, Dartmouth, Canada.

Khattab, M. and Merkel, B. (2013). Application of Landsat 5 and Landsat 7 images data for water quality mapping in Mosul Dam Lake, Northern Iraq. Arabian Journal of Geosciences, 7(9), pp.3557-3573.

Manzo, C., Bresciani, M., Giardino, C., Braga, F. and Bassani, C. (2015). Sensitivity analysis of a bio-optical model for Italian lakes focused on Landsat-8, Sentinel-2 and Sentinel-3. European Journal of Remote Sensing, 48(1), pp.17-32.

Mobley, C.D. (2001). Radiative transfer in the ocean. In: Steele, J.H. (Ed.). Encyclopedia of Ocean Sciences. Academic Press Elsevier. London, UK. Pp. 2321-2330

Morel, A., (2001). Bio-optical models. In: Steele, J.H. (Ed.). Encyclopedia of Ocean Sciences. Academic Press Elsevier. London, UK. pp. 317-326

Nauta, T., Bongco, A. and Santos-Borja, A. (2003). Set-up of a decision support system to support sustainable development of the Laguna de Bay, Philippines. Marine Pollution Bulletin, 47(1-6), pp.211-218.

Ogashawara, I., Deepak, R.M., Gitelson, A.A. (2017) Chapter 1: Radiative Transfer Theory for Inland Waters. Bio-optical Modeling and Remote Sensing for Inland Waters. pp. 1-24.

Palmer, S., Kutser, T. and Hunter, P. (2015). Remote sensing of inland waters: Challenges, progress and future directions. Remote Sensing of Environment, 157, pp.1-8.

Preisendorfer, R.W. (1976). Hydrologic Optics (2). U.S. Dept. of Commerce. Washington, DC, USA.

Schaeffer, B., Schaeffer, K., Keith, D., Lunetta, R., Conmy, R. and Gould, R. (2013). Barriers to adopting satellite remote sensing for water quality management. International Journal of Remote Sensing, 34(21), pp.7534-7544.

Tamayo-Zafaralla, M., Santos, R., Orozco, R. and Elegado, G. (2002). The ecological status of Lake Laguna de Bay, Philippines. Aquatic Ecosystem Health & Management, 5(2), pp.127-138.

MODELING EVAPOTRANSPIRATION FOR C4 and C3 CROPS IN THE WESTERN LAKE ERIE BASIN USING REMOTE SENSING DATA

Y. Marambe[1], A. Simic Milas[1,*]

[1]Bowling Green State University, School of Earth, Environment and Society, Department of Geology, Bowling Green, Ohio, USA

KEY WORDS: Evapotranspiration, Photosynthesis, Crops, C3/C4 plants

ABSTRACT:

Growing monoculture impacts not just soil properties and biodiversity but also local hydrology including evapotranspiration (ET). The Midwest region of the U.S. is known for its monoculture trend by growing and producing corn, which commonly replaces other crop types. In addition to large areas covered with corn, the photosynthetic adaptations of corn, being the C4 crop, affects ET differently than other C3 crops such as soybean, wheat, and alfalfa. This study aims to model and compare ET for C3 and C4 crops using remote sensing (Sentinel-2 data) and the Boreal Ecosystem Productivity Simulator (BEPS) model, modified to consider C3 and C4 crops. The study explores the ET rate trend for corn and soybean in an agriculture area situated in the Western Lake Erie Basin, where the balance between evapotranspiration, groundwater level, and surface runoff may play a role in agricultural runoff and Lake Erie's algal blooms caused by runoff pollution. The results suggest that the monthly average ET rates for both soybean (C3) and corn (C4) reach its maximum at the mid-to-late growing season. However, the ET rate for corn is higher than for soybean in the early season (June) (ET=121 mm month^{-1} for corn; ET=105 mm month^{-1} for soybean), while the ET rate for soybean becomes higher than for corn soon after (July) and becomes considerably higher in August (ET=181 mm month^{-1} for corn; ET=218 mm month^{-1} for soybean). It is surmised that the higher ET rate for corn in the early growing season is due to nitrogen-based fertilizer commonly applied to corn parcels at that time, whereas soybean growth is based on biological nitrogen fixation.

1. MANUSCRIPT

1.1 Introduction

Evapotranspiration (ET) is a process of water movement through plants generated by water potential gradients between the rhizosphere, plants, and the atmosphere (Landsberg and Waring, 2016). ET affects the photosynthesis rate and, together with the carbon dioxide (CO_2) uptake, controls the primary production of crops, their vigor and yield (Kool et al., 2014; Fisher et al., 2017; Irmak, 2017). Through interactions with soil moisture, as function of soil texture and soil water holding capacity, ET controls nutrient absorption and crop growth. Nowadays, when crop management practices are designed to ensure the maximum yield to satisfy the food and energy (biofuel) requirements of the growing population, understanding and predicting crop ET is critical (Black et al., 2012; FAO, 2017). Water resource management and irrigation processes are being constantly adjusted to balance ET (Liu et al., 2003; FAO, 2016), due to the constantly changing weather and crop management practices.

The recent trend of monoculture farming affects not just soil properties and biodiversity but also local hydrology including ET. The Midwest region of the U.S. is known for its high production of corn and soybean where the monoculture trend, by growing and producing corn that replaces other crop types, has been observed in recent years. The trend is observed not just in the Midwest region but across the whole U.S. (Mladenoff et al., 2016; Gao et al., 2017). The production of corn plays an important role in the U.S. economy. The U.S. Department of Agriculture (USDA) related the trend of high corn production to ethanol, in addition to human and livestock production (USDA, 2017; USDA, 2018). The USDA estimated that corn planted in 2017 covered an area of 90.9 million acres and that the daily biofuel production in 2017 was nearly 1000 barrels per day.

In addition to larger areas covered with corn, the photosynthetic adaptations of corn, as the most widely cultivated C4 plant, affects ET differently than other, more abundant, C3 crops (e.g., soybean, wheat, and alfalfa). It is well known that C4 plants have their photosynthetic mechanism developed to reduce photorespiration and save water (Ajao et al., 2017) and that they are more photosynthetically active than C3 plants (Ueno, 2001; Garner et al., 2016; Ajao et al., 2017). Lundgren and Christin (2016) summarized the findings showing how the anatomical variations in C4 plants also exhibit more efficient uptake of CO_2. Known as the ratio between primary production and transpiration, the Water Use Efficiency (WUE), which is higher in C4 plants, helps them to survive better in harsh environments (Cernusak et al., 2007). Overall, C4 plants perform better in conditions of high temperatures, with a wide range of CO_2 concentrations, and in soil with poor water conditions (Powell and Still, 2009; Black et al., 2012). By contrast, C3 crops increase water consumption under the same conditions but their yield decreases (Leakey et al., 2009; Kumar et al., 2017).

In the present study, the Boreal Ecosystem Productivity Simulator (BEPS) model (Liu et al., 1997), modified for C3 and C4 crops in this study, is used to estimate subtle ET differences between corn (C4) and soybean (C3) using Sentinel-2 satellite data for the growing season of 2016. The aim is to explore the ET temporal trends for the two crops in an agriculture area situated within the

* Corresponding author.
E-mail addresses: asimic@bgsu.edu (A. Simic Milas), ymaramb@bgsu.edu (Y.Marambe).

Western Lake Erie Basin where the balance between evapotranspiration, groundwater level, and surface runoff may play an important role in agricultural runoff and Lake Erie's algal blooms caused by runoff pollution.

2. METHODS

2.1 Study Site

The study area is located in northwest Ohio and southeast Michigan within the Western Lake Erie Basin (WLEB) (latitude 41.78°, longitude -83.92). The area partially covers Lucas and Fulton counties of Ohio, and Lenawee and Monroe counties of Michigan (Figure 1). In 2016, the agricultural lands covered around 75% of the study area (USDA, 2016). The USDA's cropland classified map measured 37% of soybean, 28% of corn, 7% of wheat, and 2% of alfalfa for the same year. Besides the agricultural lands, there were some deciduous trees and grasslands. Irrigation is not a common agricultural practice in the study area and it is not considered in the present study.

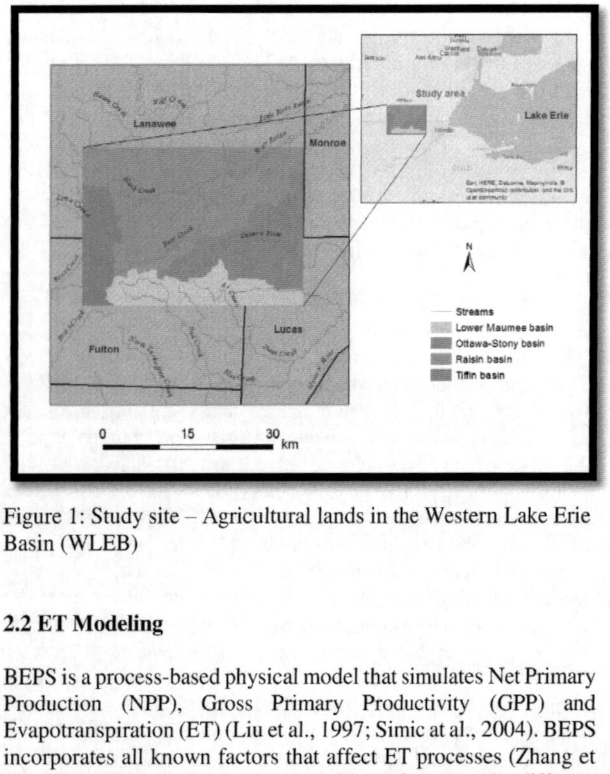

Figure 1: Study site – Agricultural lands in the Western Lake Erie Basin (WLEB)

2.2 ET Modeling

BEPS is a process-based physical model that simulates Net Primary Production (NPP), Gross Primary Productivity (GPP) and Evapotranspiration (ET) (Liu et al., 1997; Simic at al., 2004). BEPS incorporates all known factors that affect ET processes (Zhang et al., 2012). The model has proved its performance in different studies using various ecosystems worldwide (Liu et al., 2002; Govind et al., 2009; Zhang et al., 2012). BEPS incorporates meteorological, soil, and remote sensing data to estimate ET. Precipitation, solar radiation, maximum temperature, minimum temperature, and average relative humidity are required for the daily model simulation. Land cover (LC) data are necessary to simulate ET because of specific plant physiological parameters for each LC or crop types. Leaf area index (*LAI*) is another important input parameter to BEPS, generated from satellite images (Liu et al., 2002; Govind et al., 2009). BEPS considers sunlit and shaded leaf areas in the photosynthesis related calculations, plant canopy structure, *LAI*, clumping index, plant physiological parameters, stomatal resistance/conductance, and plant type (Liu et al., 2002). BEPS facilitates analyses at different scales from local and regional to global levels (Liu et al., 2002). To run BEPS in this study, two land-use classes, one for C3 crops and another for C4 crops, were added and their parameters appropriately modified. Daily ET values, calculated in each run, were averaged and represented as monthly values.

2.3 Data

GridMET daily meteorological data sets were used in the present study (http://www.climatologylab.org/gridmet.html) with the resolution of ~4-km / 1/24th degree. The gridMET data sets are a combination of NLDAS-2 (National Land Data Assimilation System-2) data with hourly measurements and 1/8th-degree grid spacing, and PRISM (Parameter-elevation Regression on Independent Slope Model) data sets with the spatial resolution of 800 m, and monthly temporal resolution (Abatzoglou, 2011).

Available Water Holding Capacity (AWHC) data sets were collected from the Soil Survey Geographic (SSURGO) database (USDA, 2017).

Sentinel-2 satellite data were used to produce *LAI* and LC, as input data to BEPS. Sentinel-2 images were obtained from the European Space Agency (ESA) Copernicus data hub (https://scihub.copernicus.eu/). Atmospheric corrections were performed using the Fast Line of Sight Atmospheric Analysis of Spectral Hypercubes (FLAASH), an advanced atmospheric correction model embedded in the ENVI 5.4 software (Harris Geospatial, 2017). The Sentinel-2 data were resampled to 10 m.

LAI images were generated in accordance with the temporal resolution of Sentinel-2 data, for 10-day intervals. An empirical algorithm was used to calculate *LAI*, as given by Boegh et al. (2002). The algorithm is a widely used in-built spectral function of ENVI 5.4 (Boegh et al., 2002):

$$LAI = (3.618 \times EVI - 0.118) \quad (1)$$

where *EVI* = Enhanced Vegetation Index

According to Boegh et al. (2002), the performance of this algorithm was successful when tested over the agricultural land covers, especially over corn, winter wheat, and grassland covers.

EVI is developed by Huete et al. (2002):

$$EVI = 2.5 \times (NIR - Red) / (NIR + 6 \times Red - 7.5 \times Blue + 1) \quad (2)$$

EVI is commonly used to improve separation of vegetation signal from background signal. It also performs well in non-vegetation areas as well as in areas with high biomass. The *EVI* algorithm uses the blue spectral band, which is sensitive to the atmosphere (Huete et al., 2002). Sentinel 2 images were classified using the maximum likelihood image classifier, a widely used parametric classification method.

3. RESULTS AND DISCUSSION

Land cover classification images have high overall accuracy (OA=95%, Kappa coefficient=0.92) when compared with the USDA cropland classification reference map (USDA, 2016). The accuracy for soybean and corn are 97% and 98%, respectively (Figure 2). The accuracy is also high for deciduous trees and grassland as well as for other crops, which include winter wheat in June, cover crop to replace winter wheat from July - October, and perennial alfalfa.

The LAI images derived using *EVI* closely follow the pattern of *EVI* (Figure 2). Zonal statistics was used to examine *LAI* for different land covers. *LAI* increases gradually for corn and more rapidly for soybean before the peak growing season in August.

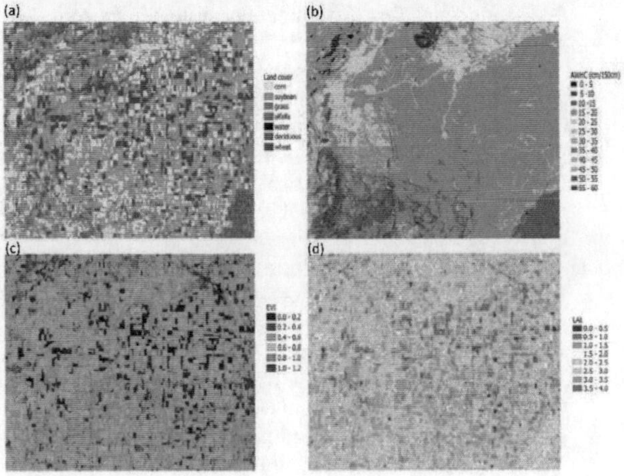

Figure 2. Input data to BEPS: a) Land cover; b) AWHC c) *EVI* d) *LAI*

The BEPS outputs (daily ET) averaged for each month reach their peaks in August for C4 and all C3 crops (Figures 3 and 4).

In June, the highest ET rate is observed for corn (ET=121 mm month^{-1}). Soybean exhibits lower ET for the same month (ET=105 mm month^{-1}). While the ET rate for corn is higher in the early growing season, the ET rate for soybean becomes higher in July and stays that way until the end of the season. The ET rate for soybean increases by 47% from July to August, while the rate for corn increases 24% between the two months (Figures 3 and 4). The cover crop on the winter wheat plots has the lowest ET rates for all months. Alfalfa is the perennial crop and has the highest ET rate in June. The lowest ET rate is observed in October for all crops, as expected.

It can be expected that corn grown as monoculture may increase the ET rate but only in the early growing season. Overall, it would decrease the ET rate under given conditions. In this preliminary study, the BEPS parameters such as stomata resistance were kept equal for soybean and corn.

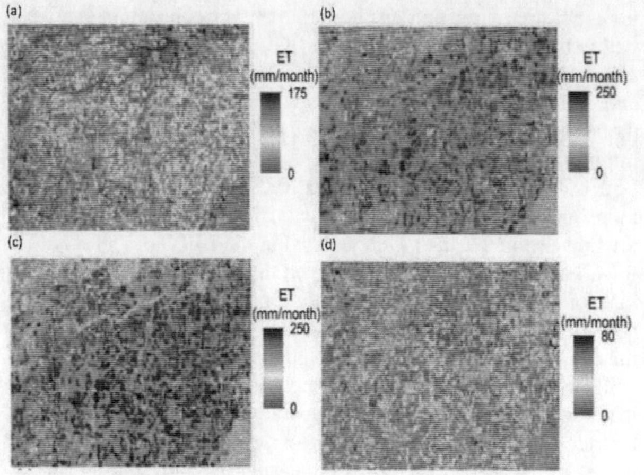

Figure 3. Spatial monthly average ET rate for: a) June; b) July; c) August; d) October 2016.

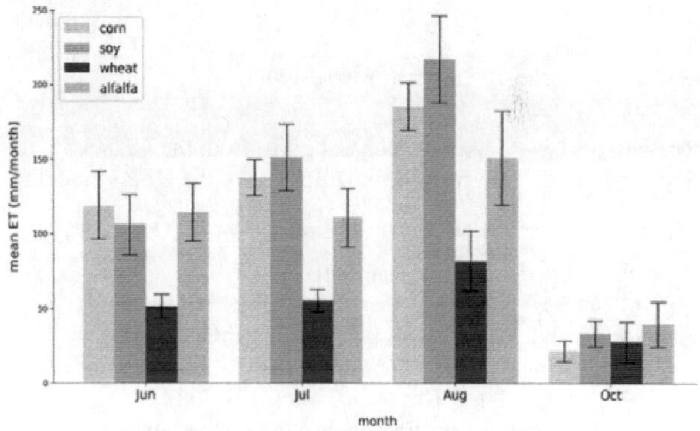

Figure 4. Monthly average ET rate for: a) June; b) July; c) August; d) October 2016.

Corn monoculture may affect the overall ET rate but most likely not considerably. The ET trend can be related to the properties of C4 vs. C3 plants as it is in accordance with the photosynthetic activities of corn being C4 and soybean being C3 plants. As C4 plants have photosynthetic mechanisms developed to reduce photorespiration and save water, it is expected that they have lower ET rate. The possible impact of the agricultural treatments on ET has not been explicitly considered in the study, but it is indirectly captured in the remote sensing derived data. It is surmised that the higher ET rate for corn in the early growing season is due to nitrogen-based fertilizer commonly applied to corn parcels at that time, whereas soybean growth is based on biological nitrogen fixation. This is in accordance with the study of Simic Milas and Vincent, 2016.

4. CONCLUSION

Corn has much higher yield than any other crop and it grows as a monoculture in many parts of America, especially in the Midwest. Corn is a versatile crop and it can be used for food for humans and animals, and converted into ethanol among other products. This preliminary study explored the possible impact of corn monoculture on evapotranspiration in an agricultural area situated within the Western Lake Erie Basin. Given that corn is the C4 crop and that other crops in the areas, such as soybean, winter wheat, and alfalfa, are C3 crops, it was hypothesized that corn monoculture might affect local evapotranspiration, which is an important hydrological parameter.

Soil properties and meteorological data, together with leaf area index and C3/C4 land-use spatial data, derived using remote sensing Sentinel-2 data, were utilized in the BEPS model modified for C3 and C4 crops. It was observed that the monthly average ET rate for corn was higher in the early growing season, while the ET rate for soybean became higher in July and stayed that way until the end of the season. It is anticipated that the findings of this study could also be considered for different agricultural settings where irrigation is included in the farming practices. Global warming may restrict irrigation water resources for farmlands in the future and farmers may rely more on C4 crops such as corn considering its better WUE and high yield. Fu et al. (2009) and Hobbins et al. (2004) suggested the importance of soybean and corn in food production, but they also emphasized the possible 'ecohydro' impacts of these major crops, including the impact on ET.

Acknowledgments

The authors would like to thank Gang Mo from the University of Toronto, for his help with the modification of the BEPS model.

References

Abatzoglou, J.T., 2011. Development of Gridded Surface Meteorological Data for Ecological Applications and Modelling. Int. J. Climatol. 33 (1), 121-131.

Ajao, A.A., Jimoh, M.A., Saheed, S.A., 2017. Studies on Anatomical Characters Indicating C3 and C4 Photosynthetic Metabolism in the Genus Boerhavia L. (Nyctaginaceae). Taiwania. 62 (3), 265-271.

Black, E., Vidale, P.L., Verhoef, A., Santiago Vianna Cuadra, S.V., Osborne, T., Hoof, C.V.D., 2012. Cultivating C4 Crops in a Changing Climate: Sugarcane in Ghana. Environ. Res. Lett. 7 (4), 044027.

Boegh E., Soegaard H., Broge N., Hasager C., Jensen N., Schelde K.,Thomsen A., 2002. Airborne multi-spectral data for quantifying leaf area index, nitrogen concentration and photosynthetic efficiency in agriculture. Remote. Sens. Environ. 81 (2-3), 179-193.

Cernusak, L.A., Aranda, J., Marshall J.D., Winter, K., 2007. Large Variation in Whole-Plant Water-Use Efficiency among Tropical Tree Species. New Phytol. 173, 294–305.

Food and Agriculture Organization (FAO). Introduction to Evapotranspiration. Url: http://www.fao.org/docrep/X0490E/x0490e04.htm (visited on March 15 2017).

Food and Agriculture Organization (FAO), 2017. The future of food and agriculture - Trends and challenges. Url: http://www.fao.org/3/a-i6583e.pdf (visited on April 12 2017).

Fisher, J.B., Melton, F., Middleton, E., Hain, C., Anderson, M., Allen, R., McCabe, Wood, E.F., 2017. The Future of Evapotranspiration: Global Requirements for Ecosystem Functioning, Carbon and Climate Feedbacks, Agricultural Management, and Water Resources. Water. Resour. Res. 53 (4), 2618-2626.

Fu, G., Charles, S.P., Yu, J., 2009. A Critical Overview of Pan Evaporation Trends over the Last 50 Years. Clim. Change. 97, 193-214.

Gao, J., Sheshukov, A.Y., Yen, H., Kastens, J.H., Peterson, D.L., 2017. Impacts of Incorporating Dominant Crop Rotation Patterns as Primary Land Use Change on Hydrologic Model Performance. Agric. Ecosyst. Environ. 247, 33-42.

Garner, D.M.G., Mure, C.M., Yerramsetty, P., Berry, J.O., 2016. Kranz Anatomy and the C4 Pathway. DOI: https://doi.org/10.1002/9780470015902.a0001295.pub3

Govind, A., Chen, J.M., Margolis, H., Ju, W., Sonnentag, O, Giasson, M., 2009. A Spatially Explicit Hydro-Ecological Modeling Framework (BEPS-Terrainlab V2.0): Model Description and Test in a Boreal Ecosystem in Eastern North America. J. Hydrol. 367, 200–216.

Harris Geospatial, 2017. Url: https://www.harris.com/what-we-do/geospatial-solutions (visited on April 4 2017).

Hobbins, M.T., Ramirez, J.A., Brown, T.C., 2004. Trends in Pan Evaporation and Actual Evapotranspiration across the Conterminous U.S.: Paradoxical or Complementary? Geophys. Res. Lett. 31, L13503, doi:10.1029/2004GL019846.

Huete, A., Didan, K., Miura, T., Rodriguez, E.P., Gao, X., Ferreira, L.G., 2002. Overview of the Radiometric and Biophysical Performance of the MODIS Vegetation Indices. Remote Sens. Environ. 83, 195–213.

Irmak, S., 2017. Evapotranspiration Basics and Estimating Actual Crop Evapotranspiration from Reference Evapotranspiration and Crop-Specific Coefficients. Nebraska Extension-Crops and Irrigation Engineering NebGuide. Url: http://extensionpublications.unl.edu/assets/pdf/g1994.pdf

Kool, D., Agam, N., Lazarovitch, N., Heitman, J.L., Sauer, T.J., Ben-Gal, A., 2014. A Review of Approaches for Evapotranspiration Partitioning. Agric. For. Meteorol. 184, 56-70.

Kumar, V., Sharma, A., Soni, J.K., Pawar, N., 2017. Physiological Response of C3, C4 and CAM Plants in Changeable Climate. Pharma Innov. 6 (9), 70-79.

Landsberg, J., Waring, R., 2016. Water Relations In Tree Physiology: Where To From Here?. Tree Physiol, 37 (1), 18-32.

Leakey, A.D.B., Ainsworth, E.A., Bernacchi, C.J., Rogers, A., Long, S.P., Ort, D.R., 2009. Elevated CO_2 Effects on Plant Carbon, Nitrogen, and Water Relations: Six Important Lessons from FACE. J. Exp. Bot. 60, 2859-2876.

Liu, J., Chen, J.M., Chillar, J., Park, W.M., 1997. A Process-Based Boreal Ecosystem Productivity Simulator Using Remote Sensing Inputs. Remote Sens. Environ., 62, 158-175.

Liu, J., Chen, J.M., Chillar, J., Chen, W., 2002. Net Primary Productivity Mapped for Canada at 1-km Resolution. Glob. Ecol. Biogeogr. 11 (2), 15-129.

Liu, J., Chen, J.M., Chillar, J., Park, W.M.. 2003. Mapping Evapotranspiration Based On Remote Sensing: An Application to Canada Landmass. Water Resour., 39 (7), 1189.

Lundgren, M.R., Christin, P.A., 2016. Despite Phylogenetic Effects, C_3–C_4 Lineages Bridge the Ecological Gap to C_4 Photosynthesis. J. Exp. Bot. 68 (2), 241–254.

Mladenoff, D.J., Sahajpal, R., Johnson, C.P., Rothstein, D.E., 2016. Recent Land Use Change to Agriculture in the U.S. Lake States: Impacts on Cellulosic Biomass Potential and Natural Lands. PLoS ONE, 11 (2). https://doi.org/10.1371/journal.pone.0148566.

Powell, R.L., Still, C.J., 2009. Biogeography of C3 and C4 vegetation in South America. Anais XIV Simpósio Brasileiro de Sensoriamento Remoto, Natal, Brasil, INPE, 2935-2942.

Simic A., Chen J.M., Liu J., Csillag F., 2004. Spatial scaling of net primary productivity using subpixel information. Remote Sens. Environ. 93, 246-258.

Simic Milas, A, Vincent, K., R., 2016. Monitoring Landsat vegetation indices for different crop treatments and soil chemistry. Int. J. Remote Sens. 38 (1), 141-160. DOI: 10.1080/01431161.2016.1259680.

Ueno, O., 2001. Environmental Regulation of C3 and C4 Differentiation in The Amphibious Sedge Eleocharisvivipara1. Plant Physiol. 127, 1524-1532.

USDA. (2016). CropScape- Cropland Data Layer. Url: https://nassgeodata.gmu.edu/CropScape/ (visited on March 3 2017).

USDA. (2017). National Agriculture Statistic Service. Url: https://www.nass.usda.gov/Newsroom/2018/01_12_2018.php (visited on April 3 2017).

Zhang, F, Chen, J.M., Chen, J., Gough, C.M., Martin, T.A., Dragoni, D., 2012. Evaluating Spatial and Temporal Patterns of MODIS GPP over the Conterminous U.S. against Flux Measurements and a Process Model. Remote Sens. Environ. 124, 717–729.

A MULTI-SCALE SEGMENTATION APPROACH TO FILLING GAPS IN LANDSAT ETM+ SLC-OFF IMAGES THROUGH PIXEL WEIGHTING

R. F. B. Marujo [a], L. M. G. Fonseca [a], T. S. Körting [a], H. N. Bendini [a]

[a]National Institute for Space Research, Image Processing Division, São José dos Campos, Brazil
(rennan.marujo, leila.fonseca, thales.korting, hugo.bendini)@inpe.br

KEY WORDS: Landsat, SLC-off, Segmentation, Gap-filling, Image processing

ABSTRACT:

Monitoring changes on Earth's surface is a difficult task commonly performed using multi-spectral remote sensing images. The absence of surface information in optical images due to the presence of cloud, low temporal resolution and sensors defects interfere in analyses. In this context, we present an approach for filling gaps in imagery mainly caused by small clouds and sensor defects. Our method consists of an adaptation from an existing method that uses spatial context of close-in-time images through the use of the most frequent value obtained using multiscale segmentation. Our method uses the pixel proportion contained in each segment to fill missing values. We applied the gap-filling methodology on three dates containing simulated images from Landsat7 using Landsat8 images. We validated the method by introducing and filling artificial gaps, and comparing the original data with model predictions. The developed approach surpassed Maxwell et al. (2007) gap-filling method for all bands, presenting a minimal R^2 of 0.78. Our method proved to enhance the Maxwell et al. (2007) gap-filling method, while also asymptotically maintaining the algorithm cost. It also allowed image texture to be conserved on reconstructed images. This characteristic enables narrow features, e.g., as roads, riparian areas, and small streams capable of being detected on the filled images. Based on that, further object-based approaches can be used on images filled using this methodology, demonstrating its capacity to estimate Earth's surface data.

1. INTRODUCTION

Satellite remote sensing instruments are one of the most valuable tools available for understanding and monitoring changes occurring in land, coastal and oceanic ecosystems. Several remote sensing applications, such as crop mapping and agricultural monitoring, need a great amount of images acquired in different times and free of clouds to map dynamic changes (Kuenzer et al., 2015, Bendini et al., 2017).

At the present, Landsat program consists in the most consistent historical source used when mapping Earth's surface (Wulder et al., 2012). Nowadays, Landsat-8 and Landsat-7 are active and acquiring images. However, Landsat-7, began to present hardware failure on Enhanced Thematic Plus (ETM+) Scan Line Corrector (SLC) in May 31, 2003 (Markham et al., 2004). This problem resulted acquisition loss for certain regions of the images. Images collected before such failures are called SLC-on images, and afterwards they are referred as SLC-off. This missing sections, also called gaps, correspond to a single pixel near the center of a path/row image and can reach 14 pixels in width near image borders (Storey et al., 2005), as illustrated in Figure 1.

Considering the possibility of data absence, it is important to develop tools capable of reconstructing Earth Observation (EO) data, removing null observations and estimating close-to-reality values to fill the gaps. Reconstruction of EO data to obtain images without gaps, remains a challenge, nevertheless some studies have been developed to solve these problems (Vuolo et al., 2017, Maxwell, 2004, Maxwell et al., 2007).

Considering Tobler's first law of geography which establishes that near things are more related to each other than distant things (Tobler, 1970), interpolating Landsat-7 SLC-off missing data would fill the gaps with neighboring pixel values. This procedure may be acceptable for small gaps, however it produces less satisfactory results for large gaps. Following that, (Maxwell, 2004) developed a segmentation approach to fill Landsat-7 SLC-off missing data, pairing it with a reference image that does not contains gaps, based on three principles:

- adjacent pixels are more likely to be similar;
- groups of the same landscape unit are likely to have similar spectral values;
- most landscapes remain constant for certain periods.

This approach consists in using a segmentation to delimit homogeneous regions, in which pixels are similar. This segmentation is acquired using a close-in-time reference image from a source that does not suffer from the gaps, e. g. a cloud-free image from Landsat-5/TM (Maxwell, 2004) or Landsat-8/OLI (Marujo et al., 2017b). Once the reference image segmentation is obtained, it is overlaid on the image which contain the gaps, the target image, and each gap area is replaced by the segment mode value, calculated using the values from the target image.

The mentioned gap-filling through segmentation has been improved further by using several segmentation levels instead of just one (Maxwell et al., 2007). The method employed three hierarchical segmentation levels and a nearest neighbor approach in case the third level isn't enough. In this approach, if a segment is fully contained within a gap, a segmentation level with larger area segments is used to fill the gap instead and so on with the third segmentation level (Maxwell et al., 2007). This approach is exemplified in Figure 2. The process is similar to (Maxwell, 2004), being the difference the number of segmentation levels.

Considering that nowadays Landsat-7/ETM+, Landsat-8/OLI and similar sensors, e. g. CBERS-4/MUX are in operation and also considering that these sensors present similar spectral bands (visible and near-infrared) and similar spatial resolution (30 and 20m), with a minimal spectral alignment and spatial resampling, they

Figure 1: A subset of Landsat-7/ETM+ SLC-off image (path/row 220/75) showing the missing data gaps size, smaller at the center of the image (top image) reaching 14 pixels in width near image borders (bottom image).

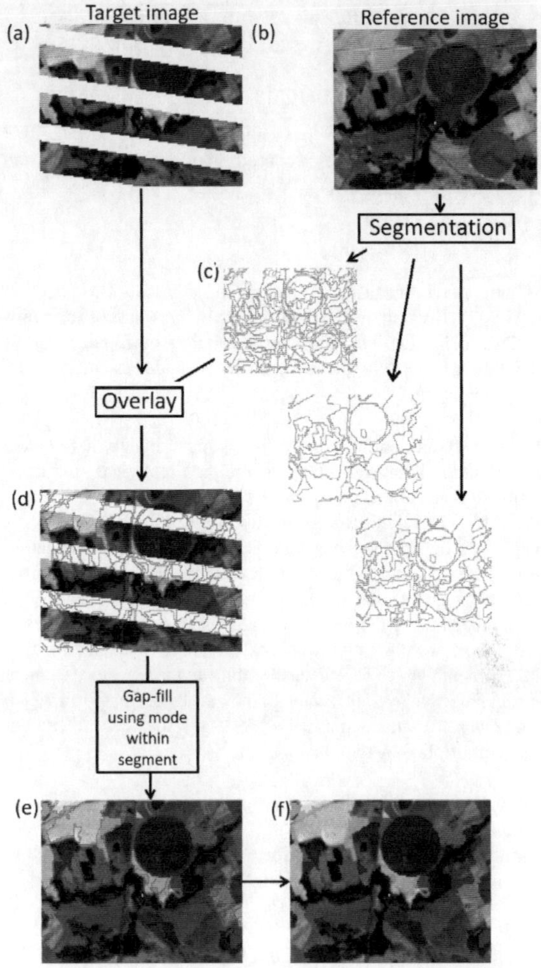

Figure 2: (a) a target image that contains gaps, in this case a Landsat-7/ETM+ SLC-off, (b) a reference image, in this case a Landsat-8/OLI acquired in a close date; (c) the reference image is segmented; (d) the segmentation is overlayed in the target image; (e) the mode value within the segment is calculated and used to fill missing data; (f) resulted synthetic image.

can be used to build a virtual constellation and increase the number of EO (Marujo et al., 2017a), helping gap-fill Landsat-7/ETM+ data. Within this perspective, the objective of this work is to propose a procedure to reconstruct Landsat-7/ETM+ gaps based on a multi-scale segmentation approach that considers pixel weights, to maintain image texture. The method uses non-gapped images of close-in-time-dates as reference to obtain homogeneous regions through segmentation. For each segment an internal distribution of the pixel values is calculated and the segments are overlaid on the gapped image. The filling values are then extracted from the gapped image, from within the segments and the pixel distribution is applied. To validate the method, artificial gaps were simulated in images and the filled result was compared to the reference.

2. DATASETS AND METHODOLOGY

2.1 Datasets

In this study, Landsat-7/ETM+ SLC-off effect was simulated using Landsat-8/OLI images. Landsat-8/OLI images were acquired, as shown in Table 1. The images were selected considering their overlap areas, which enables images from the same sensor with a difference between 7-9 days. The images were obtained as surface reflectance products, processed by Center Science Processing Architecture (ESPA) (UNITED STATES GEOLOGICAL SURVEY - USGS, 2017) using the Landsat 8 surface reflectance code (LaSRC), (UNITED STATES GEOLOGICAL SURVEY - USGS, 2019).

The SLC-off effect was simulated using a Landsat-7/ETM+ image from the same orbits (Path/Row 219/075 and Path/Row 220/075) by replicating gap pixels to the Landsat-8/OLI images. These areas were stored for further use as validation. Since the objective of this paper was to evaluate the feasibility of the proposed gap-filling algorithm, data from Landsat-7/ETM+ and Landsat-8/OLI was not used combined due to the need of harmonize the data from the different sensors (Bendini et al., 2017, Marujo et al., 2017a), which means that the only spectral information being used is derived from Landsat-8/OLI with synthetic SLC-off effect.

Image Pairs	Path/Row 219/075	Path/Row 220/075
Pair 1	02 Aug 2015	11 Aug 2015
Pair 2	28 Jul 2016	04 Aug 2016
Pair 3	31 Jul 2017	07 Aug 2017

Table 1: Acquisition dates for OLI images on Path/Row 223/072 and 222/072 to test the proposed gap-filling approach and Maxwell 2007 method

2.2 Methodology

As mentioned, our proposed algorithm, is based on Maxwell et al. (2007). Both algorithms are similar in most part of their processing, being expected that the algorithm's asymptotic cost is maintained since only linear operations are added to calculate the pixel weighting.

As input, the algorithm requires one gapped image, here called target, one non-gapped image, here called reference, and three segmentations, containing segments of three different scales, obtained by a hierarchical segmentation procedure. Considering a bottom-up approach, the method starts using the segmentation level with smaller segments, here called level-1 segmentation. For each segment that contains a gap, the segment mean value is calculated using the reference image values. Also on the reference image, a weighting factor is calculated for each pixel within the segment, which is the pixel value divided by the segment mean. This pixel weight is then used on the target image, in which gap values are estimated as the segment mean value of the target image multiplied by the pixel weight.

The validation procedure was performed by comparing the gap-filled values and the original values, which were initially removed from the image. The Pearson Correlation Coefficient (R^2), Root Mean Square Deviation ($RMSE$), Mean Absolute Error (MAE), Universal Image Quality Index ($UIQI$) (Wang and Bovik, 2002) and Visual Information Fidelity (VIF) (Sheikh and Bovik, 2006) were used to compare the results. The Pearson Correlation Coefficient indicates how the results are correlated with the expected results (the reference). However correlation does not imply causality. Based on that the visual indices were used to evaluate the results and the RMSE and MAE were used to measure the errors. RMSE and MAE both measure the results associated error, being MAE less affected by outlier values (Chai and Draxler, 2014). UIQI and VIF both measure compare the result image with the reference and are indicators of image quality. However, VIF is capable of detecting changes more similarly to the human eye when compared to UIQI (Sheikh and Bovik, 2006), while UIQI is more susceptible to be masked by the image areas that haven't changed.

3. RESULTS

In order to gap-fill the Landsat SLC-off images, reference segmentations were needed. Based on that, using a hierarchical segmentation, which is required for the proposed method, a multiscale segmentation (Baatz and Schäpe, 2000) algorithm was adopted and several segmentation parameters were tested. The scale parameter (the algorithm merge criteria) was empirically adjusted to 50, 100, 200, respectively for levels 1, 2 and 3, whereas the compactness and shape parameters were maintained in 0.5 (which implies 0.5 for smoothness) and 0.1 (which implies in 0.9 for color), respectively. These parameters were adopted for all segmentation levels and were estimated by visual interpretation, since the human eye is a strong and experienced evaluator of segmentation techniques (Gamanya et al., 2007).

A common effect through (Maxwell et al., 2007) gap-filling approach is the loss of narrow features such as roads, riparian areas, and small streams. In our method these features remain detectable, since texture is maintained. This is mainly due to the pixel weighting step. In larger gaps, like the ones found closer to the Landsat Path/Row borders, (Maxwell et al., 2007) approach fills the areas as homogeneous regions due to the repetition of the same spectral value across missing values, while in our method the filled areas look visually more realistic due to the textured filling. These visual differences can be seen in Figure 3, which illustrates a target Landsat image, presenting SLC-off gaps (upper image) and the same image filled by Maxwell et al. (2007) (central image) and by our method (bottom image).

Filling small gaps with an homogeneous value may be unnoticed in homogeneous target, as the central pivot at the bottom right corner of 3. However, in heterogeneous areas, e.g. the gallery forests and mixed fields, the single value gap-filling may not represent the reality distribution of that area, which may imply affect analysis depending on the adopted strategy. Although it is also an estimation of a non-observed area, gap-filling using the weighting pixel strategy conserves the texture, which may be a valid option for certain methods, e. g. Geographic Object-Based Image Analysis (GEOBIA).

To quantitatively evaluate both methods, Table 2 and Table 3 shows the agreement between (Maxwell et al. 2007) and our proposed method when comparing the results to a reference. The R^2, RMSE, MAE, UIQI and VIF were used to evaluate the results.

Band	R^2	RMSE	MAE	UIQI	VIF
Blue	0.82	92.17	55.07	0.99	0.56
Green	0.81	139.72	79.47	0.99	0.54
Red	0.86	207.93	111.22	0.99	0.50
NIR	0.78	319.39	199.97	0.99	0.50
SWIR1	0.85	344.98	199.80	0.99	0.49
SWIR2	0.87	246.04	144.77	0.99	0.50

Table 2: Validation agreement of Maxwell et al. (2007) multiscale segmentation gap-filling method on bands Blue, Green, Red, NIR, SWIR 1 and SWIR 2, obtained by simulating SLC-off effect using Landsat-8/OLI images.

Band	R^2	RMSE	MAE	UIQI	VIF
Blue	0.86	81.26	47.57	0.99	0.61
Green	0.85	120.91	65.64	0.99	0.60
Red	0.89	178.82	89.56	0.99	0.56
NIR	0.82	283.47	171.55	0.99	0.55
SWIR1	0.90	283.99	156.69	0.99	0.54
SWIR2	0.91	199.92	112.94	0.99	0.56

Table 3: Validation agreement of our pixel weighting multiscale segmentation gap-filling method on bands Blue, Green, Red, NIR, SWIR 1 and SWIR 2, obtained by simulating SLC-off effect using Landsat-8/OLI images.

For all spectral bands, the R^2 obtained using our method presented was more correlated to the original images than the R^2 obtained when using the original method, being the least R^2 obtained 0.78. Based on the same approach can also be noted that the residues RMSE and MAE are lower using our method. Considering that our data is scaled from 0 to 10,000, as provided by (UNITED STATES GEOLOGICAL SURVEY - USGS, 2017), the worse RMSE was obtained at the SWIR 1 band presenting values of 344.98 and 283.99, for Maxwell et al. (2007) gap-filling method and our proposed adaptation, while MAE worse results were obtained for the NIR band, presenting 199.97 and 171.55 respectively for both methods. As stated by Chai et al. (2004),

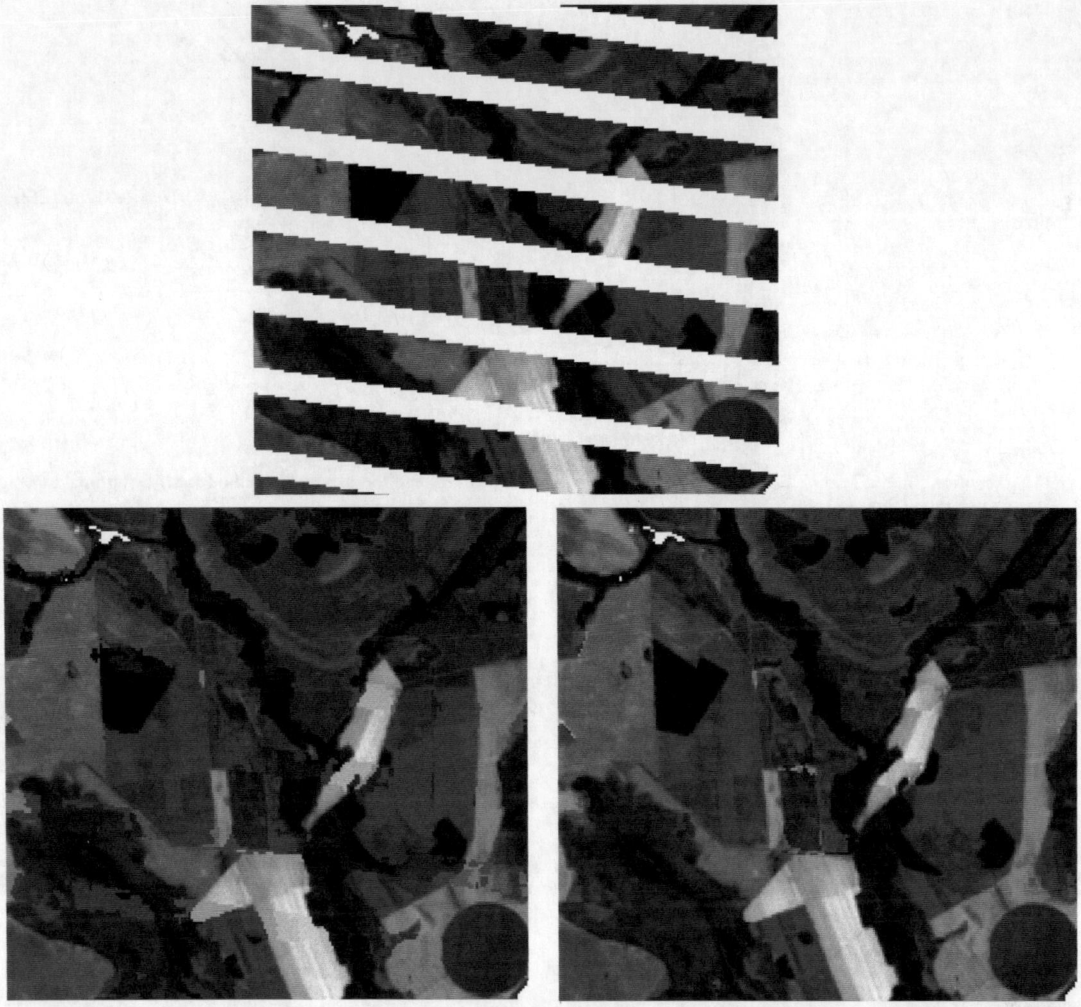

Figure 3: SLC-off Landsat-7/ETM+ image (upper image); image filled through Maxwell et al. (2007) hierarchical multi-scale segmentation method (bottom left image) and the image filled through the proposed method, which uses segmentation pixel weighting (bottom right image).

MAE presented lower values than RMSE due to this index deal better with outliers.

Analysing the image indices, UIQI and VIF, UIQI was close to 1 (0.99) at all image bands. This effect was due to UIQI been affected by the values that have not changed from the reference and the filled images. Based on that, the VIF describe better the visual differences between the images (Sheikh and Bovik, 2006). Based on the obtained VIF values, the gap-filled image quality assessment is closer to the reference when using our method to fill the gaps rather than the original proposition, since for all bands the VIF presented better values when using the proposed approach.

In Table 2 and Table 3 is also possible to note that the blue and green spectral bands presented lower correlation results when compared to the other spectral bands. The almost crescent correlation results occurred probably due to atmospheric interference in shorter wavelength bands, also known as Rayleigh Scattering (Jensen, 2007), that is not completely suppressed even with atmosphere correction.

Another characteristic that we noticed is that the algorithm's asymptotic cost was maintained since only linear operations were added to calculate the pixel weighting. We tested the algorithm performance using a 16GB of memory ram and an Intel processor i7-6500U of 2.5GHz computer. Figure 4 shows the execution time for both algorithms using images ranging from 100x100 to 2000x2000 pixels. Implementations of the methods, in Python and R are being developed as an Earth Observation fill package and can be found in *https://github.com/marujore/*.

4. CONCLUSIONS

Our multiscale segmentation gap-filling method enhanced the original Maxwell et al. (2007) method, while asymptotically maintaining the algorithm cost. Our approach allowed image texture to be conserved on reconstructed images, allowing object based analysis to be further applied on Landsat-7/ETM+ images. Another improvement was that narrow features, e.g., roads, riparian areas, and small streams, remained detectable in reconstructed images.

Pragmatically, it is expected that the approach developed in this

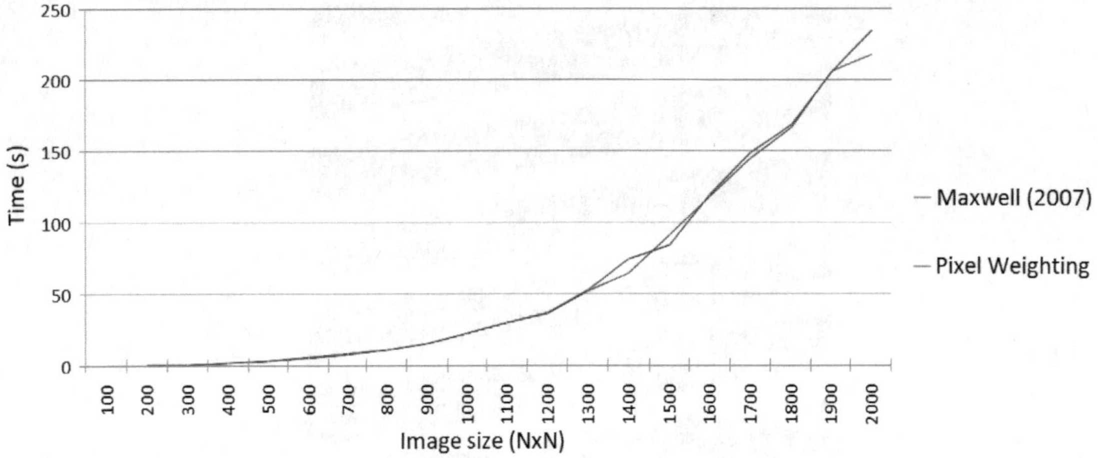

Figure 4: Execution time of Maxwell et al. (2007) gap-filling algorithm (blue) and the proposed method (red) using images ranging from 100x100 to 2000x2000 pixels.

thesis can be used effectively in academic and industrial environments and henceforth contribute to improves overall multi-source harmonization and Earth observation data reconstruction. The developed codes are open source, free and published so it can be used or adapted, as is all its dependencies.

ACKNOWLEDGEMENTS

The authors would like to thank the Brazilian National Institute for Space Research (INPE) and FAPESP (Project # 2014/08398-6 and Process # 2016/08719-2), São Paulo Research Foundation for funding this research.

REFERENCES

Baatz, M. and Schäpe, A., 2000. Multiresolution segmentation: an optimization approach for high quality multi-scale image segmentation. In: Proceedings..., Angewandte Geographische Informationsverarbeitung, Heidelberg, pp. 12–23.

Bendini, H. N., Fonseca, L. M., Körting, T. S., Marujo, R. F. B., Sanches, I. D. and Arcanjo, J. S., 2017. Assessment of a multisensor approach for noise removal on Landsat-8 OLI time series using CBERS-4 MUX data to improve crop classification based on phenological features. Revista Brasileira de Cartografia 69(5), pp. 240–251.

Chai, T. and Draxler, R. R., 2014. Root mean square error (rmse) or mean absolute error (mae)? arguments against avoiding rmse in the literature. Geoscientific Model Development 7(3), pp. 1247–1250.

Gamanya, R., De Maeyer, P. and De Dapper, M., 2007. An automated satellite image classification design using object-oriented segmentation algorithms: A move towards standardization. Expert Systems with Applications 32(2), pp. 616–624.

Jensen, J., 2007. Remote sensing of the environment: an Earth resource perspective. Prentice Hall series in geographic information science, Pearson Prentice Hall, New Jersey.

Kuenzer, C., Dech, S. and Wagner, W., 2015. Remote sensing time series. Remote Sensing and Digital Image Processing, Vol. 22, Springer International, Cham.

Markham, B. L., Storey, J. C., Williams, D. L. and Irons, J. R., 2004. Landsat sensor performance: history and current status. IEEE Transactions on Geoscience and Remote Sensing 42(12), pp. 2691–2694.

Marujo, R. F. B., Fonseca, L. M. and Körting, T. S., 2017a. Spectral normalization between Landsat-8 / OLI , Landsat- 7 / ETM + and CBERS-4 / MUX bands through linear regression and spectral unmixing. In: Anais..., Brazilian Symposium on Geoinformatics, INPE, São José dos Campos, p. 10.

Marujo, R. F. B., Fonseca, L. M. G., Körting, T., Bendini, H., Queiroz, G., Vinhas, L. and Ferreira, K., 2017b. Remote sensing image processing functions in Lua language. Journal of Computational Interdisciplinary Sciences 8(3), pp. 1–14.

Maxwell, S. K., 2004. Filling Landsat ETM+ SLC-off gaps using a segmentation model approach. Photogrammetric Engineering & Remote Sensing 70(10), pp. 1109–1111.

Maxwell, S. K., Schmidt, G. L. and Storey, J. C., 2007. A multiscale segmentation approach to filling gaps in Landsat ETM+ SLC-off images. International Journal of Remote Sensing 28(23), pp. 5339–5356.

Sheikh, H. R. and Bovik, A. C., 2006. Image information and visual quality. IEEE Transactions on Image Processing 15(2), pp. 430–444.

Storey, J., Scaramuzza, P. and Schmidt, G., 2005. Landsat 7 scan line corrector-off gap-filled product development. In: Proceedings..., PECORA GLOBAL PRIORITIES IN LAND REMOTE SENSING, ASPRS, South Dakota, pp. 23–27.

Tobler, W. R., 1970. A computer movie simulating urban growth in the Detroit region. Economic Geography 46(277), pp. 234.

UNITED STATES GEOLOGICAL SURVEY - USGS, 2017. User guide: Earth resources observation and science (EROS) Center Science Processing Architecture (ESPA) on demand interface. Technical report, Sioux Falls. Access in: 18 Nov. 2017.

UNITED STATES GEOLOGICAL SURVEY - USGS, 2019. Product guide: Landsat 8 surface reflectance code (LaSRC) Product. Technical report, Sioux Falls.

Vuolo, F., Ng, W.-T. and Atzberger, C., 2017. Smoothing and gap-filling of high resolution multi-spectral time series: example of Landsat data. International Journal of Applied Earth Observation and Geoinformation 57, pp. 202–213.

Wang, Z. and Bovik, A. C., 2002. A universal image quality index. IEEE Signal Processing Letters 9(3), pp. 81–84.

Wulder, M. A., Masek, J. G., Cohen, W. B., Loveland, T. R. and Woodcock, C. E., 2012. Opening the archive: how free data has enabled the science and monitoring promise of Landsat. Remote Sensing of Environment 122, pp. 2–10.

LAND COVER MONITORING OF LAGUNA LAKE WATERSHED USING MODIS NDVI DATA

J.M. Medina[1,*], A.C. Blanco[1,2], C.G. Candido[1]

[1]Training Center for Applied Geodesy and Photogrammetry, University of the Philippines Diliman, 1101 Quezon City, Philippines - jmmedina@up.edu.ph, chriscandido93@gmail.com

[2] Department of Geodetic Engineering, University of the Philippines Diliman, 1101 Quezon City, Philippines - acblanco@up.edu.ph

KEY WORDS: LULC, vegetation index, ISODATA, spacetime cube, hot spot analysis, MAD, canonical correlation

ABSTRACT:

Land use and land cover monitoring is an important component in the management of Laguna Lake watershed due to its impacts on the lake's water quality. Due to limitations caused by cloud cover, satellite systems with limited revisit capability fail to provide sufficient data to more effectively monitor the land surface. Normalized difference vegetation index (NDVI) derived from MODIS image data were used to generate land cover maps for the years 2001, 2005, 2009, 2013, and 2017. These were produced by classifying ISODATA classes using annual NDVI profiles, which resulted in land cover classes, namely, agricultural land, built-up, forest, rangeland, water, and wetland. The resulting maps were post-processed using multi-variate alteration detection (MAD), resulting in multi-temporal land cover maps with improved overall accuracies and kappa coefficients that indicate moderate agreement with ground truth data. Spatiotemporal hot spot analysis was also performed using NDVI data from 2001 to 2017 to identify vegetation hot spot areas, where clustering of low NDVI values were observed over the years. Results showed an increasing trend in built-up areas accompanied by decreasing trends in water and wetland areas, indicating impacts caused by land reclamation and expansion of residential subdivisions near the lakeshore. The decrease in total vegetation area from 2001 to 2017 could be attributed to conversion of land to built-up surface. Vegetated areas in identified hot spots decreased from 41% in 2001 to 19% in 2017. This suggests that vegetation cover in these hot spots was converted to non-vegetated surface during the time period studied.

1. INTRODUCTION

Remote sensing satellites acquire tremendous amount of data which have been useful in monitoring the Earth's surface. Time-series data coming from high temporal resolution sensors such as the Moderate Resolution Imaging Spectroradiometer (MODIS) have been used in numerous studies to monitor and assess land cover using vegetation datasets such as the normalized difference vegetation index (NDVI). The common finding is that using MODIS time-series data for land cover studies is cost- and time-efficient for mapping large areas (Nyamekye, et al., 2014).

Laguna Lake is the largest inland body of water in the Philippines. It is situated in the heart of the CALABARZON-Metro Manila region, which consists of six provinces and is one of the most densely populated areas in the country. These geographical and demographic characteristics make land cover monitoring a vital component in resource management of the lake as land use and land cover have an impact on the lake's overall quality, among other concerns.

Nationwide land cover maps, which are commonly used in different agencies, are released every five years in the Philippines. The long time between successive releases makes monitoring of land cover changes difficult especially for purposes of resource management. This concern is important when dealing with highly dynamic and fast-changing systems such as the Laguna Lake watershed. Because of the large extent of land area covered and the many different human activities happening in it, there is a need to consider other datasets which can provide information faster at an acceptable level of accuracy. Time-series MODIS NDVI offer an alternative to address this issue.

2. MATERIALS AND METHODS

2.1 Study Area

Laguna Lake has a surface area of about 900 sq km, making it the largest inland waterbody in the Philippines and third largest in Southeast Asia (Laguna Lake Development Authority, 2017). Otherwise known as Laguna de Bay, the lake has an average depth of about 2.5 m (Santos-Borja, Nepomuceno, 2003). It is used for various purposes such as aquaculture, navigation, flood control, and source of water for irrigation, industrial cooling, and drinking.

With a shoreline of about 285 km, the lake is divided into four regions namely West, Central, South, and East Bays, each with unique water quality characteristics as a result of effluents coming from more than 100 tributaries draining into the lake from the surrounding area. Its only outlet is the Pasig River which drains into Manila Bay. Because of such hydrologic set-up, backflow occurs during the dry season when lake water level is lower than the mean sea level. This results in the intrusion of saline water which enters the West Bay through the Pasig River making the lake water brackish (Santos-Borja, Nepomuceno, 2003).

The watershed area of Laguna Lake is about 2,980 sq km and geographically covers, in part or in full, 59 towns in Metro Manila and the Provinces of Laguna, Rizal, Cavite, Batangas, and Quezon (Santos-Borja, Nepomuceno, 2003). The urban towns in Laguna and Metro Manila are situated in the west portion of the watershed while the north and east are characterized by a mixture of built-up and vegetation cover. Agricultural lands are mostly found in the east and south areas. For management purposes, Laguna Lake watershed is divided into 24 sub-watersheds. Figure 1 shows a map of the watershed area.

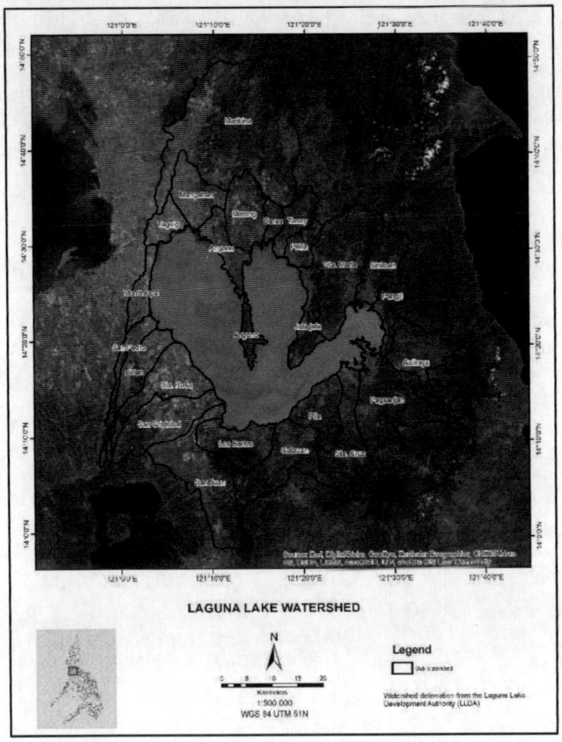

Figure 1. Laguna Lake watershed area

2.2 MODIS NDVI Data

Moderate Resolution Imaging Spectroradiometer (MODIS) is a remote sensing instrument onboard the Terra and Aqua satellites. These two satellites acquire complete coverage of Earth observation data every one to two days in sun-synchronous near-polar orbits, with Terra making an overpass from north to south across the equator in the morning and Aqua making an overpass in reverse direction in the afternoon (MODIS Moderate Resolution Imaging Spectroradiometer, n.d.)

The MODIS sensor captures data in 36 spectral bands with wavelength ranging from 0.4 to 14.4 µm available in three spatial resolutions. Bands 1 (red; 620-670 nm) and 2 (near-infrared; 841-876 nm) are at 250-m resolution. Bands 3 to 7 (459-479 nm, 545-565 nm, 1230-1250 nm, 1628-1652 nm, 2105-2155 nm, respectively) are at 500-m resolution. Finally, Bands 8 to 36 have 1-km resolution (MODIS Moderate Resolution Imaging Spectroradiometer, n.d.). Bands 1 to 7 are commonly referred to as land bands as these are used for remote sensing applications of the land surface (Carroll, et al., 2010).

The normalized difference vegetation index (NDVI) is a remote sensing product used to measure vegetation health and density. It has a value ranging from -1 to 1, where lower values indicate sparse and/or less healthy vegetation while higher values denote denser and/or healthier vegetation. It is based on measured reflectance from red and near-infrared bands of a satellite image and is defined by the following equation:

$$NDVI = \frac{NIR - Red}{NIR + Red} \quad (1)$$

MOD13Q1 Version 6 data are a product derived from image data acquired by Terra MODIS sensor. The dataset is a Level 3 MODIS product generated as 16-day composites at 250-m spatial resolution. To produce the composite data, the algorithm chooses the best pixel from all acquisitions within the 16-day period based on the following criteria: (i) minimum cloud contamination; (ii) low view angle; and (iii) highest NDVI value. The dataset is in sinusoidal projection and is available from February 2000 up to present (Didan, 2015). Accuracy of MODIS NDVI data is within ±0.025. This value represents the accurate retrieval of top-of-canopy and nadir VI values for high quality observations (MODIS Land Team Validation, n.d.).

2.3 Pre-processing

MOD13Q1 data covering Luzon island from 2001 to 2017 were downloaded from NASA's Earthdata site. Each year consists of 23 granules (or scenes), with each granule representing 16-day composite data consisting of 12 layers including VI layers, surface reflectance in MODIS Bands 1, 2, 3, and 7, and pixel quality layers, among others.

The NDVI layer was extracted from each of the downloaded MODIS data and re-projected to WGS 84 UTM Zone 51N. The NDVI layers acquired during the same year were stacked in order from Julian date-of-year 001 (January 01-16) to 353 (December 19-31) to produce the NDVI dataset for each year. Subsets of the datasets were generated following the extent of Laguna Lake watershed with 500-m buffer.

As the data are already a Level 3 product, these have already undergone pre-processing to minimize effects of the atmospheric elements, such as clouds, in the data. However, to further minimize cloud contamination effects, Savitzky-Golay (SavGol) filtering was done on each of the NDVI stacks. The SavGol filter is based on the study of Savitzky and Golay (1964) which presented an approach to smoothen a set of consecutive values, or a spectrum, by means of a simplified convolution using least squares fitting within a moving window. Weights are given such that the observations within the filter window are fitted to a polynomial whose degree is equal to the weight (Chen, et al., 2004).

Half-width smoothing window m and degree of polynomial d are required parameters for SavGol filter. For this study, the value of (m, d) used was $(4, 6)$, as it is the optimal pair of parameters determined by Chen, et al. (2014) to provide the best fit. For each pixel location in the stacked NDVI layers, the values were taken by SavGol filter and fitted to generate the smoothened dataset for that location. The results of this filtering are smoothened NDVI datasets for Laguna Lake watershed from 2001 to 2017.

2.4 Land Cover Classification

Smoothened NDVI datasets were used to generate land cover classes using Iterative Self-Organizing Data Analysis Techniques (ISODATA) algorithm. ISODATA is an unsupervised classification technique in which pixels are clustered based on mean distances between pixels and class centroids and standard deviations. Splitting of clusters is done when the maximum standard deviation exceeds a user-specified value and when the number of pixels in the cluster is twice a specified minimum number. Clusters are merged if the distances between their centroids are less than the minimum distance specified (Ball, Hall, 1965). New class distances and standard deviation for each cluster are calculated after each iteration and new clustering is performed. The iteration stops when the maximum number of iterations is reached, the maximum number of ISODATA classes specified is achieved, and/or no classes can be further split or merged (Dhodhi, et al., 1999).

The threshold values for class standard deviation and class distance were determined by calculating these statistics from annual average NDVI. Land cover and land use data for the study area released by the National Mapping and Resource Information Authority (NAMRIA), the Philippines' national mapping agency, as of 2010 were used as arbitrary classes for the computation of zonal statistics. The mean values and standard deviations for the zones were compared and used to derive the minimum class difference (distance) and maximum standard deviation used as input parameters in ISODATA classification. The maximum number of classes was set to 40 and the maximum iteration set was 100.

The classification was done on time-series NDVI datasets for the years 2001, 2005, 2009, 2013, and 2017. The result for each are 40 ISODATA classes with no specific land cover class. Due to the coarse resolution of the data used, only Level 1 classification of the Anderson Classification Scheme was considered. These land cover classes are agricultural land, built-up, forest, rangeland, water, and wetland (Anderson, et al., 1976).

Assignment of ISODATA classes to each of the six land cover classes were done by NDVI profiling. Three sample pixels were randomly selected from each ISODATA class. Time-series NDVI values from Julian date-of-year 001 to 353 were determined for each sample pixel and the average value per date-of-year of the three samples was calculated and plotted. This process resulted in an NDVI profile for each ISODATA class. Ground truth data from Google Earth images were used to match an NDVI profile with a certain land cover.

2.5 Post-processing

A common issue with generating multitemporal land cover maps is the presence of illogical land cover transitions. An example of this is the change from agricultural land to built-up and then back to agricultural land. While this may not necessarily be impossible, this becomes less likely to happen for shorter time intervals. To address this, multivariate alteration detection (MAD) was implemented. This transformation algorithm is based on canonical correlation analysis (CCA) which investigates the relationship between two groups of variables such as spectral bands from two remote sensing datasets. This approach was applied in this study to detect significant changes between two NDVI datasets. Given two NDVI datasets expressed as vectors:

$$X = [X_1 \cdots X_k]^T \text{ and } Y = [Y_1 \cdots Y_k]^T \tag{2}$$

where k is the number of NDVI values in a dataset, the transformation change detection is given by:

$$X - Y = [X_1 - Y_1 \cdots X_k - Y_k]^T \tag{3}$$

Optimizing the measure of change in a simple difference requires maximizing the variance which measures the quality of being divergent:

$$Var\{v_1(X_1 - Y_1) + \cdots + v_k(X_k - Y_k)\} = Var\{v^T(X - Y)\} \tag{4}$$

High absolute values of $v^T(X - Y)$ indicate pixel locations with maximum change. Hotelling (1936) standardized the approach in determining the optimum value for v through CCA. In a standard CCA, linear transformation is performed on each set of variables resulting in linear combinations arranged by correlation or strength of similarity. The highest mutual correlation is called the first canonical variable, the next highest is the second canonical variable, and so on. Given the canonical variables U and V, and $\rho = Corr(U, V)$ is the correlation between them, the problem can be generalized by:

$$\sum XY \sum^{-1} YY \sum^{T} YX\, a = \rho^2 \sum XXa \tag{5}$$

$$\sum^{T} YX \sum^{-1} XX \sum XY\, b = \rho^2 \sum XXb \tag{6}$$

where $\sum^{-1}XX$ and $\sum^{-1}YY$ are the covariance matrices of the two images, and $\sum XY$ and $\sum^T YX$ are the inter-image covariance matrices which are equal. The system of Equations (5) and (6) generates new images $U = [U_1 \cdots U_N]^T$ and $V = [V_1 \cdots V_N]^T$ which are called the canonical variates. The pair of elements (U_1, V_1) is the maximally correlated pair, and (U_2, V_2) is maximally correlated which is orthogonal to, or uncorrelated with, (U_1, V_1). Pair differencing is performed after CCA generating a sequence of transformed difference images:

$$\left\{ \begin{array}{l} U = \begin{pmatrix} U_1 \\ \vdots \\ U_N \end{pmatrix} \\ V = \begin{pmatrix} V_1 \\ \vdots \\ V_N \end{pmatrix} \end{array} \right. \rightarrow MAD = \begin{pmatrix} MAD_1 \\ \vdots \\ MAD_N \end{pmatrix} = \begin{pmatrix} U_1 - V_1 \\ \vdots \\ U_N - V_N \end{pmatrix} \tag{7}$$

Each element of the MAD image generated using Equation (7) is called a MAD variate, which can be used for change detection analysis. For the NDVI datasets used in this study, 23 MAD variates were generated ($N = 23$), each representing one date-of-year since the comparison was done between two successive years.

Statistical thresholding was done using the MAD variates of each pixel to identify whether it changed or not. MAD variates indicating significant change tend to deviate strongly from a multivariate normal distribution. Furthermore, assuming independence of orthogonal MAD variates, one may expect the sum of the square MAD variates after standardization to unit variance approximately follow the chi-squared (χ^2) distribution with N degrees of freedom. The MAD variate standardization function is given by:

$$z = \sum_{i=1}^{N} \left(\frac{MAD_i}{\sigma MAD_i} \right)^2 \tag{8}$$

Each pixel, having a corresponding value for z, can be given weights according to the:

$$P(unchanged) = 1 - P_{\chi^2;N}(z) \tag{9}$$

A small value of z typically implies greater probability that the pixel is unchanged. A statistical threshold was selected such that $P(unchanged) > t$, where t is the threshold value taken from the χ^2 distribution at $\alpha = 0.05$. The degree of freedom (df) was taken to be equal to N. Figure 2 shows an example of a χ^2 distribution plot. The broken line marks the threshold value determined using the χ^2 distribution table. A pixel whose z value is greater than, or lying to the right of, t is classified as unchanged. Otherwise, the pixel is marked as changed pixel.

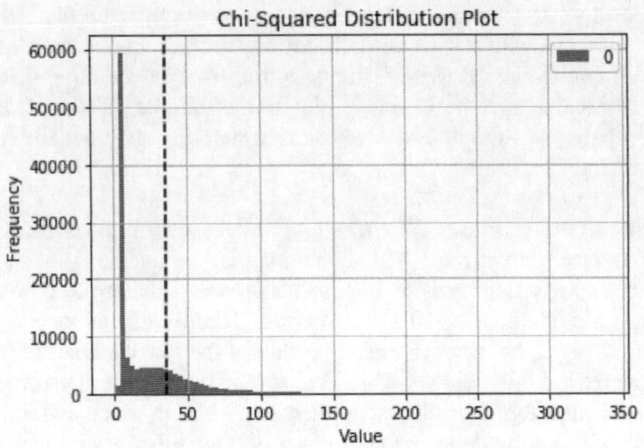

Figure 2. Sample χ^2 distribution plot

MAD was implemented to generate a mask for unchanged and a mask for changed land cover pixels based on NDVI values. Refinements in the initial land cover maps were done pairwise starting with a reference map (i.e., land cover at time t_0), which was the one with the highest overall accuracy. Land cover classification for unchanged pixels were retained. For changed pixels, the classification in the map being refined (i.e., land cover at time t_1) is adopted. The land cover at t_1 was then used as reference to refine the land cover map at time t_2, and so on.

2.6 Accuracy Assessment

To generate a validation data, the watershed area was divided into 1 km by 1 km grids within each of which a validation point was chosen. The land cover of this point was noted based on ground truth data from Google Earth images. This process generated a total of 1,296 validation points used for checking the accuracy of each land cover map produced. A set of validation dataset was generated based on image data for each of the years 2005, 2010, and 2015 used for assessing the accuracy of 2001, 2005 and 2009, 2013, and 2017 land cover maps, respectively. The five-year interval follows the same period for updating official land cover and land use maps by the national mapping agency.

2.7 Vegetation Hot Spot Analysis

Vegetation hot spot mapping based on NDVI time-series data involved two stages: (i) creating space-time cube and (ii) emerging hot spot analysis. A space-time cube is three-dimensional representation of a time-series dataset whose base denote the spatial component or location (X, Y) and whose height indicate the temporal component or timestamp of the data (Kraak, 2003). Each element of a space-time cube is a bin, which represents a specific instance of the entire dataset at a given location and time. Bins that share the same location represents a bin time series, which is the dataset for that specific location through time. Bins with different locations but share the same temporal extent make up a time slice (Esri, n.d.). In the NDVI dataset, each bin time series corresponds to a unique pixel location while a time slice corresponds to a time-step interval of one year.

Emerging hot spot analysis uses the space-time cube as input data along with parameters such as neighborhood distance and neighborhood time step. The neighborhood distance indicates the spatial extent of the analysis to be considered for each location while the neighborhood time step gives the number of time-step interval to be included in the analysis. The analysis first calculates the Getis-Ord Gi* statistic for each bin to test whether there is significant clustering of high or low values at each location considering spatial neighbors. Each bin is given a hot spot classification based on the results. Mann-Kendall test is then performed to identify whether there is a significant trend in each bin time series to determine the final hot spot classification for a specific location (Esri, n.d.).

Annual average NDVI data from 2001 to 2017 were used as input to create the space-time cube. Inland water bodies such as lakes were masked out so that the analysis focused only on the watershed's land area. The NDVI raster data were converted to polygon grid features with the same dimensions as the raster pixels and with object ID, date, and NDVI as attributes. The object ID was set to be the same for a unique pixel location regardless of the year while the date was set to January 01 for all pixels belonging to the same year. NDVI values (range 0.1-0.8) were negated so that originally low values would be taken as greater in numerical value in the hot spot analysis. This was done because the statistical calculations tag locations where clustering of high values occur as hot spots. The negation ensured that locations of decreasing vegetation (low NDVI in the actual data) over time were correctly identified as hot spots.

Vegetation hot spot analysis was done using Esri's Space Time Mining Toolbox in ArcGIS Pro. Cold spots were not considered as these indicate locations of healthy and dense vegetation and are not a primary concern of this study. A spatial neighbor distance equal to one pixel and neighborhood time step of 2 were used as input parameters.

3. RESULTS

3.1 NDVI Profiles of Land Cover Classes

The NDVI profiles generated for each of the 40 ISODATA classes were grouped together based on similarity in trends and land cover class. Different trends were observed within the same land cover. This is due to the varying characteristics of the surface. Figure 3 shows the NDVI profiles of the ISODATA classes categorized under each land cover type.

Water 1, with NDVI value ranging from -0.1 to 0.0, shows a stable trend throughout the year, representing water areas such as most portions of Laguna Lake. Water 2 has the same NDVI and trend as Water 1 during the first half of the year but with spikes towards the latter part. These spikes represent increases in NDVI due to relatively large areas covered by water hyacinths which accumulate for several days near aquaculture structures in the lake or close to the lake shore while moving very slowly around the lake. Water 3 represent areas very close to the shore where water hyacinths accumulate more frequently with longer retention period than they do in areas far from the shore, hence.

The profile for wetland areas show increasing NDVI during the dry season and decreasing as the wet season starts. These areas, which are located along the lake's shoreline, are lands that get exposed when the lake water level is lower than average during the dry season months. Patches of vegetation, particularly grasses, grow on the exposed land, which explains the NDVI values that are as high as 0.6. As the water level increases again during the wet season, these lands get submerged again giving lower NDVI values similar to those of water.

All NDVI profiles for ISODATA classes identified as built-up areas show a similar trend that is relatively steady for the entire year, having only changes within ± 0.1. These trends indicate land cover that barely changes, which is true for built-up areas. The difference in NDVI values between different built-up ISODATA classes are influenced by the amount of sparse vegetation present. One class, for example, is characterized by areas dominated by built-up environment with more sparse vegetation than another class, which represents high density urban areas.

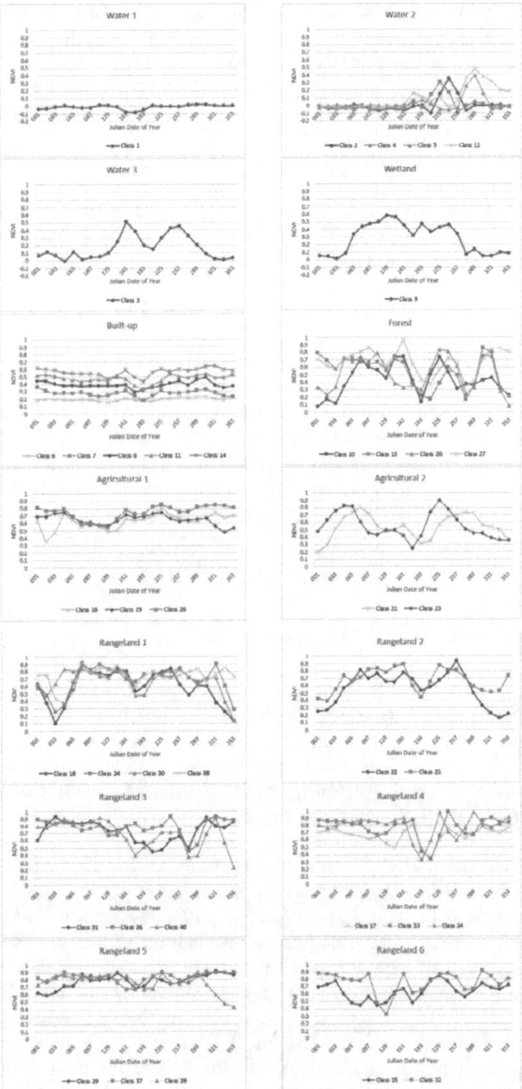

Figure 3. NDVI profiles of different land cover classes

Two types of agricultural land cover classes were identified based on their NDVI profiles. Agricultural 1 correspond to perennial cropland areas, which are for plants with longer agricultural cycles such as fruit trees. The profile shows a steadily high NDVI values which may drop after several months or even more than a year. Agricultural 2 shows the profile of croplands with agricultural cycles of five to six months as shown by two peaks within a year. One cycle shows clearly the phenology of the crops in which the increasing NDVI marks the growth period, the peak indicates the time when the crops are fully grown, and the decreasing NDVI happens during the harvest season. The minimum in the plot marks the post-harvest period when the soil is almost bare until new crops are planted, and a new cycle starts. Agricultural 3 also represents croplands with two cycles. However, the period is likely to be more than six months and starts and ends in the middle of the year compared to Agricultural 2 whose first planting season starts in January.

The NDVI profiles of ISODATA classes identified as forest cover are more erratic. This behavior may be attributed to the fact that forested areas in Laguna Lake watershed are found in high altitude areas such as mountains. Because of this location and topography, a considerable portion of the forests are usually covered by thick clouds which are too difficult to be corrected even with the corrections applied on MODIS data and the SavGol filtering done on the time-series NDVI. However, image interpretation and validation showed that these classes belong to forestlands. In general, however, the four ISODATA classes show agreement in the trend during some portions of the year, such as high values during the dry season and drop in NDVI during some wet season months when persistent cloud cover is present.

Finally, six different profiles for rangeland areas were identified. The behavior of these different profiles may be attributed to the different phenological characteristics of grass and shrub species which cover the rangeland areas. For example, grassland areas are almost bare during the summer months when precipitation is low, while some types of bushes may shed leaves on certain seasons while other species may have abundant leaves throughout the year. Separation of specific vegetation types and plant species was not done in this study.

3.2 Multi-temporal Land Cover Mapping

Land cover maps were generated using the NDVI datasets for years 2001, 2005, 2019, 2013, and 2017. Table 1 shows that post-processed land cover maps showed improved overall accuracy and kappa coefficient compared to the initial maps. The map for 2005 was not post-processed because it was used as reference in the MAD implementation for change detection. Kappa coefficient values show moderate agreement with ground truth data (Congalton, 2001).

Year	Overall Accuracy		Kappa Coefficient	
	Initial	Post-processed	Initial	Post-processed
2001	0.75	0.79	0.57	0.64
2005	0.79	0.79	0.67	0.67
2009	0.69	0.72	0.52	0.55
2013	0.67	0.69	0.49	0.51
2017	0.69	0.71	0.51	0.54

Table 1. Accuracy assessment of MODIS NDVI-derived land cover maps

Table 2 shows the user's (UA) and producer's accuracies (PA) for each class in the post-processed land cover maps. Accuracy values range from 0.50 to 1.00, with the low values calculated for vegetation classes (agricultural land, forest, and rangeland) and the high values for the water and wetland classes. Confusion among vegetation classes are due to lack of more refined separation between the three vegetation land cover types as an effect of using a 250-m resolution data. Figure 4 shows the final land cover maps.

Class	2001		2005		2009		2013		2017	
	UA	PA	UA	PA	UA	PA	UA	PA	UA	PA
Agricultural	0.63	0.63	0.59	0.66	0.53	0.67	0.50	0.54	0.51	0.62
Built-up	0.84	0.74	0.83	0.79	0.79	0.74	0.68	0.65	0.63	0.73
Forest	0.80	0.51	0.69	0.72	0.58	0.53	0.56	0.62	0.73	0.62
Rangeland	0.82	0.89	0.86	0.84	0.95	0.79	0.78	0.76	0.81	0.74
Water	0.71	0.88	1.00	0.88	0.95	0.88	1.00	0.89	0.95	0.77
Wetland	0.71	0.63	0.80	1.00	0.63	0.83	0.75	1.00	0.83	0.83

Table 2. Individual accuracies of different land cover classes

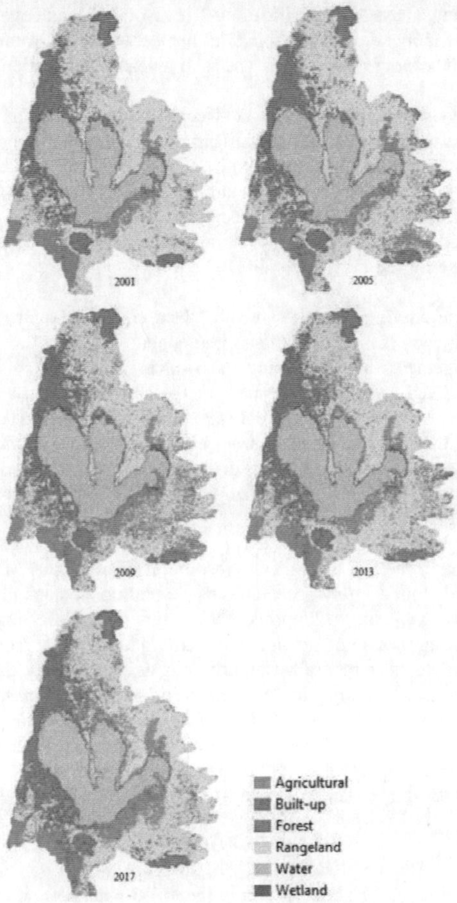

Figure 4. MODIS NDVI-derived land cover maps

Table 3 shows the percent area coverage of each land cover class per year. Rangeland, which includes grasslands, shrublands, and bushlands, covers most of the watershed at 39%-48%. This is followed by waterbodies, which includes Laguna Lake, at around 20%. Built-up environment makes up 12%-15% of land cover. Forested areas account for 4%-7% of the watershed, while wetlands cover 1% of the area. The total area of the watershed based on the land cover maps generated is 4,087.90 sq. km.

	2001	2005	2009	2013	2017
Agricultural	12.62	14.45	18.47	15.10	18.76
Built-up	12.43	13.29	12.22	13.15	15.43
Forest	4.22	7.35	6.67	6.72	5.64
Rangeland	48.65	43.58	40.02	43.09	39.25
Water	21.20	20.66	21.50	21.07	20.50
Wetland	0.88	0.68	1.12	0.86	0.42

Table 3. Percent area coverage per land cover class

3.3 Vegetation Hot Spot Analysis

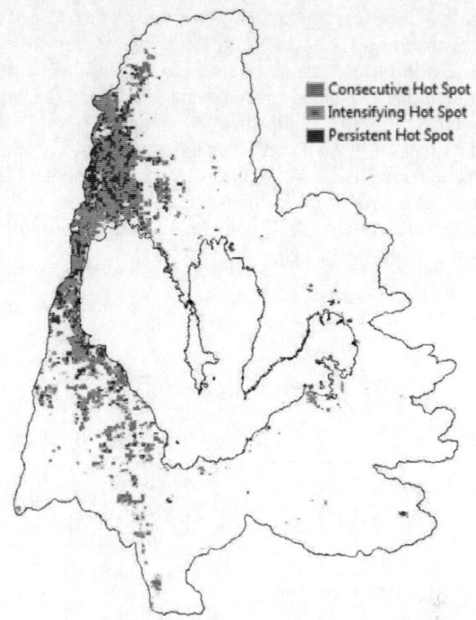

Figure 5. Vegetation hot spot map of Laguna Lake watershed

Figure 5 shows the distribution of vegetation hot spots based on the analysis of 2001-2017 NDVI data for Laguna Lake watershed. Hot spot areas cover 12.36% of the watershed. Of these hot spots, 39.55% are intensifying, 25.63% are persistent, and 20.70% are consecutive hot spots. Majority of the identified hot spots, particularly intensifying and persistent ones, are in the northwest side of the watershed, which is the high-density urban area of Metro Manila. Consecutive hot spots dominate the low-density urban towns of Laguna in the southwest. Hot spot areas can also be found along shoreline areas and rural town centers where clusters of built-up structures are found. These environments are characterized by sparse vegetation.

4. DISCUSSION

Analysis of the changes in built-up area coverage showed an overall increasing trend from 2001 to 2017, with as much 2.28% or 93 sq km increase in area happening between 2013 and 2017. Despite the very little fluctuations in area (less than 1% or 40 sq km), water and wetlands area coverage have been generally decreasing. This change may be attributed to land reclamation which has been one of the challenges faced by the lake's monitoring agency in managing the lake and watershed resources. Reclamation of land within the lake's shoreline area is allowed only after securing a clearance from the agency. However, some illegal reclamation activities have still occurred especially in the West Bay shoreline areas due to its proximity to highly urban Metro Manila. The reclaimed lands are commonly used for residential, commercial, and industrial purposes. Figure 6 shows the plot of built-up, water, and wetland area coverages from 2001 to 2017.

Figure 7 shows the percent area coverage of agricultural land, forest, and rangeland. agricultural land showed an overall decreasing trend while rangeland and forest cover were decreasing. One possible scenario is the conversion of rangelands to croplands to help in the demand for food supply in rural areas. In general, vegetation cover in Laguna Lake watershed steadily declined from 65.49% to 64.91% (change of 0.11%-0.24% or 4-10 sq km) between 2001 and 2013. The decrease in area was highest, however, between 2013 and 2017, with as much as 1.27% or 51 sq km of vegetation loss. The increase in built-up area along with the decrease of vegetation cover can be explained by conversion of land to built-up environment. In the Philippines, a common example is the development of agricultural lands into mixed residential and commercial land use, which is the usual case for Cavite, Laguna, and Batangas. Industrial facilities are also common, such as in the case of western and southern Laguna where different industrial parks are located.

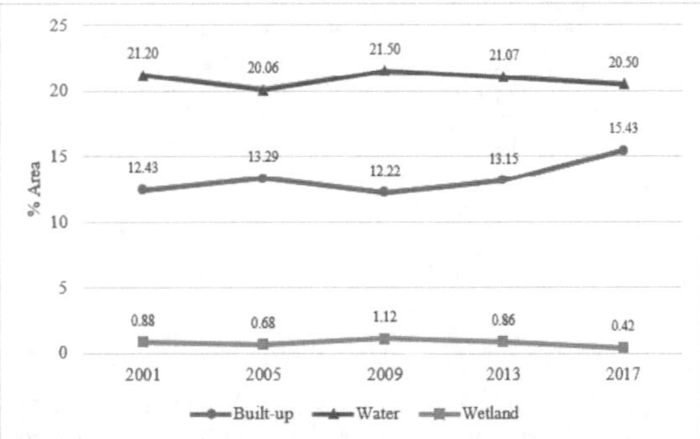

Figure 6. Percent area coverage of built-up, water, and wetland

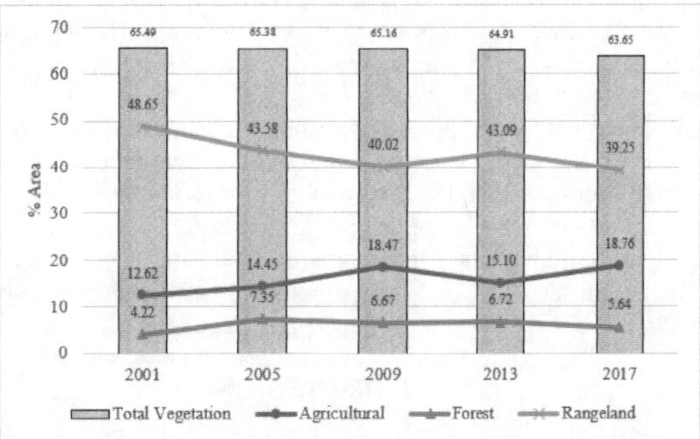

Figure 7. Percent area coverage of vegetation cover types

This finding is supported by the result of hot spot analysis. Intersecting the hot spot map with the vegetation cover per year revealed that 41.57% of the identified hot spots were vegetated areas in 2001. This decreased to 35.98% in 2005, 32.92% in 2009, 27.29% in 2013, and 19.27% in 2017. The decreasing trend confirmed that vegetation cover in hot spot areas were removed and the lands were converted to non-vegetated surface.

Persistent hot spots are characterized by long-term low to absent vegetation cover. These are found in high-density urban areas such as Metro Manila, in which built-up environment has been in place for at least 15 years The presence of intensifying hot spots indicates already low but continuously declining vegetation cover, which happens in low-density urban areas experiencing continuous development, for the past 15 years including 2017. Consecutive hot spots appear to be areas of concern, as these are locations where low vegetation have been observed in the recent years. If declining vegetation persists, these hot spots may intensify.

5. CONCLUSIONS

The use of MODIS NDVI data to monitor land cover in a watershed such as the Laguna de Bay region provides an alternative to remote sensing data acquired by optical sensors with limited overpass period. Using time-series NDVI allows the identification of different land cover classes based only on statistical values and trends in the data and without the need for training dataset for classification. The result is a land cover map that is moderately acceptable with respect to ground truth data, which is enough for certain types and levels of applications.

Analysis of multi-temporal land cover maps of Laguna Lake watershed revealed changes in land cover types that are driven by anthropogenic actions as shown by the overall increase in built-up areas along with declining vegetation cover. This is also reflected in the observed decrease in vegetation cover in identified hot spot areas, indicating conversion to non-vegetated surface for the 17-year period studied.

The monitoring and assessment of land cover in Laguna Lake watershed is an important component in lake resource management. While findings shown in this study may be considered an initial assessment of the area, these can already provide insights on the current state of the watershed and on the steps that need to be taken not only by the lake's monitoring agency but by all stakeholders in order to mitigate the impacts of land cover in the lake water quality, which is the primary concern in Laguna Lake.

Although the use of moderate resolution data such as MODIS provided moderately acceptable maps, additional steps may be taken, however, to improve the results. This may include exploring the use of satellite data with higher spatial resolution but moderate temporal resolution and using ancillary data to improve the classification. Implementing other algorithms, such as Markov chain models, to address inconsistencies in multi-temporal land cover maps may also be considered. With these, more specific land cover classification may lead to better understanding of land cover transformation occurring in the area.

ACKNOWLEDGEMENTS

This research was done as part of the Multi-platform and Cross-sensor Water Quality Monitoring (MCWQM) Project. The Project was implemented by the University of the Philippines Training Center for Applied Geodesy and Photogrammetry (TCAGP), through the support of the Department of Science and Technology (DOST) of the Republic of the Philippines and the Philippine Council for Industry, Energy, and Emerging Technology Research and Development (PCIEERD). The Project was also done in collaboration with the Laguna Lake Development Authority (LLDA).

REFERENCES

(n.d.). Retrieved October 25, 2019, from MODIS Land Team Validation: https://landval.gsfc.nasa.gov/ProductStatus.php?ProductID=MOD13&_ga=2.62252712.526663290.1572839604-536403554.1566955206

(n.d.). Retrieved October 25, 2019, from MODIS Moderate Resolution Imaging Spectroradiometer: https://modis.gsfc.nasa.gov/

Anderson, J. R., Hardy, E. T., Roach, J. T., & Witmer, R. E. (1976). *A land use and land cover classification system for use with remote sensing data.* United States Geological Survey. doi.org/10.3133/pp964

Ball, G. H., & Hall, D. J. (1965). *ISODATA, A Novel Method of Data Analysis and Pattern Classification.* Menlo Park: Stanford Research Institute. Retrieved from https://apps.dtic.mil/dtic/tr/fulltext/u2/699616.pdf

Carroll, M., Townsend, J., Hansen, M., DiMicelli, C., Sohlberg, R., & Wurster, K. (2010). MODIS Vegetative Cover Conversion and Vegetation Continuous Fields. In *Land Remote Sensing and Global Environmental Change* (p. 726). New York, New York: Springer. doi.org/10.1007/978-1-4419-6749-7_32

Chen, J., Jonsson, P., Tamura, M., Gu, Z., Matsuhita, B., & Eklundh, L. (2004). A simple method for reconstructing high-quality NDVI time-series data set based on the Savitzky-Golay filter. *Remote Sensing of Environment*, 322-344. doi.org/10.1016/j.rse.2004.03.014

Congalton, R. G. (2001). Accuracy assessment and validation of remotely sensed spatial information. *International Journal of Wildland Fire*, 321-328. doi.org/10.1071/WF01031

Dhodhi, M. K., Saghri, J. A., Ahmad, I., & Ul-Mustafa, R. (1999). D-ISODATA: A Distributed Algorithm for Unsupervised Classification of Remotely Sensed Data on Network of Workstations. *Journal of Parallel and Distirbuted Computing* , 280-301. doi.org/10.1006/jpdc.1999.1573

Didan, K. (2015). MOD13Q1 MODIS/Terra Vegetation Indices 16-Day L3 Global 250m SIN Grid V006 [Data set]. NASA EOSDIS Land Processes DAAC. doi.org/10.5067/MODIS/MOD13Q1.006

Esri. (n.d.). *Create Space Time Cube From Defined Locations*. Retrieved from ArcGIS Pro: https://pro.arcgis.com/en/pro-app/tool-reference/space-time-pattern-mining

Esri. (n.d.). *How Emerging Hot Spot Analysis Works*. Retrieved from ArcGIS Pro: https://pro.arcgis.com/en/pro-app/tool-reference/space-time-pattern-mining/learnmoreemerging.htm

Kraak, M. (2003). The Space-time Cube Revisited from a Geovisualization Perspective. *21st International Cartographic Conference*. 1988-1996.

Laguna Lake Development Authority. (2017). *2017 Annual Report.*

Nyamekye, C., Osei Jnr, E., Oseitutu, A. (2014). *Classification of time series NDVI for the assessment of land cover using NOAA/AVHRR data*. Journal of Geomatics, 34-39.

Santos-Borja, A., Nepomuceno, D. S. (2003). Experience and Lessons Learned Brief for Laguna de Bay, Philippines. *Lake Basin Management Initiative ILEC/LakeNet Regional Workshop for Asia: Sharing Experience and Lessons Learned in Lake Basin Management.*

Savitzky, A., & Golay, M. (1964, July). Smoothing and Differentiation of Data by Simplified Least Squares Procedures. *Analytical Chemistry*, 1627-1639. doi.org/10.1021/ac60214a047

United States Geological Survey. (n.d.). doi.org/10.5067/MODIS/MOD13Q1.006

ASSESSING POPULATION SENSITIVITY TO URBAN AIR POLLUTION USING GOOGLE TRENDS AND REMOTE SENSING DATASETS

Prakhar Misra[1,*], Wataru Takeuchi[1]

[1] Institute of Industrial Science, The University of Tokyo Tokyo - 153-8505, Japan - (mprakhar, wataru)@iis.u-tokyo.ac.jp

Commission VI, WG VI/4

KEY WORDS: MODIS AOD, AirRGB, time series, sensitivity, perception of air pollution, developing countries

ABSTRACT:

This study demonstrates relationship between remote sensing satellite retrieved fien aerosol concentration and web-based search volumes of air quality related keywords. People's perception of urban air pollution can verify policy effectiveness and gauge acceptability of policies. As a serious health issue in Asian cities, population may express concern or uncertainty for air pollution risk by performing search on the web to seek answers. A 'social sensing' approach that monitors such search queries, may assess people' perception about air pollution as a risk. We hypothesize that trend and volume of searches show impact of air pollution on general population. The objectives of this research are to identify those atmospheric conditions under which relative search volume (RSV) obtained from Google Trends shows correlation with measured fine aerosol concentration, and to compare search volume sensitivity to rise in aerosol concentration in seven Asian megacities. We considered weekly relative search volumes from Google Trends (GT) for a four year period from January, 2015 to December, 2018 representing diverse $PM_{2.5}$ concentrations. Search volumes for keywords corresponding to perception of air quality ('air pollution') and health effects ('cough' and 'asthma') were considered. To represent $PM_{2.5}$ we used fine aerosol indicator developed in an earlier research. The results suggest that tendency to search for 'air pollution' and 'cough' occurs when AirRGB R is in excess and temperature is below the baseline values. Consistent with this, in cities with high baseline concentrations, sensitivity to rise in AirRGB R is also comparatively lower. The result of this study can used as an indirect measure of awareness in the form of perception and sensitivity of population to air quality. Such an analysis could be useful for forecasting health risks specially in cities lacking dedicated services.

1. INTRODUCTION

1.1 Background

In most Asian cities outdoor fine particulate matter air pollution are increasing (Misra et al., 2017). Studies have shown that fine particulate matter results in severe health impacts that include pulmonary disorders, irritation in eyes as well as cardiac issues. As per WHO, particulate matter has well documented and strongest evidence of health risks that leads to harmful impacts on the cardiovascular, respiratory and pulmonary functionality. In India alone, outdoor air pollution was responsible for more than 670,000 deaths in 2016. To deal with these health effects, several countries have launched large-scale policies, e.g. China's Action Plan on Prevention and Control of Air Pollution and India's National Clean Air Program. In other countries, like Thailand, there is a growing demand for 'Clean-air Act' to overcome the issue of polluted air. However to move in the direction for achieving air quality and environmental management objectives, public perception about air pollution as a risk needs to be studied (Bickerstaff, Walker, 2001). An understanding of what levels of air pollution are perceived as acceptable among the public, can verify the effectiveness of policies as well as help improve the communication framework for achieving the desired change in attitude and behavior (Eden, 1996, Elliott et al., 1999). Perception of people regarding exposure and health effects is critical to gauge response of people and acceptability of related policies (Egondi et al., 2013). Although it is known environmental factors affect the health impact, (Elliott et al., 1999) suggested that perception of air pollution affects the relation between air pollution and health.

*Corresponding author

Since epidemiological datasets are not publicly available for many cities, web-based social sensing datasets may be useful in reflecting impact of pollution on human heath. However it is not clear to what extent information derived social sensing dataset, such as Google Trend, is related to air pollution concentration. Furthermore, as cities differ in baseline concentrations and social parameters, whether a unit rise in pollutant concentration results in a similar response on social sensing datasets across all cities is also unclear.

1.2 Objective of this research

The objective in this paper is to: 1) assess whether Google Trends for air quality related search queries shows correlation and seasonality with measured fine aerosol concentration, and 2) estimate health sensitivity of general population to measured fine aerosol concentration.

2. METHODOLOGY

The flowchart of dataset and methodology employed for studying the relationship between Google Trends based search activity and pollution concentration is shown in Figure 1. Data location consist of 7 Asian megacities.

Seven Asian megacities were chosen to study their perception and sensitivity. In order of their increasing per capita gross national income (GNIpc), these cities are: Dhaka ($3,677), Karachi ($ 5,311), Delhi ($ 6,353), Bangkok ($ 12,376) Taipei ($ 24,318), Seoul ($ 35,945) and Tokyo ($ 38,986). These cities were chosen based on their population size, known air pollution issues and popularity of Google as the dominant search engine.

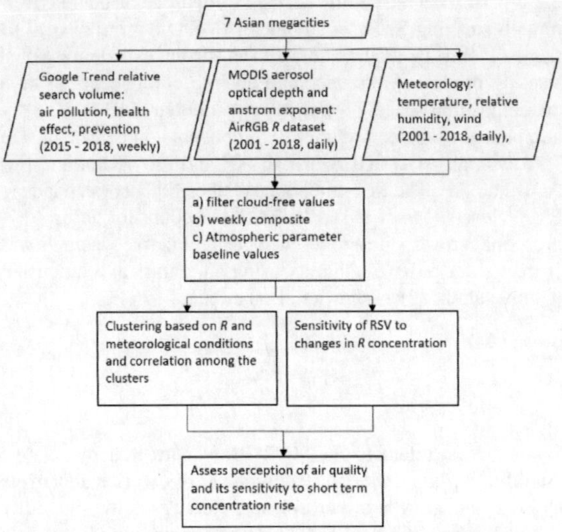

Figure 1. Flowchart studying relationship between Google Trends based search activity (RSV) and fine aerosol concentration R

Each of them (except Tokyo) sees PM2.5 concentrations above the WHO specified 24 hour limits although to a much variable degree.

2.1 Theoretical background

Within social sciences, risk is defined as (Rosa, 2003) "a situation or an event where something of human value (including humans themselves) is at stake and where the outcome is uncertain" while perception of risk is defined as 'the subjective assessment of the probability of a specified type of accident happening and how concerned we are with the consequences' (Sjöberg et al., 2004). There are many studies that have studied public perception of air pollution, but few have probed the correlation between perception and measured air pollutant concentrations. Even fewer studies have focused on developing countries (Saksena, 2011). This is important to explain why there is a gap between the perception of general public and an expert who relies on physical information. The first such study found that in Los Angeles County air quality was perceived as 'smoggy' when visibility (as a surrogate indicator of pollution), O_3 and SO_2 concentrations exceeded their baseline values (Flachsbart, Phillips, 1980).

Measuring perception is difficult as it is a subjective phenomenon and lacks any related open data. In absence of real-time data, social sensing and remote sensing data could be used to understand perception and sensitivity of humans to air pollution to some extent. Social sensing is defined as 'any source of information that can be identified in modern social networking and Web tools that expresses some situation or fact about users (e.g., their preferences or scheduled activities) and their social environments' (Rosi et al., 2011). Popular data sources include Google Trends, Twitter API and Sina Weibo. Researches have shown impact of air pollution on expressed happiness on Sina Weibo (Kahn et al., 2019). Google Trends has been used to assess the impact of meteorological environment on human behavior. Although search data is not significantly better than autoregressive model, however in absence of other data sources search data could be useful (Goel et al., 2010). COPD symptoms are triggered with chronic particulate air pollution (Schikowski et al., 2014) as well as short-term exposures (Li et al., 2016). In recent years, use of Google Trends in health care research as been increasing (Nuti et al., 2014), such as for COPD symptoms (Boehm et al., 2019), asthma (Bousquet et al., 2017), influenza (Polgreen et al., 2008) as well as flu (Eysenbach, 2006). (Maurer, Holbach, 2016) showed that search requests represent not only the concern for the issue concern (also known as salience) but also the public uncertainty regarding the issue, where uncertainty can be defined as absence of knowledge on an issue. Since cough is a common symptom of COPD (Smith, Woodcock, 2006), it was chosen as a search keyword indicating health impact. Based on this we hypothesized that among these keywords, search for 'air pollution' corresponds to perception of air quality, while 'cough' and 'asthma' correspond to short term health-effect.

2.1.1 Perception of fine aerosol concentration Perception of air quality is subjective to each individual. Perception of indoor air quality is established on the basis of temperature, humidity and velocity of the air as well as the quotient of the amount of air pollution and fresh air supplied (Roelofsen, 2018). Since perception of air quality is ultimately related with health effects (Elliott et al., 1999), search keywords RSV keywords may inform how varying outdoor pollutant concentration affects the health. Perception of air quality was assessed by identifying correlation of GT search volume relative to R. Clustering methods were used to group observations with similar atmospheric conditions. Correlation between mean observation of each cluster was computed to examine whether variation in atmospheric parameters influences RSV. Following the perception index of (Roelofsen, 2018), R, $temp$, and $wind$ were considered for clustering the data. As past studies have suggested the role of atmospheric visibility in perception of air pollution (Flachsbart, Phillips, 1980), AOD values were also for cluster correlation included as inverse indicator of visibility (Kaufman, Fraser, 1983). It should be noted however that if the light attenuation is high, retrieved AOD will be high irrespective of the particle being coarse or fine. Hierarchical Agglomerative Clustering (HAC) algorithm was used to generate the clusters. HAC is a bottom-up algorithm where each object is initially a single-element cluster and at each step, similar clusters are combined to progressively bigger cluster. Two set of clusters, C_1 and C_2, were generated using HAC, to obtain the correlation between among cluster components as shown in Equation 1.

$$C_1 : HAC(R, temp, wind)$$
$$C_2 : HAC(\Delta R, \Delta temp, \Delta wind) \quad (1)$$

R, $temp$, $wind$ and RSV for each location was standardized from 0 to 1 and pooled into a single dataset. Cluster set, C_1, were computed over this dataset. Another cluster set, C_2 was computed over ΔR, $\Delta temp$ and $\Delta wind$ such that, ΔX refers to $X - \bar{X}_{2015-2018}$, where $\bar{X}_{2015-2018}$ is the baseline obtained as mean X from 2015 to 2018 .

2.1.2 Sensitivity of perception We are interested in identifying how sensitive is RSV in responding to rise in R. Sensitivity is commonly defined as a measure of the proportion of actual positives that are identified as such. This increase or decrease is quantified through a relative change metric (Preis et al., 2013). The relative change metric $\Delta' n$ at time t within a time-frame of δt for R or RSV is defined as Equation 2.

$$\Delta' n(t,\delta t) = n(t) - N(t-1,\delta t) \quad (2)$$

where, $N(t-1,\delta t) = (n(t-1)+n(t-2)+..+n(t-\delta t))/\delta t$. For this research, N was computed at δt of both 1 week and 2 weeks, which means that relative change of current search volume or concentration was computed in comparison with its mean over the previous 2 and 3 weeks respectively. Based on this, we define the time-frame sensitivity, s, of relative change in RSV, $\Delta' RSV$, to rise in R, $\Delta' R$, as Equation 3. Only those cases were considered when $\Delta' R$ was greater than 10 units so as to ensure $\Delta' R$ is large enough to be perceivable.

$$s(\delta t) = \frac{\Sigma TP_{\delta t}}{\Sigma TP_{\delta t} + \Sigma FN_{\delta t}} \quad (3)$$

where, $TP_{\delta t}$ (true positive) is considered 1 when within a time-frame δt, both $\Delta' R$ and the co-occurring $\Delta' RSV$ are positive, $FN_{\delta t}$ (false negative) is considered 1 when $\Delta' R$ is positive but co-occurring $\Delta' RSV$ is negative. We also define weighted sensitivity $sw(\delta t)$, that takes into not only the direction of change, but also the magnitude of relative change as shown in Equation 4. A high value (≥ 1) of $\frac{\Delta' RSV}{\Delta' R}$ is a result of larger change in RSV relative to R. This could be on account media-generated hype or signify the high sensitivity of population.

$$sw(\delta t) = \frac{\Sigma(TP_{\delta t}.\frac{\Delta' RSV}{\Delta' R})}{\Sigma(TP_{\delta t}.\frac{\Delta' RSV}{\Delta' R}) + \Sigma(FN_{\delta t}.\frac{\Delta' RSV}{\Delta' R})} \quad (4)$$

2.2 Data used

2.2.1 Google Trends (GT) Relative search volume (RSV) for any search term is available from the web search engine Google, since 2004 at daily scale on a geographical basis of a metropolitan region or a state. If the specified duration is bigger than 9 months then the RSV is available at weekly scale. These relative search volumes are available as integers on a relative scale of 0 to 100, 100 being the maximum RSV within the specified time period duration and 0 being the minimum. RSV was considered between January, 2015 to December to obtain data free of Google's previous algorithmic adjustments. The RSV values are affected by any new relative maxima or minima in search volume. Due to this, in some locations a very high RSV for a few dates along with very low values (≤ 5) for other dates was obtained, masking the subtler daily changes in RSV. To partly overcome this, smaller overlapping periods were downloaded and then joined by inter-calibration (Challet, Bel Hadj Ayed, 2014). In addition RSV was further scaled between 0 percentile and 95 percentile to remove extreme value RSV. The keywords probed were the search terms: 'air pollution', 'cough' and 'asthma'. These keywords are collectively referred to as GT henceforth. For Bangkok, Tokyo, Taipei and Seoul corresponding translated terms in Thai, Japanese, Chinese and Korean were used. The Google Trend time series for GT were batch downloaded using a Python module 'pytrends' available from GitHub. The RSV also exist in the form search 'Topics' in Google Trends, which allow search volume of all words related to that topic but since the full list of related words is not disclosed, this option was not considered.

2.2.2 MODIS based fine aerosol indicator Moderate Resolution Imaging Spectro-radiometer (MODIS) retrieved AOD (aerosol optical depth) and AE (angstrom exponent) global satellite products were decomposed into components using a decomposition scheme following (Misra et al., 2017). The decomposition scheme provides 3 components on a scale of 0 to 100 for high AOD and AE, high AE and low AOD, and low AOD and AE. The first component (labeled R) corresponds to $PM_{2.5}$ levels (Misra et al., 2017). To account for aerosol hygroscopic growth during presence of precipitable water, R was corrected for relative humidity using an empirical relationship (Chin et al., 2002) as shown in Equation 5.

$$R = \frac{R_{obs}}{\sigma_{rhum}} \quad (5)$$

Where, R_{obs} refers to the AirRGB R corrected by relative humidity. Mass extinction coefficient (σ_{rhum}) characterizes hygroscopic growth in aerosol at different relative humidity ($rhum$) values (Chin et al., 2002). Relative humidity values for the locations were obtained from the reanalysis weather data, described below in 2.2.3. This dataset was composited at weekly level from 2011 to 2018 and was gridded at 10 km resolution.

2.2.3 Reanalysis weather data Daily local weather climatological information for each location was obtained between 2012 to 2019 from the NCEP/NCAR Reanalysis 1 data set from NOAA Earth System Research Laboratory (NOAA ESRL, n.d.). It is globally available at 2.5 degree mesh grid. Minimum temperature, maximum temperature, wind speed vectors and relative humidity were obtained at daily scales. Daily mean temperature ($temp$) was calculated as average of minimum and maximum temperature, which along with relative humidity ($rhum$) and wind speed magnitude (wnd) was averaged to weekly level for further analysis.

3. RESULTS AND DISCUSSION

3.1 Trend of R and RSV

Figure 2 shows difference of R from 2015 and 2001 highlighting the study locations. Detailed four-year weekly trend for AirRGB R and RSV of 'air pollution', 'cough' and 'asthma' over some study locations is shown in Figure 6 in Appendix (Section 4). Taipei has the highest missing R due to frequent cloud cover (80.08% missing) followed by Bangkok (67.69% missing). Within the observable days, the highest R baseline values between 2010 and 2018 are seen in Bangkok (42.94), followed by Delhi (41.58) and Dhaka (39.13). In Delhi, Dhaka and Karachi, the baseline R between 2010 to 2018 is higher than the baseline between 2001 to 2010 by 1.92, 1.12 and 0.92 respectively and shows existence of seasonal trends. This clearly shows a rising fine mode aerosol in these locations. During the same period, R decreased by in Taipei, Seoul, Bangkok and Tokyo respectively. The factors affecting the seasonality are meteorological conditions like boundary layer inversion and rainfall and anthropogenic factors like annual crop-burning cycles. It appears that 'air pollution' searches also follow a seasonal regime and are only partially dependent on R values as higher value of R do not always translate into higher searches. At the same time there is a very clear but non-overlapping seasonality in 'cough' and 'asthma' RSV for Karachi, Delhi, Taipei, Seoul and Tokyo. The coefficient of variation (defined

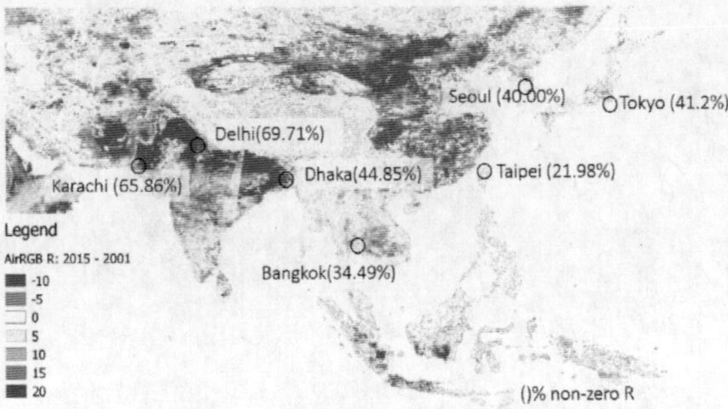

Figure 2. Difference of R from 2015 and 2001 highlighting the study locations and annual percentage cloud-free observable days using MODIS.

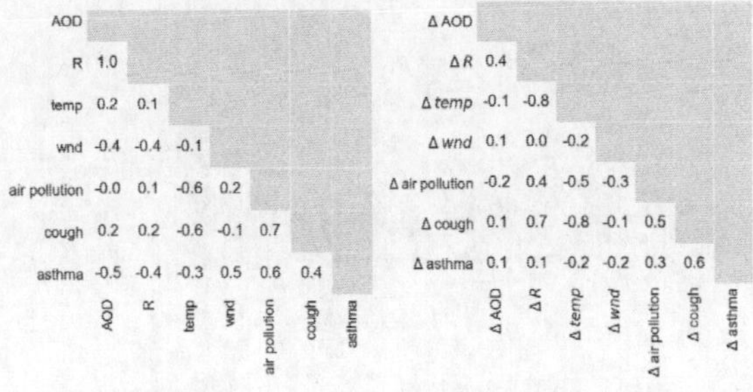

(a) Clusters based on absolute values

(b) Clusters based on difference from baseline values

Figure 3. Heat map showing correlation between mean values of the clusters based on the (a) absolute values and (b) and difference from mean(Δ) for concentration R, temperature $temp$ and wind speed wnd.

as ratio of σ and μ) indicates the standardized dispersion of frequency distribution, the high value of which can suggest strong presence of seasonality. Coefficient of variation for 'cough' RSV is highest in Delhi and for 'asthma' RSV it is highest in Tokyo. In some cities like Delhi there is overlap in seasonality of 'cough' and R. Specially for Delhi 'cough' shows seasonality with high R values in months of November, December and January. Also in Delhi, three exceptional peaks were noted in the December of 2016, 2017 and 2018 where 'air pollution' searches coincided very well with high R. This is likely a result of rising mass media news articles (Negi et al., 2017) and social campaigns since 2015. This could correspond to the 'issue-attention cycle' (Downs, 1996) wherein public attention is transformed from pre-discovery of problem stage to alarmed discovery.

3.2 Correlation between R and Google Trends RSV

Figure 3 shows the result of correlation between mean value of clusters derived from absolute values and deviance from baseline values of R, $temp$, and $wind$. Correlation among cluster set C_2 (comprising difference from baseline values) is higher than the cluster based only on absolute values (C_1 in Equation 1). In the C_2 case, correlation of ΔR with 'air pollution' ΔRSV (0.4), 'cough' ΔRSV (0.7) and 'asthma' ΔRSV (0.1) is higher than the C_1 case of correlation of R with 'air pollution' RSV (0.1), 'cough' RSV (0.2) and 'asthma' RSV (-0.4). Similar but smaller improvements are seen for correlation of RSV with $temp$ and $wind$. It is interesting to note that compared to C_1, C_2 results in more correlation improvement in 'cough' and 'asthma' RSV than 'air pollution'. It is also seen that AOD has lower correlation compared to R with RSV. This implies that although visibility of the air could be a visual indicator of perception, it is the presence of fine aerosol that governs perception.

There also exists an inverse relationship between absolute values of R and $wind$ speed as well as ΔR and $\Delta temp$. This is a well-studied phenomena where high wind speed leads to pollutant dispersion and dilution. Higher than usual temperatures cause to surface-to-air convection currents also leading to dilution. However it is interesting to note that both low $temp$ (Figure 3 (a)) as well as negative $\Delta temp$ (Figure 3 (b)) lead to high 'air pollution' RSV. This correlation of $temp$ with 'air pollution' RSV is stronger than that of R with 'air pollution' RSV. It suggests absolute and relative differences of temperatures govern the perception of air quality more than the R. This is shown more clearly in Figure 4. It is seen that a $\Delta temp$ lower than -3° and a ΔR larger than -2 than leads to positive 'air pollution' ΔRSV. For both 'air pollution' and 'cough', very high positive RSV also occur when $\Delta temp$ is lower than -7° and ΔR is higher than 11. For 'asthma', although positive ΔRSV occur with lower $\Delta temp$ and higher ΔR, it is not as high as 'air pollution' and 'cough'. It could be because apart from air pollution, asthma is also triggered by allergens. Further asthma affects children more than adults which could be another reason.

3.3 Sensitivity of GT to short-term rise in R

The sensitivity and weighted sensitivity in terms of relative change in search volume of keywords to rise in R in the 7 cities between 2015 and 2018 is shown in Figure 5. Compared to other cities in Figure 5 (a), in Tokyo 'air pollution' $\Delta' RSV$ has the highest 1-week sensitivity ($s(1) = 0.60$) to $\Delta' R$. Also, a lower sensitivity to 2-week ($s(2) = 0.50$) suggests that 'air pollution' RSV is lower in the second week (compared to 'air pollution' RSV in the first week) when continuous rise of R takes place. While in Seoul and Bangkok, 'air pollution' $\Delta' RSV$ has the lowest 1-week sensitivity ($s(1) = 0.45$). Dhaka has the highest sensitivity of 'air pollution' $\Delta' RSV$ to 2-week $\Delta' R$ ($s(2) = 0.65$), showing that continuous 2-week $\Delta' R$ lead to more positive $\Delta' RSV$ than just 1-week rise. There is a large gap between $s(1)$ (0.46) and $s(2)$ (0.64) in Seoul and between $s(2)$ (0.44) and $sw(2)$ (0.65) in Delhi, signifying that less than 50% positive $\Delta' R$ events are noticed, but when noticed they lead to larger than usual 'air pollution' $\Delta' RSV$.

A high 1-week sensitivity of 'cough' $\Delta' RSV$ ($s(1) > 0.55$) is found in Dhaka, Karachi and Delhi in Figure 5 (b). 2-week sensitivity is highest in Tokyo ($s(2) = 0.66$) followed by Delhi ($s(2) = 0.6$). Low weighted sensitivity ($sw(1) = 0.37$, $sw(2) = 0.35$) in Delhi compared to non-weighted sensitivity ($sw(1) = 0.54$, $sw(2) = 0.60$) suggest that positive 'cough' $\Delta' RSV$ takes place during positive $\Delta' R$, however there is low correspondence between their magnitudes. In Figure 5 (c), Karachi has the highest 1-week sensitivity of asthma $\Delta' RSV$ ($s(1) = 0.66$) followed by Seoul ($s(1) = 0.56$) and the lowest in Delhi ($s(1) = 0.34$) and Tokyo ($s(1) = 0.34$). Apart from Karachi and Seoul, $s(1)$ is lower than 0.5 for other cities, which means less than 50% of positive $\Delta' R$ correlates with increase in asthma RSV. Sensitivity of 'cough' to $\Delta' R$ is more sensitive than 'asthma' in Dhaka, Delhi and Tokyo in contrast to 'asthma' which is more sensitive than 'cough' in Karachi and Seoul. This could be due to difference in the constituents of the local air pollutants, leading to differences in health impacts.

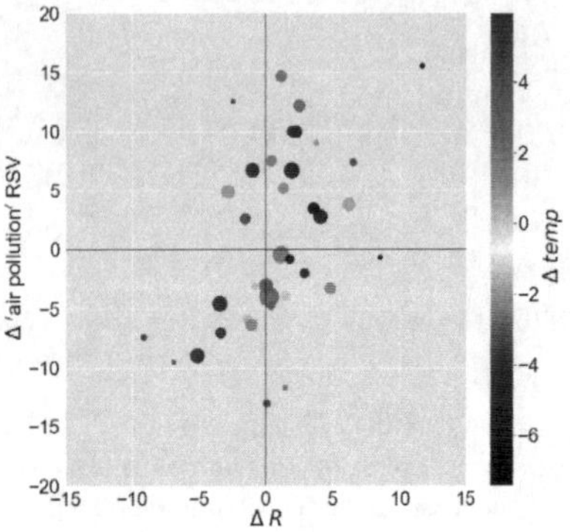

(a) 'air pollution'

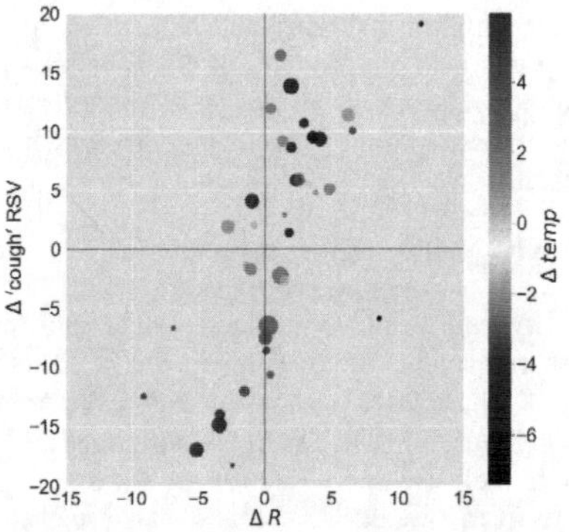

(b) 'cough'

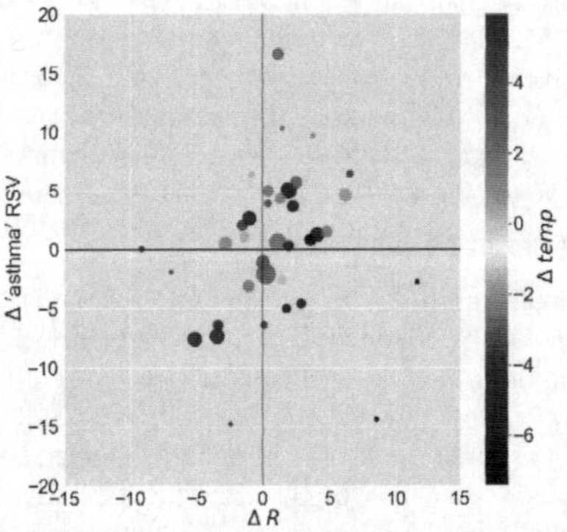

(c) 'asthma'

Figure 4. Plot of mean cluster values based on difference from baseline of concentration indicator (ΔR) and temperature ($\Delta temp$) showing difference from mean relative search volume (ΔRSV) values of (a) 'air pollution', (b) 'cough' and (c) 'asthma'. Generally positive ΔR and negative $\Delta temp$ leads to positive ΔRSV of 'air pollution', 'cough' and 'asthma'. Bubble size represents relative cluster size.

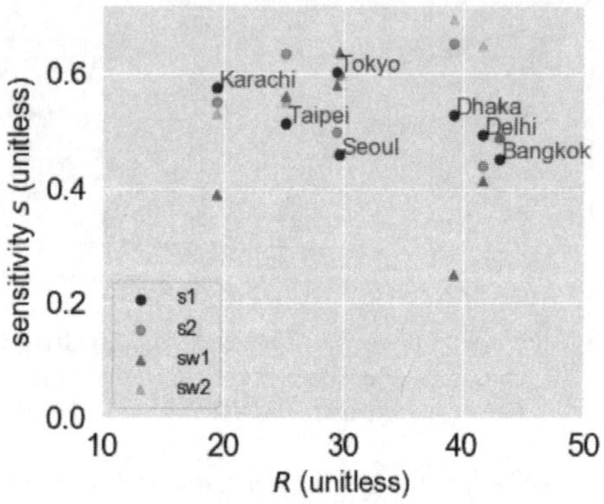

(a) 'air pollution'

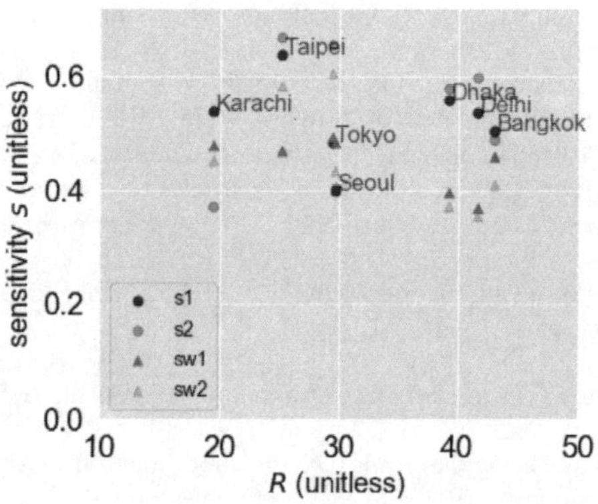

(b) 'cough'

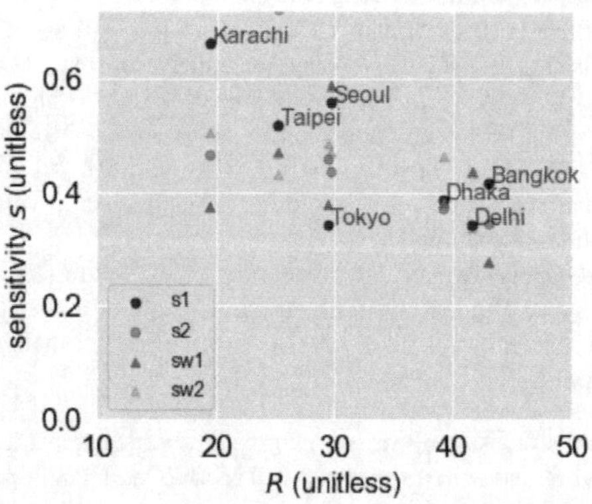

(c) 'asthma'

Figure 5. Mean sensitivity of the change in (a) 'air pollution', (b) 'cough' and (c) 'asthma' $\Delta' RSV$ with respect to change in AirRGB $\Delta' R$ for the 7 mega cities. $s1$ and $s2$ refer to sensitivity for the period of 1 and 2 weeks while $sw1$ and $sw2$ refer to their respective weighted sensitivity.

As asthma is also affected by biogenic particles such as pollen grains, it is possible that the pollen season and high pollutant period coincides in Karachi and Seoul, leading to high sensitivity in 'asthma'.

3.4 Discussion

In this study we established that people's perception of air quality and its health risk, cough is related with difference of fine mode aerosol level from its baseline. At same R but a different $temp$, perception as shown by RSV could be different. Since COPD diseases, whose common symptom is cough, could be caused to presence of certain viruses which spike in winter season, this confounds the influence of air pollution on health. A similar observation was also regarding COPD search terms (Boehm et al., 2019). Policy implication of this finding could be that for the success of policies that target people behavior, current approaches should also contains different provisions for awareness campaigns based on weather conditions. In fact, warmer seasons may needs extra measures and campaigns to spread awareness regarding pollution. This also implies that in places with climate warming, perception of air pollution as a risk may decrease. These findings are partially similar to that oc climate change perception where frequent anomalies lead to lowered expectations and less chances of the events being perceived as remarkable (Moore et al., 2019).

From the perspective remote sensing this result has some bias due to consideration of only non-cloudy observations. A known limitation is that compared to surveys which provide a direct indication of public concern based in individual response, internet search data can only provide active collective public response (Bromley-Trujillo, Poe, 2018), and by extension cannot inform individualistic inclinations that may govern the reason for concern. Further internet searches represents only active interest, so when air pollution may not be searched for on a clean air day, it si unclear whether perception regarding air pollution has changed. Although studies using Google Trends dataset are increasing, currently set standards for data collection and an optimal search strategy is lacking (Nuti et al., 2014). Our analysis was limited by a narrow selection of Google Trend keywords. There are two issues here. First, a broader selection of keywords is needed that to identify effect of polluted air on people, e.g. 'smog', 'sick' 'mask', etc. However these words are also carry several other connotations which make can confound the interpretation of search volume leading to lower content validity (Mellon, 2014). Second, when there is a preference to search in a local informal language, the interpretation of the word may not always coincide with the direct English translation. e.g. in Hindi language, the same word is used to inquire about wind and air quality. A systematic approach could be scraping the most frequently used words in air pollution related news articles in the local language and obtaining search volumes corresponding to those terms. Another limitation is with using Google Trends unavailability of daily-level data and absolute search frequencies (Maurer, Holbach, 2016). So far we assumed that the most popular method of obtaining information is by searching for 'air pollution' or 'asthma'. However with rise in smartphone app based alert systems in many places, the apparent concern for these topic as measured on Google Trends may decrease leading to erroneous conclusions. Further, this analysis was performed at the scale of one week with a dataset of 4 years. For a much smaller time period, Google Trends is also available at daily and hourly scales. Temporally denser data can be obtained and studied with dense concentration readings, e.g. Aeronet retrieved AOD or personal PM2.5 measurements. If behavior of other emission sources like real time traffic congestion status can be obtained, its inclusion can reveal impact of emission sources and rising concentration on human health. This could be useful for preparation of medical facilities and stocking medicinal supplies. Nighttime light from VIIRS is also an important predictor of human activities which shows discrimination with air pollution (Misra, Takeuchi, 2016) and can be included as independent variable to separate perception among diverse study locations.

4. CONCLUSION

In this paper we have shown that geo-located social sensing data can combine with traditional remote sensing data to reveal perception and sensitivity of web-based search volumes with respect to outdoor air pollution. This is the first study to demonstrate relationship between air quality related search volumes and the air quality concentration. The results suggest that tendency to search for 'air pollution' and 'cough' occurs when Air-RGB R is in excess and temperature is below the baseline values. For the Asian megacities, the highest sensitivity to rise in pollutant concentration is shown by 'air pollution' in Delhi, 'asthma' and 'cough' in Dhaka, 'headache' in Seoul and 'mask' in Tokyo. Further the level of R that results in increased RSV of GT varied for the cities. It was found that rise in RSV of GT occurred at higher R in low GNIpc cities than the R levels of high GNIpc cities. RSV and remote sensing datasets could be beneficial is assessing perception and health effects during high pollution episodes.

ACKNOWLEDGEMENTS

NCEP Reanalysis data provided by the NOAA/OAR/ESRL PSD, Boulder, Colorado, USA, from their web site at https://www.esrl. noaa.gov/psd/. The first author is thankful to Tateisi Foundation for providing international conference travel grant.

REFERENCES

Bickerstaff, K., Walker, G., 2001. Public understandings of air pollution: The 'localisation' of environmental risk. *Global Environmental Change*, 11(2), 133–145.

Boehm, A., Pizzini, A., Sonnweber, T., Loeffler-Ragg, J., Lamina, C., Weiss, G., Tancevski, I., 2019. Assessing global COPD awareness with Google Trends . *European Respiratory Journal*, 1900351.

Bousquet, J., O'Hehir, R. E., Anto, J. M., D'Amato, G., Mösges, R., Hellings, P. W., Van Eerd, M., Sheikh, A., 2017. Assessment of thunderstorm-induced asthma using Google Trends. *Journal of Allergy and Clinical Immunology*, 140(3), 891–893.e7.

Bromley-Trujillo, R., Poe, J., 2018. The importance of salience: public opinion and state policy action on climate change. *Journal of Public Policy*, 1–25.

Challet, D., Bel Hadj Ayed, A., 2014. Do Google Trend Data Contain More Predictability than Price Returns? *SSRN Electronic Journal*, 1–17. http://www.ssrn.com/abstract=2405804.

Chin, M., Ginoux, P., Kinne, S., Torres, O., Holben, B. N., Duncan, B. N., Martin, R. V., Logan, J. a., Higurashi, A., Nakajima, T., 2002. Tropospheric Aerosol Optical Thickness from the GOCART Model and Comparisons with Satellite and Sun Photometer Measurements. *Journal of the Atmospheric Sciences*, 59(3), 461–483.

Downs, A., 1996. Up and down with ecology– the " issue-attention cycle". P. Peretz (ed.), *The politics of american economic policy making*, M.E. Sharpe, New York, 48.

Eden, S., 1996. Public participation in environmental policy: Considering scientific, counter-scientific and non-scientific contributions. *Public Understanding of Science*.

Egondi, T., Kyobutungi, C., Ng, N., Muindi, K., Oti, S., 2013. Community Perceptions of Air Pollution and Related Health Risks in Nairobi Slums. *International Journal of Environmental Research and Public Health*, 4851–4868.

Elliott, S. J., Cole, D. C., Krueger, P., Voorberg, N., Wakefield, S., 1999. The Power of Perception: Health Risk Attributed to Air Pollution in an Urban Industrial Neighbourhood. *Risk Analysis*, 19(4), 621–634. http://doi.wiley.com/10.1111/j.1539-6924.1999.tb00433.x.

Eysenbach, G., 2006. Infodemiology: tracking flu-related searches on the web for syndromic surveillance. *Annual Symposium proceedings. AMIA Symposium*, 244–8.

Flachsbart, P. G., Phillips, S., 1980. An Index and Model of Human Response to Air Quality. *Journal of the Air Pollution Control Association*, 30(7), 759–768.

Goel, S., Hofman, J. M., Lahaie, S., Pennock, D. M., Watts, D. J., 2010. Predicting consumer behavior with Web search. *Proceedings of the National Academy of Sciences*, 107(41), 17486–17490. http://www.pnas.org/lookup/doi/10.1073/pnas.1005962107.

Kahn, M. E., Zhang, X., Wang, J., Sun, C., Zheng, S., 2019. Air pollution lowers Chinese urbanites' expressed happiness on social media. *Nature Human Behaviour*. http://dx.doi.org/10.1038/s41562-018-0521-2.

Kaufman, Y. J., Fraser, R. S., 1983. Light Extinction by Aerosols during Summer Air Pollution. *Journal of Climate and Applied Meteorology*, 22(10), 1694–1706.

Li, J., Sun, S., Tang, R., Qiu, H., Huang, Q., Mason, T., Tian, L., 2016. Major air pollutants and risk of COPD exacerbations: a systematic review and meta-analysis. *International Journal of Chronic Obstructive Pulmonary Disease*, Volume 11, 3079–3091. https://www.dovepress.com/major-air-pollutants-and-risk-of-copd-exacerbations-a-systematic-revie-peer-reviewed-article-COPD.

Maurer, M., Holbach, T., 2016. Taking Online search queries as an indicator of the public agenda: The role of public uncertainty. *Journalism and Mass Communication Quarterly*, 93(3), 572–586.

Mellon, J., 2014. Internet Search Data and Issue Salience: The Properties of Google Trends as a Measure of Issue Salience. *Journal of Elections, Public Opinion and Parties*, 24(1), 45–72. http://dx.doi.org/10.1080/17457289.2013.846346.

Misra, P., Fujikawa, A., Takeuchi, W., 2017. Novel decomposition scheme for characterizing urban air quality with MODIS. *Remote Sensing*, 9(8), 1–19. http://www.mdpi.com/2072-4292/9/8/812.

Misra, P., Takeuchi, W., 2016. Analysis of air quality and nighttime light for Indian urban regions. *IOP Conference Series: Earth and Environmental Science*, 37(2), 012077.

Moore, F. C., Obradovich, N., Lehner, F., Baylis, P., 2019. Rapidly declining remarkability of temperature anomalies may obscure public perception of climate change. *Proceedings of the National Academy of Sciences*, 116(11), 4905–4910. http://www.pnas.org/lookup/doi/10.1073/pnas.1816541116.

Negi, N., Onyon, L., Murukutla, N., Puri, P., Mullin, S., 2017. Online media coverage of air pollution risks and current policies in India: A content analysis. *WHO South-East Asia Journal of Public Health*, 6(2), 41.

NOAA ESRL, n.d. NCEP/NCAR Reanalysis 1. https://www.esrl.noaa.gov/psd/data/gridded/data.ncep.reanalysis.pressure.html.

Nuti, S. V., Wayda, B., Ranasinghe, I., Wang, S., Dreyer, R. P., Chen, S. I., Murugiah, K., 2014. The use of google trends in health care research: A systematic review. *PLoS ONE*, 9(10).

Polgreen, P., Chen, Y., Pennock, D., Nelson, F., 2008. Using Internet Searches for Influenza Surveillance. *Clinical Infectious Diseases*, 47(11), 1443–1448.

Preis, T., Moat, H. S., Stanley, H. E., 2013. Quantifying Trading Behavior in Financial Markets Using Google Trends. *Scientific Reports*, 3(1), 1684. http://www.nature.com/articles/srep01684.

Roelofsen, P., 2018. A new methodology for the evaluation of the perceived air quality depending on the air pollution , caused by human bioeffluents , the temperature , the humidity as well as the air velocity. 8975.

Rosa, E. A., 2003. The logical structure of the social amplification of risk framework (SARF): Aferatheoretical foundations and policy implications. *The Social Amplification of Risk*.

Rosi, A., Mamei, M., Zambonelli, F., Dobson, S., Stevenson, G., Ye, J., 2011. Social sensors and pervasive services: Approaches and perspectives. *2011 IEEE International Conference on Pervasive Computing and Communications Workshops (PERCOM Workshops)*, 17number 1, IEEE, 525–530.

Saksena, S., 2011. Public Perceptions of Urban Air Pollution Risks. *Risk, Hazards & Crisis in Public Policy*, 2(1), 19–37. http://doi.wiley.com/10.2202/1944-4079.1075.

Schikowski, T., Mills, I. C., Anderson, H. R., Cohen, A., Hansell, A., Kauffmann, F., Kramer, U., Marcon, A., Perez, L., Sunyer, J., Probst-Hensch, N., Kunzli, N., 2014. Ambient air pollution: a cause of COPD? *European Respiratory Journal*, 43(1), 250–263. http://erj.ersjournals.com/cgi/doi/10.1183/09031936.00100112.

Sjöberg, L., Moen, B., Rundmo, T., 2004. Explaining risk perception. An evaluation of the psychometric paradigm in risk perception research. Technical report, Trondheim.

Smith, J., Woodcock, A., 2006. Cough and its importance in COPD. *International Journal of COPD*, 1(3), 305–314.

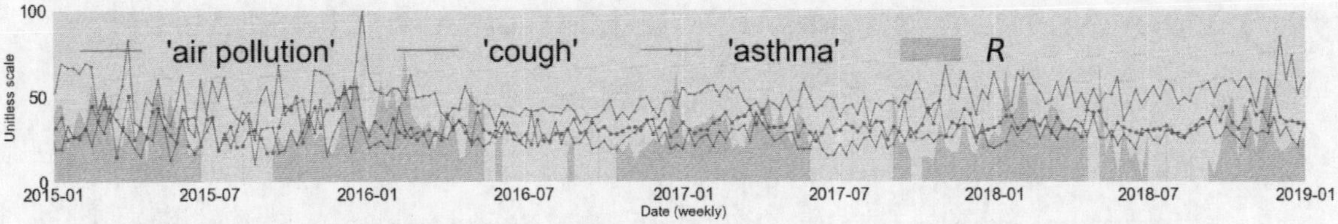

(a) Dhaka

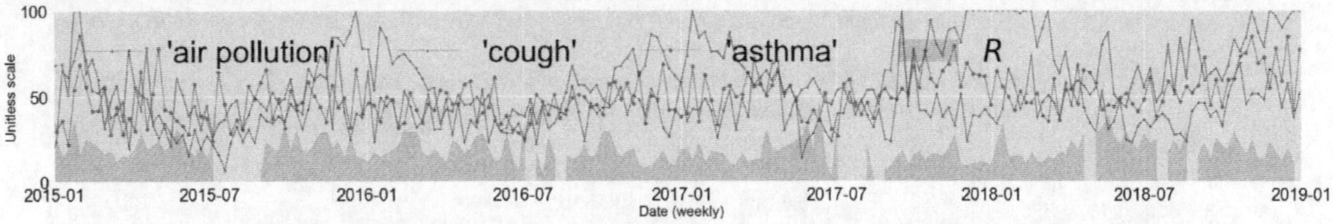

(b) Karachi

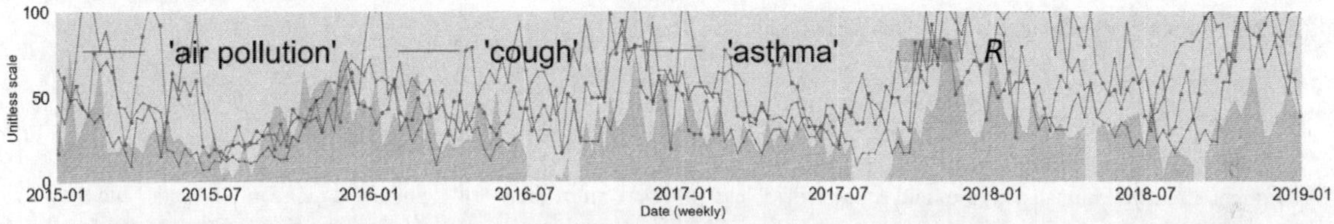

(c) Delhi

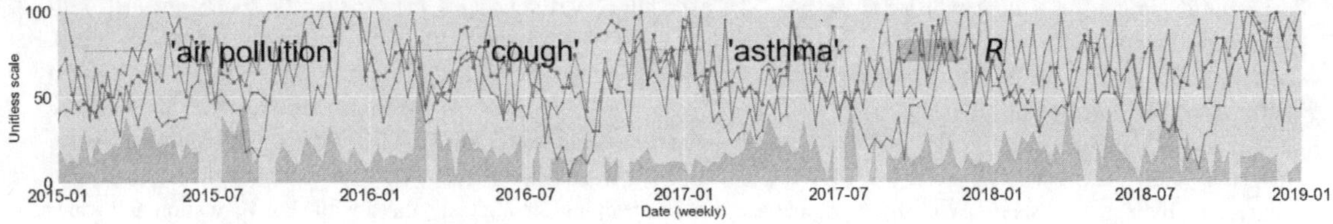

(d) Tokyo

Figure 6. Weekly relative search volume (RSV) of the search terms: 'air pollution' (black), 'cough' (red) and 'asthma' (green) are shown along with the fine aerosol component R (shaded gray) for (a)Dhaka, (b) Karachi, (c) Delhi and (d) Tokyo under study.

APPENDIX

Weekly relative search volume (RSV) of the search terms is shown in Figure 6.

IMPROVING THE LAND SURFACE GENERAL DROUGHT INDEX MODEL

Alfred Homère Ngandam Mfondoum*[1, 2], Paul Gérard Gbetkom[3] Ryan Cooper [4], Sofia Hakdaoui[5], Mamane Barkawi Mansour Badamassi[6]

[1] StatsN'Maps, Private Consulting Firm, Dallas, Texas 75287, USA. stats.n.maps.expertise@gmail.com
[2] Department of Geography, University of Yaoundé I, Yaoundé, Cameroon. Alfred.homere@gmail.com
[3]Department of Geography, University of Aix-Marseille, Marseille, France. Paul-gerard.gbetkom@etu.univ-amu.fr
[4] Erik Jonsson School of Engineering and Computer Science, University of Texas in Dallas, Dallas, Texas 75080, USA. Rdc190001@utdallas.edu
[5] Earth Observation department, Geo-Biodiversity and Natural Patrimony Laboratory, Geophysics, Natural Patrimony and Green Chemistry Research Center, Scientific Institute, Mohamed V University, Rabat, Morocco. s.hakdaoui@gmail.com
[6]Laboratory of Botanic, Mycology and Environment, Mohammed V University, Rabat, Morocco. barkawimansour@yahoo.fr

Keywords: Drought, Remote Sensing, Spectral index, Improved Land Surface General Drought Index, Vegetation Moisture Index, Normalized Difference Soil Drought Index, Landsat, Semi-arid regions.

ABSTRACT

Drought affects all human activities and ecosystems. Nearly 40 percent of the world's population inhabit Drylands, and they depend on agriculture for their food, security and livelihoods. Among the remote sensing indices developed, the Land Surface General Drought Index (LSGDI) was recently proposed. This paper proposes an improved model of LSGDI to face the issue of drought in semi-arid and arid regions. The experiment was conducted for the Maga's floodplain, in North-Cameroon. The method uses satellite images of Landsat in 1987, 2003 and 2018, for January and March or April, corresponding to the middle and the end of the dry season. A Vegetation Moisture Index (VMI) and a Normalized Difference Soil Drought Index (NDSoDI) are both developed. On an orthogonal plan, their projections give a drought line that expresses the improved LSGDI (LSGDI2) as the root sum square of the NDSoDI and the VMI. The LSGDI2 results are ranged in [0.09 – 0.14] interval, which is used to define the threshold and ease the qualifiers for drought classes. The visual patterns easily match the sandy areas of the original Landsat images with the highest values, while the vegetation and water areas match the lowest values. Compared with the LSGDI and Second Modified Perpendicular drought Index (MPDI1), the new index reflectance values are higher. Finally, although LSGDI2 curve's evolution follows the NDSoDI one at 94%, the new spectral index values depends on the both components, helping to map highest values of drought and moisture in Maga's floodplain, for a sustainable rice culture expansion.

* Corresponding author (stats.n.maps.expertise@gmail.com)

1. INTRODUCTION

Drought refers to an extended period, i.e. a season, a year, or several years, of deficient rainfall, relative to the statistical multi-year average for a region (Graham, S. 2000). Although there are dozens of definitions covering meteorological and hydrological aspects, this study addresses the agricultural effects of drought, in terms of the risk of low soil moisture and crop needs in the semi-arid regions. This affects all human activities and ecosystems, and more than 2 billion people, nearly 40 percent of the world's population (White and Nackoney, 2003), inhabit Drylands. These populations depend on agriculture for their food, security and livelihoods. Over the years, Earth Observation (EO) sciences have been efficient in spatializing drought areas. Spectral indices are among the most important approaches.

From the vegetation moisture view, the Normalized Difference Vegetation Index (NDVI) and its derivatives such as the Soil Adjusted Vegetation Index (SAVI) and the Modified Adjusted Vegetation Index (MSAVI), based on the red and near infrared (NIR) wavelengths, were developed to monitor the vegetation cover and stage of growth. The Normalized Difference Water Index (NDWI) (Gao, 1996), the Anomaly Vegetation Index (AVI), the Vegetation Condition Index (VCI) (Kogan, F., 1995), and the Vegetation Health Index (VHI) (Kogan, F., 1997), help with assessing vegetation stress and health. Further, the Normalized Difference Drought Index (Gu, Y. et al., 2007) was used to assess the covered land surface dryness.

From the perspective of soil drought or moisture, the Normalized Multi-band Drought Index (NMDI) (Wang, L. and Qu, J.J., 2007) and the Perpendicular Drought Index (PDI) (Qin, Q. et al., 2008) were developed. The PDI was modified through the MPDI (Ghulam A., et al., 2008) and the MDPI1 (Li, Z. and Tan, D., 2014).

The main goal of this paper is to assess the land surface drought, through another spectral model that assesses the covered and the uncovered land dryness. Its gives a second technical approach of the Land Surface General Drought Index, LSGDI (Ngandam Mfondoum and Gbetkom, 2019), for sustainable agriculture expansion in arid and semi-arid regions.

2. METHODOLOGY

2.1. Research location

The research was conducted on the site of Maga's floodplain, located between latitudes 10°37' - 10°54' N, and longitudes 14°40' - 15°10'E, in northern-Cameroon (Central Africa). It is an artificial floodplain extended on 1560 hectares and more than 200000 hectares with its surroundings, where a project of irrigated rice cultivation was launched in 1985 to promote national production. According to the national bioclimatic map, Maga's floodplain belongs to the Sudano-sahelian or semi-arid zone (Figure 1). Yearly, the rainfall is around 500 mm, and temperatures vary between 15° and 41° Celsius. Natural vegetation is made of savannah that exposes the soil. The soil is loamy, clay and alluvial, characterized by sediments deposits. Several types of soil degradation have been identified, modelled and mapped (Ngandam Mfondoum, A.H. et al., 2016; Ngandam Mfondoum, A.H. and Gbetkom, P.G., 2019; Gbetkom, P.G. et al., 2019). However, there is still a need for EO support to anticipate the risk of increased silting, salinization and sodisation during the irrigated rice project's expansion planned by Cameroon's government.

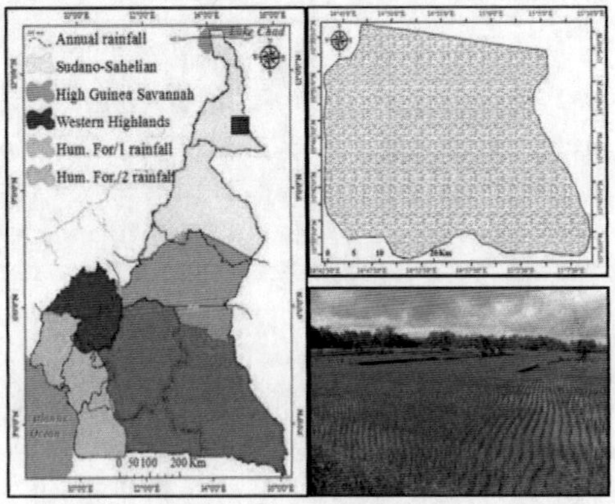

Figure 1. Maga's Floodplain, the study area

2.2. Data acquisition and pre-processing

The data used is from the Landsat sensors Thematic Mapper (TM) of 1987, Enhanced Thematic Mapper plus (ETM+) of 2003 and Operational Land Imager (OLI) of 2018 (Table1). They were downloaded from the United States Geological Survey website and displayed under false colour composite SWIR2-NIR-Red (Figure 2).

Table 1. Landsat images information

Year	Month	Scene ID
1987	January	LT51840531987024XXX03
	March	LT51840531987072XXX03
2003	January	LE71840532003012SGS00
	March	LE71840532003060SGS00
2018	January	LC81840532018013LGN00
	April	LC81840532018109LGN00

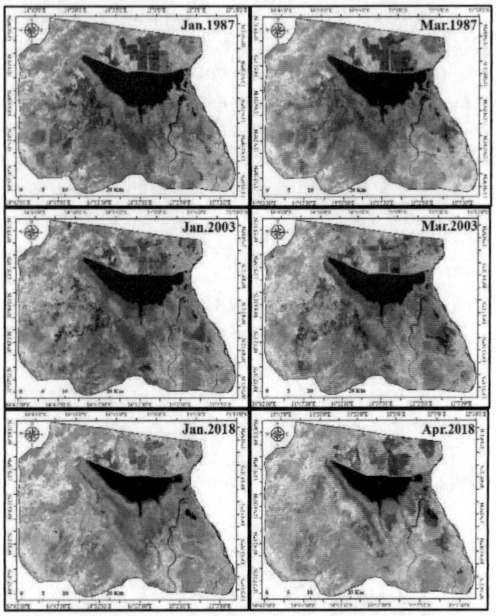

Figure 2. Landsat images in SWIR2-NIR-Red colour

Pre-processing help to minimize the effects of sensor, solar, atmospheric and topographic distortions. The images used are Level-1 products, which are delivered as digital numbers, DNs. Bands blue, green, red, near infrared, and both shortwave infrared were stacked. Applying the Cosine Solar TAUZ (COST) radiometric calibration model to the stacked image, the DNs were converted from at-sensor radiance to top-of-atmosphere (TOA) reflectance via solar correction, and rescaled from 64-bit to 8-bit. Atmospheric corrections and haze reduction have helped to remove other noises and then approximate values of surface reflectance. The last step concerned the topographic correction to lower altitude artifacts.

2.3. Experiment design

Ngandam Mfondoum and Gbetkom (2019) assume that balancing the moisture of the vegetated area with the dryness of the bare soil, is necessary to learn about the general drought retrieved from remote sensing techniques. They propose the Land Surface General Drought Index (LSGDI) as a normalized difference written as follows:

$$\mathrm{LSGDI} = 1 - \frac{\mathrm{NDSoDI - VMI}}{\mathrm{NDSoDI + VMI}} \quad (1)$$

Where the NDSoDI is the Normalized Difference Soil Drought Index and the VMI is the Vegetation Moisture Index. The subtraction of the normalized difference from 1 is suggested in order to: ***i)*** invert the values of soils from lower to higher values in the VMI; ***ii)*** keep highest values of the soil drought in the NDSoDI.

2.3.1 The Vegetation Moisture Index (VMI)

This index utilizes as entries the NDWI (Gao, B-C., 1996) and the SAVI (Huete, A.R., 1988).

$$\mathrm{NDWI} = \frac{\mathrm{NIR - SWIR1}}{\mathrm{NIR + SWIR1}} \quad (2)$$

$$\mathrm{SAVI} = \frac{(\mathrm{NIR - Red})}{(\mathrm{NIR + Red} + L)} * (1 + L) \quad (3)$$

The NDWI is sensitive to changes in liquid water content and in spongy mesophyll of vegetation canopies (Gao, 1996). Its values usually exhibit a quicker response to drought conditions than NDVI (Gu et al., 2007). In sparse vegetation cover conditions characterizing semi-arid zones, the SAVI is most suitable to assess vegetation cover and stages of growth and for adjusting the background soil reflectance through constant *L* value.

Spectral space plot between the NDWI and the SAVI for the main land cover categories, i.e. sandy soil, other soil, water body and rice crops, shows that they are highly correlated by the vegetation (Figure3). However, while the other objects values are decreasing for the SAVI in the order [***Sandy soil-Other soil-Water body***], for the NDWI, the order is [***Water body-Sandy soil-Other soil***].

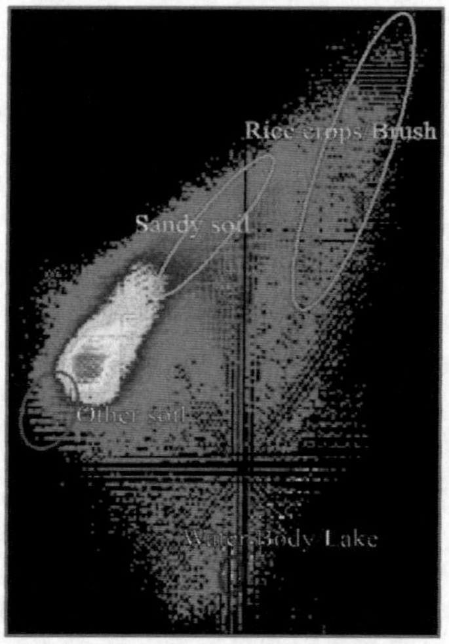

Figure 3. The spectral relation between NDWI (X axis) and SAVI (Y axis)

GAO (1996) notices that the soil background and snow cover effects are not masked in the NDWI. Therefore, to keep only the vegetation moisture as high values, the water body's values would be lower by subtracting the blue wavelength values from the sum of the SAVI and the NDWI, as follows:

$$\mathrm{VMI} = \frac{(\mathrm{NDWI + SAVI}) - \mathrm{Blue}}{(\mathrm{NDWI + SAVI}) + \mathrm{Blue}} \quad (4)$$

When visually compared to the NWDI and the SAVI, the VMI shows more efficiency in keeping vegetation moisture, lowering surface water, and being more sensitive to vegetation, soil features and sediments deposits at the lake's bottom (Figure 4).

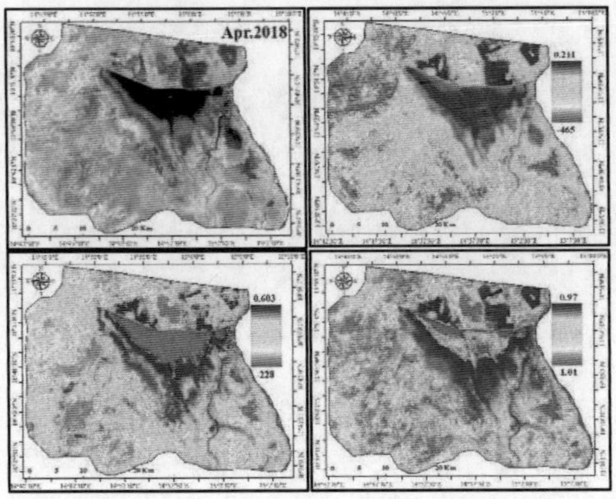

Figure 4. Visual patterns. Left-right and top-down: Landsat image, NDWI, SAVI and VMI.

2.3.2 The Normalized Difference Soil Drought Index (NDSoDI)

Its computation is based on the spectral behaviour of soils according to their moisture. Lobell and Asner (2002) have demonstrated that the soil reflectance varies according to its moisture through the following exponential model:

$$R = f \times R_{dry} + (1 - f) \times R_{dry} \exp(-c \times \theta) \qquad (5)$$

Where R is the soil reflectance at a particular wavelength, θ is the volumetric soil water content, R_{dry} is the reflectance of dry soil (at $\theta = 0.0$), c describes the rate of soil reflectance change with moisture, and f is the ratio of the saturated to dry reflectance.

A local spectral library was constructed. Figure 5 shows the land surface reflectance variation of the four training samples types of soils according to their predicted moisture.

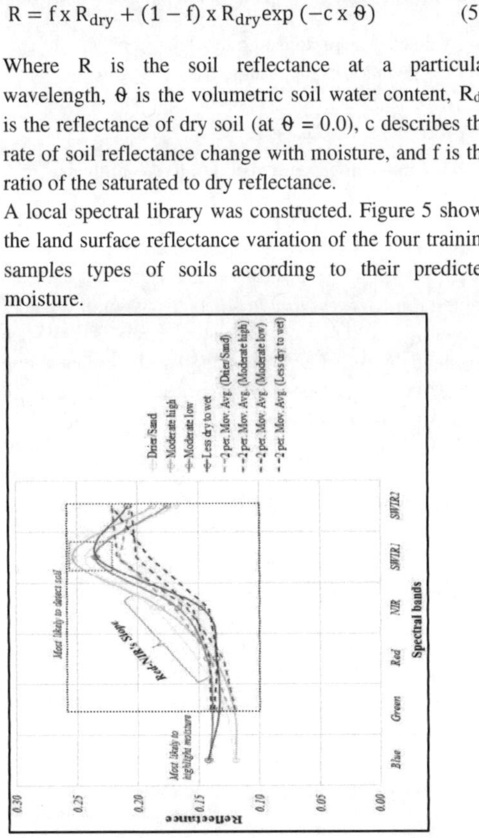

Figure 5. Reflectance of the four types of soils dryness

The evolution the "two periods" moving average trend lines built for each type of soil delimits the suitable electromagnetic spectrum wavelengths between the middles of the green and the SWIR2 (black square), while the most suitable is the SWIR1 (green square). These lines also highlight the Red-NIR's slope (green brace) as the transition between the first period going from the SWIR2 to NIR and the second period going from the NIR to the green. From the green to the blue, there is no trend change, signs of end of the slope.

Complementary analysis is made through the reflectance space between bands, sampling the four major land cover objects, i.e. sandy soil, other soils, water body and vegetation. Plots between the blue, the red and the SWIR1 bands in the six periods of the study. The example of April 2018 (Figure6) let shows that, in the blue-red feature space, the decreasing correlation order is [***Sandy soil-Other soil-Water body-Rice crops***], water body and rice crops being too close in values. The blue-SWIR1 and the red-SWIR1 plots show the same order, i.e., [***Sandy soil-Other soil-Rice crops-Water body***].

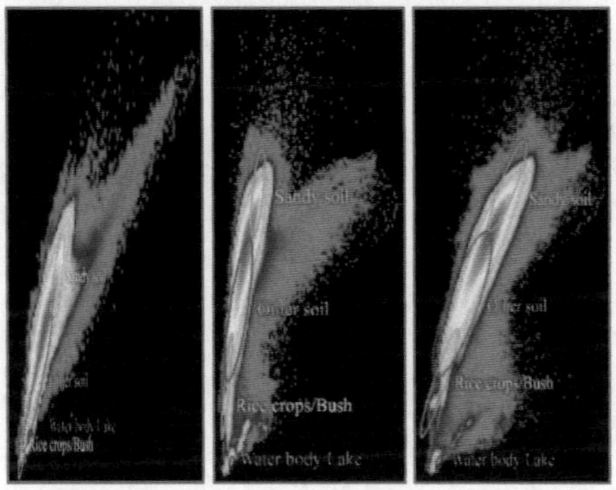

Figure 6. Features spaces. Left's plot - *Blue (X axis) Red (Y axis)*; Middle's plot - *Blue (X axis) SWIR1 (Y axis)*; Right's plot - *Red (X axis) SWIR1 (Y axis)*

The reflectance differences among water and sandy soils spectral curves are also to be noticed between in the three concerned wavelengths (Figure 7).

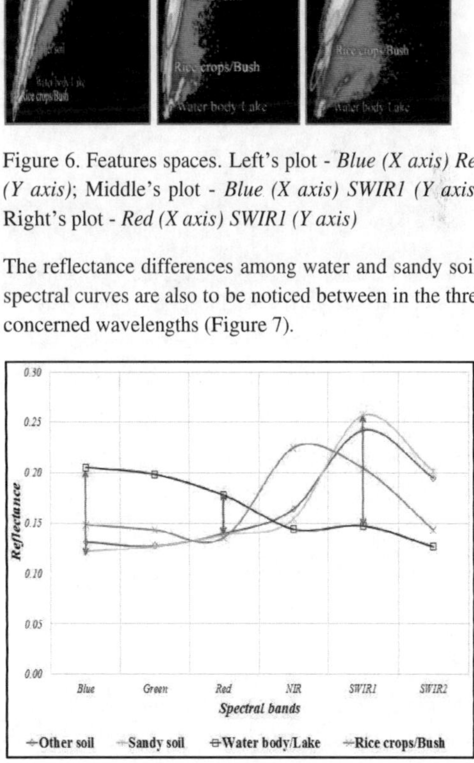

Figure 7. Reflectance curves for April 2018

From these views, it is assumed that, sand and soil features would be highlighted, when adding the red and SWIR1 bands, showing a strong positive convergence along a soil slope. The moisture would be assessed by subtracting the blue band and adjusting the water and soil reflectance values. The adjustment factor's values are settled according to the reflectance difference between the sand's spectral curve, which has the highest bare soil reflectance, and the water body's curve, in the blue, the red and SWIR1 bands for each month.

Table 2. Adjustment factor's values (*L*)

1987		**2003**		**2018**		**Average**
Jan.	*Mar.*	*Jan.*	*Mar.*	*Jan.*	*Apr.*	
0.49	0.38	0.47	0.44	0.23	0.24	0.375

The average for the six periods is 0.375≈0.4. However, considering the high difference values between years, the *L* value of each month is used. The NDSoDI is then written as follows:

$$\text{NDSoDI} = \frac{(\text{Red+SWIR1})-\text{Blue}}{(\text{Red+SWRI1})+(\text{Blue}+\sum L)}(1+\sum L) \qquad (6)$$

Where $\sum L$ is the sum of the differences between the sandy soil and the water body curves in the blue, red and SWIR1 bands. The highest values describe soil dryness and the lowest values its moisture, as shown on figure 8.

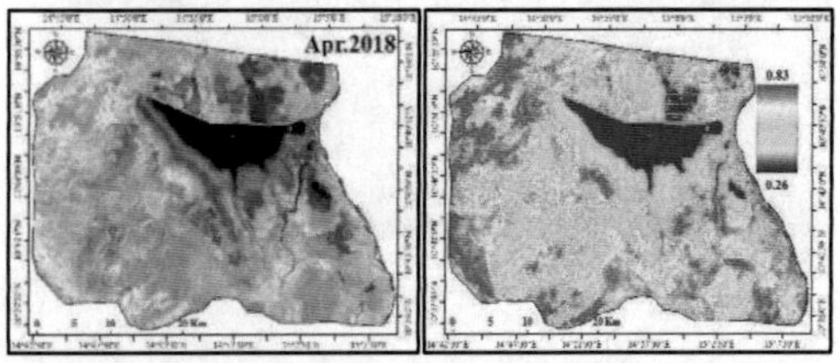

Figure 8. NDSoDI visual patterns

Accordingly, this index can be called the Soil Adjusted Moisture Index (SAMI) when the lowest values are to be considered.

2.3.3 The improved Land Surface General Drought Index (LSGDI2)

From the above reasoning, it is hypothesized that adding the drought of the land vegetated area with the one of the bare soil will provide better information regarding general droughts. When combining the VMI and NDSoDI spectral space, dry and wet surfaces can be represented on x-y orthogonal axes to locate the LSGDI2 (Figure 9).

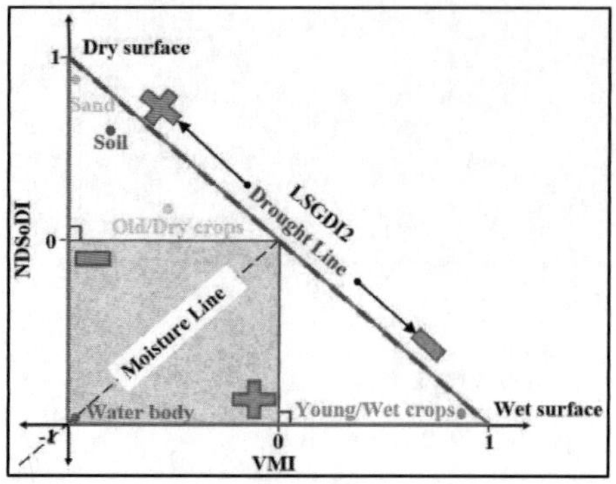

Figure 9. LSGDI2 on the VMI-NDSoDI orthogonal plan

The moisture square includes the water body in both VMI and NDSoDI lower values, and its diagonal, the moisture line, separates two triangles representing the both indices areas of influence. The value needed equals to the triangles right angles projection, named the drought line, and corresponding to the LSGDI2. According to Pythagorean Theorem, it should be expresses as the square root of sum of squares of the NDSoDI and the VMI, such as:

$$\text{LSGDI2} = \sqrt{\text{NDSoDI}^2 + \text{VMI}^2} * 0.1 \qquad (7)$$

The value 0.1 helps adjusting the values below 1, to ease the interpretation of the results.

3. RESULTS AND DISCUSSIONS

3.1 Dynamic ranges and visual patterns appraisal

The LSGDI2 global range dynamics are between 0.097 and 0.137 ≈ 0.14 for the studied period (Figure10). The difference with the LSGDI is that from the obtained values, it is easy to affect qualifiers to classes of drought. Therefore, the resulting image was reclassified and the thresholds were defined, using the five descriptors from Palmer's drought severity classification: [*Exceptional drought – Extreme drought – Severe drought – Moderate drought – Abnormally drought*]. With maximums of 0.137 and 0.134 respectively in January and March, 2003 was the drier year. It could be predicted that all values from 0.13 and above (≥ 0.13) are classified "exceptional drought", and those below 0.1 (0.9) are classified "abnormally drought" or moist areas.

Moreover, the visual patterns compared with the original Landsat images demonstrate that highest values correspond to sand and lower values to water body and rice crops (Figure 10).

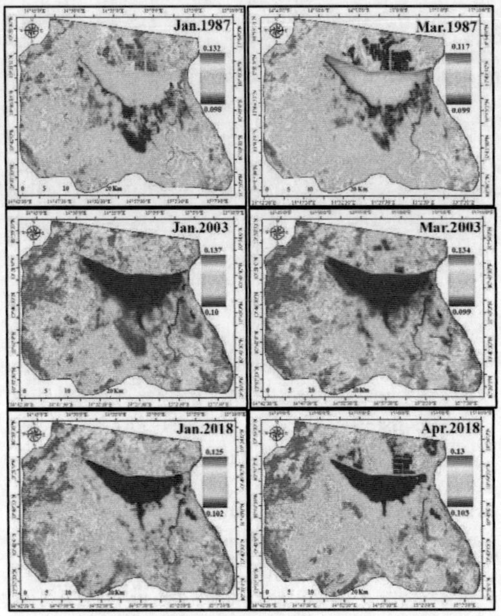

Figure 10. Visual patterns of the LSGDI2

3.2 Reflectance efficiency and its statistical impact

A comparison was run with the LSGDI and the MDPI1 (Li, Z. and Tan, D., 2014). As reminding, MDPI1 processing also assesses the soil and vegetation moisture/drought in one algorithm:

$$\mathrm{MDPI1} = \sqrt{\mathrm{PDI}^2 + \mathrm{PVI}^2} \quad (8)$$

The three indices values were resampled in the interval [0 – 1] to ease the comparison. After that, fifteen (15) pixels were sampled on each stretched image, with correspondence in all the five (05) classes of drought. It appears that LSGDI2 reflectance neatly enhances those given by the MDPI1 and LSGDI (Figure 11).

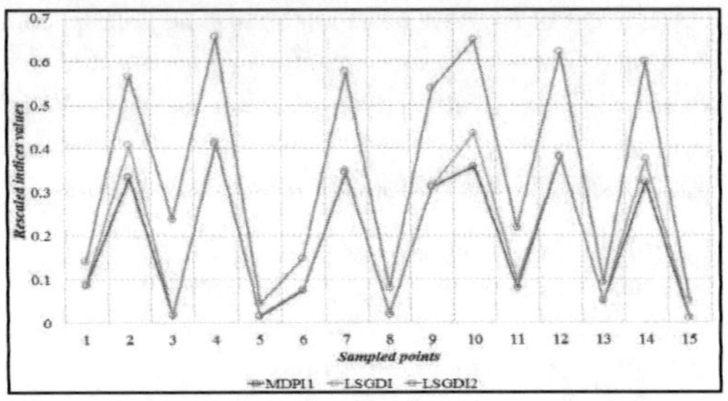

Figure 11. Reflectance of the three indices for April 2018

A statistical analysis with each class confirms the LSGDI2 efficiency in detecting drought, by varying comparatively to visual patterns interpretation and the two others indices (figure 12). For instance, its "Abnormally drought" class areas for January 2003 and April 2018, are respectively 1/6 (16716 ha / 92976 ha) and 1/7 (9658 ha/69257 ha), less than LSGDI ones. Concerning the MDPI1, its areas for the "Exceptional drought" class values has been almost the same, i.e. around 52680 hectares during the six periods (Figure 12).

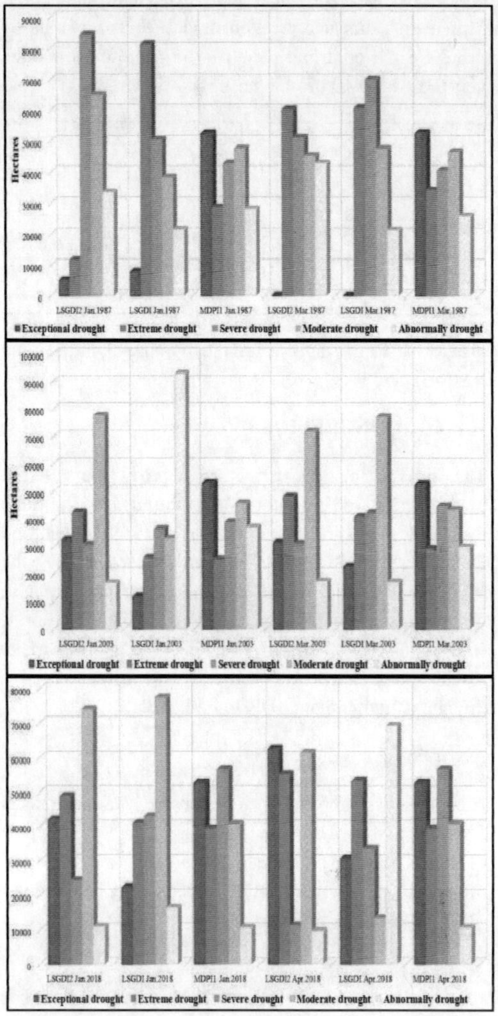

Figure 12. Compared areas statistics for the three indices

3.3 Drought and moisture informative analysis

LSGDI2 values come from both the NDSoDI and VMI. It appears on fifteen (15) sampled points that LSGDI2 curve is distinct from its two components' ones (Figure13). Therefore, they both contribute to the model.

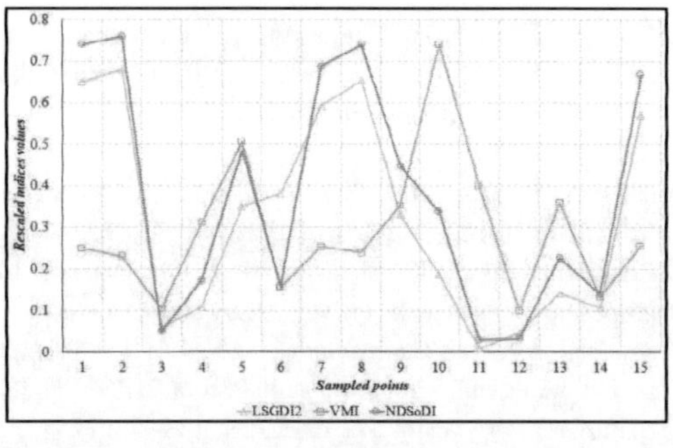

Figure13. LSGDI2, NDSoDI and VMI for April 2018

Nevertheless, the LSGDI2 and NDSoDI curves have the same evolution. On fifteen (15) pixels sampled, fourteen (14) have exactly the same behaviour, i.e. 94%. Moreover, a linear regression model (Figure 14) shows that NDSoDI contribute at 90% (R^2=90.3 to the model, versus 10% (R^2=0.07) for the VMI.

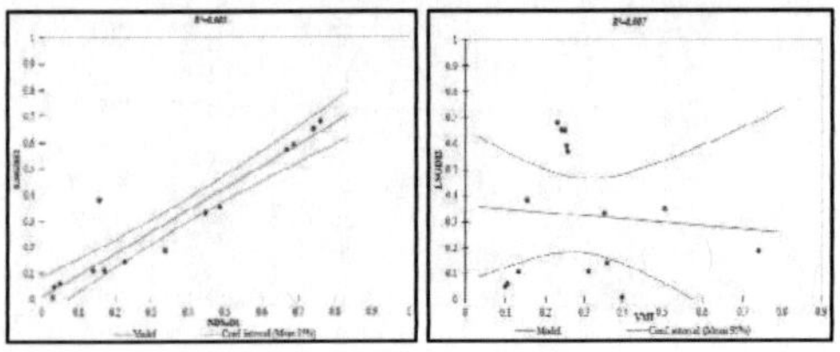

Figure 14. Regression of LSGDI by NDSoDI (left) and VMI (right)

4 CONCLUSIONS

The aim of this paper was to improve the LSGDI, formerly assessed as a normalized difference. The main methodology point has been to sum the vegetation moisture and the soil drought, both assessed through the VMI and the NDSoDI, to express the general drought. The proposed LSGDI2 appeared to be the square root of sum of squares of the two previous. The analysis of its results shows that its reflectance values in each class of drought is higher than MDPI1 and LSGDI.

AKNOWLEDGEMENTS

Authors are grateful to USGS for the satellite images freely available. They especially appreciate the support of StatsN'Maps management, for initiating and directing this work since its presentation to PECORA21/ISRSE38.

REFERENCES

Eslamian, S., Ostad-Ali-Askari, K., Singh, V.P., Dalezios, N.R., Ghane, M., Yihdego, Y., Matouq, M., 2017. A Review of Drought Indices. International *Journal of Constructive Research in Civil Engineering*, vol.3 (4), pp. 48-66.
Doi: http://dx.doi.org/10.20431/2454-8693.0304005

Gao, B., 1996. NDWI - a normalized difference water index for remote sensing of vegetation liquid water from space *Remote Sensing of Environment* 58: 257–266

Gbetkom, P.G., Gadal, S., El Aboudi, A., Ngandam Mfondoum, A.H., Mansour Badamassi, M. B., 2019. Combining spectral indices and statistics analysis to proposed a new risk of soil degradation method – application to the Cameroonians shores of Lake Chad and its hinterland. Proceedings of Pecora 21/ISRSE 38, Baltimore, Maryland, USA. HAL Id: hal-02327627

Ghulam, A., Qin, Q., Kusky, T. and Li, Z., 2008. A re-examination of perpendicular drought indices. *International Journal of Remote Sensing* **29** 6037–6044. Doi: 10.1080/01431160802235811

Graham, S., 2000. Drought: The Creeping Disaster. Portions reprinted courtesy of the NOAA National Weather Service Office of Meteorology. https://earthobservatory.nasa.gov/features/DroughtFacts, last visited on November 5, 2019.

Gu, Y., Brown,J.F., Verdin, J.P. and Wardlow, B. 2007. A five-year analysis of MODIS NDVI and NDWI for grassland drought assessment over the central Great Plains of the United States, Geophys. Res. Lett., 34, L06407, doi:10.1029/2006GL029127.

Huete, A.R., 1988. A Soil-Adjusted Vegetation Index. Remote Sensing of Environment, 25:295-309 (1988)

Kogan F 1995 Application of vegetation index and brightness temperature for drought detection *Advances in Space Research* 15:91–100

Kogan F 1997 Global drought watch from space *Bulletin of the American Meteorological Society* 78:727–636

Li, Z. and Tan, D. 2014. A modified perpendicular drought index in NIR-Red reflectance space. ISRSE35, IOP Conf. Ser.: Earth Environ. Sci. Doi: 10.1088/1755-1315/17/1/012040

Lobell, D. B., and G. P. Asner (2002), Moisture effects on soil reflectance, Soil Sci. Soc. Am. J., 66, 722– 727.

Ngandam Mfondoum, A.H., Etouna, J., Buji, K.N., Mvogo Moto, F.A., Noulaquape Deussieu, F.G., 2016. Assessment of Land Degradation Status and Its Impact in Arid and Semi-Arid Areas by Correlating Spectral and Principal Component Analysis Neo-Bands. *International Journal of Advanced Remote Sensing and GIS*, Vol.5 (2), 1539-1560. Doi: https://doi.org/10.23953/cloud.ijarsg.77

Ngandam Mfondoum, A.H., Gbetkom, P.G., 2019. Proposing a new drought assessment spectral model to anticipate desert fronts directions through sustainable agriculture expansion. Proceedings of Pecora 21/ISRSE 38, Baltimore, Maryland, USA. http://pecora.asprs.org/wp-content/uploads/2019/10/POS_004_Ngandam-Mfondoum.pdf

Qin Q, Ghulam A, Zhu L, Wang L, Li J and Nan P 2008 Evaluation of MODIS derived perpendicular drought index for estimation of surface dryness over northwestern China. *International Journal of Remote Sensing* 29:1983–1995.

Wang, L. and Qu, J.J., 2007. NMDI: A normalized multi-band drought index for monitoring soil and vegetation moisture with satellite remote sensing. Geophys. Res. Lett., 34, L20405, doi: 10.1029/2007GL031021.

White, R.P., Nackoney, J., 2003. Drylands, people, and ecosystem goods and services: a web-Based Geospatial Analysis. World Resources Institute, 58P.

THERMAL REMOTE SENSING OF URBAN CLIMATES IN SOUTH AFRICA THROUGH THE MONO-WINDOW ALGORITHM

Adeline Ngie

AQCCI: Unit for Env. Sc. and Management, School of Geo-and Spatial Sciences, Faculty of Natural and Agricultural Sciences, North – West University, South Africa: adeline.ngie@nwu.ac.za

KEY WORDS: Urban heat islands • thermal remote sensing • land surface temperature • Landsat ETM+

ABSTRACT:

Urban Heat Island (UHI) is among some of the challenges plaguing urban environments. There is increase human population within urban environments especially in the developing world, which is a need to understand the climates for their wellbeing. The use of multispectral satellite remote sensing to investigate the climatic conditions through radiation measurement is applied across the two major South African cities. The thermal remote sensing technique applied for this study is the direct determination of land surface temperatures (LST) using multispectral thermal imagery (ETM+). In addition, meteorological data which included air temperature and relative humidity for the same satellite image dates were used. The LST values obtained showed Johannesburg has many micro heat islands scattered across the metro than in Cape Town. These areas of heat islands corresponded to areas of human settlement and more so the unplanned as opposed to the planned ones. The estimated LST values and observed air temperature values with an R^2 of 0.9. It could be concluded that expansion of urban areas in South Africa has led to increased thermal radiation of land surface in densely populated areas.

1. INTRODUCTION

Urbanisation is a phenomenon of change in landuse type by humans for their well being. One of its key aspects is the reduction in vegetation cover and an increase in concreted or paved areas. Increasing urbanisation especially in the developing countries as South Africa, is continuously challenging for town planners and policy makers to incorporate knowledge about climatic conditions within urban spaces and manage towards minimal impacts on human health (Orimoloye et al., 2018). In which case it would be even more difficult for countries like South Africa where these urban spaces are invaded by unplanned settlements which make use of corrugated sheets to construct houses and have no home-garden spaces (Chikoto, 2009). Therefore, it is but essential that studies be conducted for a spatio-temporal appraisal of some climatic parameters such as surface temperature.

One of the important characteristics of urban environments is impervious paved surfaces and reduced vegetation. This characteristic causes more of the sun's incoming radiant energy to be reflected from urban areas and, less of it is converted to latent energy involved in evaporation or transpiration of moisture (Taha, 1997). The larger volume of asphalt, brick, concrete, and other materials in urban areas give it a higher thermal storage capacity than surrounding vegetated surfaces. The urban canopy also stores large amounts of energy during the day which it releases at night (Oke, 1987). Another factor exacerbating relatively high temperatures within urban areas than its surroundings is the additional heat generated by automobiles, air conditioning and other machinery.

The UHI has been described as one of the major challengers to sustainable livelihood within urban areas. The UHI is the excess temperature near the ground (canopy layer) of the central urban locations as being higher than those of nearby or surrounding areas (Voogt and Oke, 2003). This phenomenon has previously been considered as the difference between urban and rural areas but with recent developments in cities, there is no distinct borderline between 'urban' and 'rural' areas as a result of urban growth (Voogt, 2004). In recent times it has been considered as the difference between the central parts of the city and its surrounding areas (Ngie et al., 2014). The continuous urbanisation of urban areas has led to various settlement areas across metropolitan cities and is a motivation for this study.

Therefore, this study sought to assess the spatio-temporal display of surface temperature across two major metropolitan cities of South Africa with both planned and unplanned settlements. This made used of the mono-window algorithm in estimating LST (Qin *et al.,* 2001) for Johannesburg and Cape Town metros. There will be a highlight of surface temperature situation across these metropolitan cities.

2. STUDY AREA AND DATA COLLECTION

Urban areas in post-apartheid South Africa where movement of people is no longer restricted have experienced tremendous growth rates (Ndebele, 2013). According to the last national census, the country experienced a 15.5 % population growth from 2001 to 2011 meanwhile some provinces like Gauteng (seat to the city of Johannesburg) experienced a 33.7 % increase within the same 10-year period being double the national average (Stats SA, 2012). Based on their rapid rate of urbanisation the two major cities were selected for this investigation (Figure 1).

2.1 Study Area

In terms of population and economic status, Johannesburg is the biggest city in South Africa with huge urban sprawl problems influenced by highly linked highway interchanges. On a land surface area of 1 645 km^2 the city supports a population of 4 434 827 inhabitants. Cape Town is the second largest with 3 740 026 inhabitants on a total land area of 2 455 km^2 (census 2011: Stats SA, 2012).

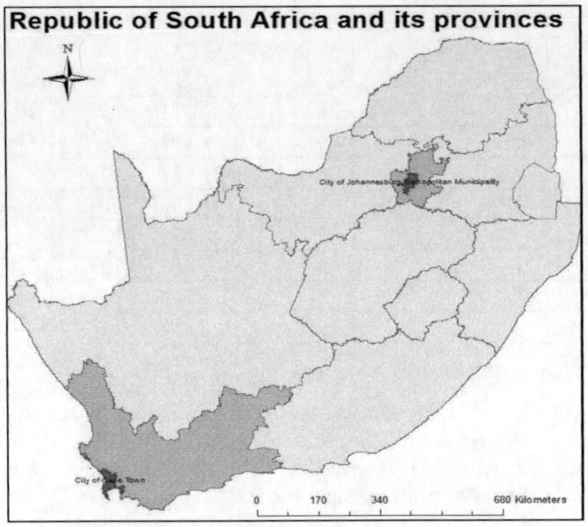

Figure 1: Map of South Africa showing the 2 cities

2.2 Data

2.2.1 Satellite Imagery: Satellite data were acquired over two different seasons: winter (June-August) and summer (January-March) during the day periods for each season for two different years. The years selected were 2002 and 2012. Where suitable 2012 images were not available (due to high cloud cover), 2011 images within the same season were acquired. For instance, summer images over Cape Town were selected in 2011. Therefore, a total of 8 LANDSAT ETM+ images were acquired for the study by downloaded from the United States Geological Society (USGS) website.

2.2.2 Other Auxiliary Data: The weather data were obtained from South African Weather Service. These data included air temperature and relative humidity (Table 1). Daily measurements for the required dates of satellite images were obtained. The air temperature and relative humidity data were used for the analysis.

2.3 Data Pre-Processing

The Enhanced Thematic Mapper Plus (ETM+) data was acquired through the USGS website after a thorough search of available and cloud free scenes for both the winter and summer seasons of the years of study. The downloaded data was unzipped then for those that have no gaps, layer stacking and clipping of study areas were done using ENVI version 5.0[1].

The Scan Line Corrector (SLC) which compensates for the forward motion of ETM+ on Landsat 7 failed on May 31, 2003 and recovery efforts were unsuccessful. ETM+ now traces a zig-zag pattern along the satellite ground track with striped images of increasing strip width towards the scene edge. An estimated 22 % of any given scene is lost because of the SLC failure and precise locations of the missing scan lines will vary from scene to scene (Landsat 7 Science Data Users Handbook[2]).

Therefore, the acquired ETM+ images post 2003 had to undergo gap filling using different scenes of the same Landsat ETM+ sensors from different adjacent dates of acquisition. These dates fall into the same season being investigated. The gap filling module developed in ERDAS Imagine® was applied. This module operates on two functions; to fetch gaps existing on the given scene for the first image, and to match the histogram for the gaps from the scene of the second image.

The separate image bands were combined into a single multispectral image file to serve as input into the algorithm. This is known as layer stacking and also commonly used to combine image derivatives with spectral bands for further analysis. The images were clipped to the desired study areas (municipal boundaries of the cities which are metropolitan).

3. METHODOLOGY

3.1 LST retrieval methods

The UHI effect has been studied through the retrieval of LST from satellite remote sensing data (Voogt & Oke, 2003; Mirzaei & Haghighat, 2010; Ngie *et al.*, 2014; Abutaleb *et al.*, 2015; Ngie *et al.*, 2016; Sharma & Bisht, 2019). Landsat TM and ETM+ have a single thermal band (band 6) which is the thermal infrared (TIR) channel that records radiation within a 10.4-12.5μm spectral range. Different LST retrieval methods have been developed according to different data sources (thermal bands on a sensor), such as the split-window, temperature/emissivity separation, mono-window, and the single-channel methods (Qin *et al.*, 2001, Sobrino *et al.*, 1996; Jimenez-Munoz & Sobrino, 2003). Among these methods three can be applied to Landsat data; these include the radiative transfer equation, mono-window algorithm and single-channel algorithm.

Although all of these methods can provide good results, some of them such as the radiative transfer equation can only yield results with *in situ* atmospheric parameters captured simultaneously as the satellite passes over the study area. In addition, the mono-window algorithm can get better results than the single-channel algorithm with a root mean square deviation of 0.9°C (Sobrino, *et. al.*, 2004). It has proven to yield better accuracy for the retrieval of LST from Landsat images as illustrated for Hong Kong with a 0.7°C error (Liu & Zhang, 2011).

This study therefore makes use of the mono-window algorithm to retrieve the LST from the ETM+ imagery for the cities of Johannesburg and Cape Town. The algorithm is based on thermal radiance transfer equation and only requires three parameters: emissivity, transmittance and effective mean atmospheric temperature to retrieve LST from Landsat multispectral imagery (Qin *et al.*, 2001). The model was designed using ERDAS Imagine® spatial model tool (Sun *et al.*, 2010; Liu & Zhang, 2011). It required two measured parameters: near-surface air temperature and relative humidity (RH) simultaneously measured as the satellite passed over the study area. These data were obtained from the local weather stations situated within the study areas (Table 1). The two parameters were then converted to atmospheric transmittance and effective mean atmospheric temperature. The third parameter (emissivity) was calculated from the normalized difference vegetation index (NDVI).

---

[1] *Exelis Visual Information Solutions. 2010. Name of the manual. Boulder, Colorado: Exelis Visual Information Solutions.*

[2] http://landsathandbook.gsf.nasa.gov/ Accessed 19/11/2016

CITY	DATE	MAX_TEMP	MIN_TEMP	MIN_RH	MAX_RH	Water Vapour content	TRANSMITTANCE			
Johannesburg	January 7th 2002	22	20	65	84	1.904	0.822	0.799	0.812	0.784
	July 2nd 2002	6	5	76	100	0.814	0.909	0.904	0.937	0.939
	January 3rd 2012	24	21	56	70	1.752	0.834	0.814	0.829	0.806
	June 27th 2012	7	6	50	65	0.575	0.928	0.927	0.965	0.972
Cape Town	March 15th 2002	26	13	89	97	2.286	0.791	0.762	0.768	0.730
	July 21st 2002	24	13	42	84	1.275	0.872	0.860	0.884	0.873
	March 26th 2012	28	15	64	93	2.033	0.811	0.787	0.797	0.766
	June 12th 2011	20	2	74	99	1.214	0.877	0.865	0.891	0.882

Table 1: Satellite and auxiliary data collected and estimated parameters (water vapour content and transmittance)

3.2 Mono-window algorithm

3.2.1 Conversion of digital number into spectral radiance: The satellite sensors measure reflectance from the earth surface as digital numbers (DN) representing every pixel of the image. The DN data from Landsat was converted to spectral radiance using Equation (1):

$$L_\lambda = L_{\lambda min} + (L_{\lambda max} - L_{\lambda min}) \times \frac{(Q_{\lambda DN} - Q_{\lambda min})}{(Q_{\lambda max} - Q_{\lambda min})} \quad (1)$$

Where L_λL is at-sensor spectral radiance;

$L_{\lambda max}$L is maximum at-sensor spectral radiance;

$L_{\lambda min}$ L is minimum at-sensor spectral radiance;

($Q_{\lambda max}$ Q= 255) and ($Q_{\lambda min}$Q= 1) are the maximum and minimum DN value of pixels respectively and, $Q_{\lambda DN}$Q is the DN value of pixel.

For Landsat ETM+ data, equation (1) can be rewritten in two dimensions depending on the low and high gain bands of the thermal band and expressed in equations 2 and 3.

1 The low gain band which provides an expanded dynamic range with lower radiometric resolution (sensitivity), has less saturation at high DN values[3] and expressed thus:

$$L_{61} = 0.0671(Q_{\lambda DN} - 1) \quad (2)$$

2 The high gain band has a higher radiometric resolution (sensitivity) although it has a more restricted dynamic range (*ibid*) and expressed thus:

$$L_{62} = 0.0372(Q_{\lambda DN} - 1) + 3.2 \quad (3)$$

3.2.2 Conversion of spectral radiance into at-sensor brightness temperature: The conversion formula is:

$$T_i = \frac{K_2}{\ln(\frac{K_1}{L_\lambda} + 1)} \quad (4)$$

Where T_i is at-sensor temperature in Kelvin, K_1 and K_2 are pre-launch calibration constants and, L_λ is the spectral radiance in watts per metre squared steradian micron (W/ (m^2*sr*μm)). For ETM+ the calibration constants are:

K_1 = 666.09 (W/ (m^2*sr*μm))

K_2=1282.71 K

3.2.3 Calculation of NDVI: NDVI is a simple graphical indicator usually used to analyze remote sensing measurements, and assess if observed target includes green vegetation or not. NDVI is one of the outputs that run into the model to retrieve LST. Therefore, eight NDVI maps were produced according to the various dates of satellite capture in this study using equation (5)

$$NDVI = \frac{NIR - R}{NIR + R} \quad (5)$$

Where NIR is Near Infra-red band and R is the red band

3.2.4 Estimation of emissivity: In this study, the land surface emissivity was derived from the NDVI (Zhang, 2006). From the NDVI results obtained above (Equation 5) corresponding land emissivity values were established (Table 2).

NDVI	Land surface emissivity (ε_i)
NDVI < -0.185	0.995
-0.185 ≤ NDVI < 0.157	0.970
0.157 ≤ NDVI ≤ 0.727	1.0094+0.0047 * Ln(NDVI)
NDVI > 0.727	0.990

Table 2: Estimation of land surface emissivity from NDVI (Adopted from Zhang, 2006)

3.2.5 Estimation of atmospheric transmittance: Atmospheric transmittance which is defined as the capacity of the atmosphere to transmit electromagnetic energy is dependent on the air mass penetrated by radiations, as well as on the amount of water vapour and dust in the air. It varies for radiations of different wavelengths; the smaller the atmospheric absorption and scattering of light, the greater the atmospheric transmittance.

3.2.6 Calculation of water vapour content: The calculation of water vapour is done through equation (6) which entails deriving relative humidity. Relative humidity can be defined as the ratio of water vapour at a given temperature to the saturation water vapour pressure at the same temperature (Wagner and Pruß, 2002).

$$RH = \frac{P_a}{P_s} \times 100 \quad (6)$$

[3]http://landsathandbook.gsfc.nasa.gov/data_properties/prog_sect6_4.html. Accessed 27/02/2014

Where RH is the relative humidity;
P_a is the actual water vapour pressure;
P_s is the saturation water vapour pressure.
Equation (6) is rewritten to obtain P_a as follows:

$$P_a = P_s \frac{RH}{100} \tag{7}$$

The saturation water vapour pressure (P_s) at a given temperature (K) can be calculated (Wagner and Pruß, 2002):

$$P_s = P_c * e^{[\frac{T_c}{T}(C_1\vartheta + C_2\vartheta^{1.5} + C_3\vartheta^3 + C_4\vartheta^{3.5} + C_5\vartheta^4 + C_6\vartheta^{7.5})]} \tag{8}$$

Where:
T = ambient temperature in K
P_s = Saturation water vapour pressure (hPa)
T_c = Water critical temperature (647.096 K)
P_c = Critical pressure (220 640 hPa)
C_1 = -7.85951783
C_2 = 1.84408259
C_3 = -11.7866497
C_4 = 22.6807411
C_5 = -15.9618719
C_6 = 1.80122502
And

$$\vartheta = 1 - \frac{T}{T_c}$$

Atmospheric transmittance is then estimated on the basis of some linear equations derived from statistical regression analysis (Qin *et al.,* 2001). These equations relate the atmospheric transmittance to the water vapour content (g/cm^2) (Equation (8)). Based on the results obtained for water vapour content in this study, corresponding atmospheric transmittance equations were selected (**Error! Reference source not found.**). The air temperature (t_a) profiles were set as t_a <15° C being low and t_a >15° C as being high.
3.2.7 Calculation of mean atmospheric temperature: Equations (9) and (10) are used to estimate the atmospheric mean temperature (T_a) based on given near surface air temperature (Qin *et al.,* 2001):
For mid-latitude summer

$$T_a = 16.0110 + 0.92621 T_0 \tag{9}$$

For mid-latitude winter

$$T_a = 19.2704 + 0.91118 T_0 \tag{10}$$

LST retrieval from the mono-window algorithm: In the mono-window algorithm, the LST value is calculated using Equation 11 but with two unknowns (C and D) being realised through Equations 12 and 13. The three required parameters (emissivity, transmittance and effective mean atmospheric temperature) are included in the equation as follows:

$$T_s = \frac{(a + bT_i)(1 - C - D) + D(T_i - T_a)}{C} + T_i \tag{11}$$

$$D = (1 - \varepsilon_i)[1 + (1 - \varepsilon_i)t] \tag{12}$$

$$C = \varepsilon_i * t \tag{13}$$

Where: T_s is LST (K); T_i is at-sensor temperature (K); ε_i is emissivity estimated from NDVI; *t* is atmospheric transmittance calculated from water vapour content; T_a is mean atmospheric temperature; *a* = −67.355351, and b = 0.458606.

4. RESULTS AND DISCUSSION

4.1 Relationship between the observed air temperature and the estimated LST

Statistical regression analysis was used to establish a relationship between the estimated LST values and the measured air temperature values from weather stations in the two cities. The analysis was done using mean temperature values for both scenarios over the cities and seasons. The results yielded an R^2 of 0.9. Therefore, results confirmed the fitness of the mono-window algorithm in estimating LST values. The difference between the estimated LST values and the measured air temperature values can be associated partly with the effects of surface roughness on surface temperature and emissivity not considered (Weng, 2001) and partly because of atmospheric impurities which can obstruct the smooth passage of radiant energy (Brooks, 2006).

4.2 Spatial distribution of surface temperature

The spatial distribution of surface temperatures from these two cities indicated that not just do the central city regions illustrate heat islands but also around the cities which correlated to the informal human settlement areas. The intensity of these heat islands varied within these cities and the various areas.

In Cape Town the city centre showed a surface temperature increase to a maximum of 5°C compared to the surroundings in summer 2012 (Figure 2). However, the geographical features surrounding this city being the Table Mountain in the south and the ocean (Atlantic and Indian Oceans) to the east and west respectively should play a role in the experience but require further research to ascertain.

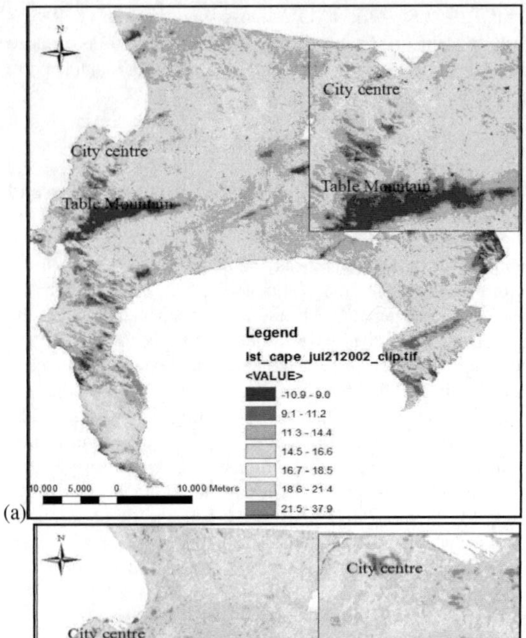

Figure 2: LST maps for Cape Town (a) Winter 2002; (b) Summer 2012

When new surface materials which include asphalt and concrete are introduced to a naturally vegetated area the thermal climate also changes with an increase in canopy temperature. The emission of heat, moisture and pollutants dramatically change radiative, thermal, moisture, roughness and emissions properties of the surface and the atmosphere above (Joshi & Bhatt, 2012). There is also the emission of a lot of radiant energy through automobiles and heavy machines within these urban centres which contributes in the higher LST values (Figure 3).

Generally, the intensity of surface temperature in Johannesburg is at micro levels due to the high density of planted trees which do not only absorb the energy but cause shaded areas that disrupt the heat island (Liu *et al.,* 2007). There are other major centres of heat islands in Johannesburg which could be linked to the expansion of urban areas with or without necessary planning. The expansion of planned areas towards the northern parts of the city, distinctively shows the Sandton, Fourways and Randburg city centres. However, the unplanned and unorganized rapid expansions of settlements such as the Alexandra Township in the north as well and SOWETO in the south western section illustrate heat island intensities slightly higher than their planned human settlements (Figure 3). For instance, Sandton which is adjacent to Alexandra but is a formally planned up-market area experiences a 2°C lower land surface temperature.

This study highlighted the fact that it is not just a challenge with development that has transformed vegetated areas to paved or concreted land covers but also that of land use. The human activities within these unplanned settlements which include burning coal for cooking as well as heating in open fires can be seen as contributors to raising the surface temperature. These unplanned settlements have limited land space for vegetation due to the high influx of people from rural to urban areas without any resources to live within decent or planned neighbourhoods (Chikoto, 2009). They end up raising structures with metal sheets and of high density as is the case with the informal sections of the south western part of Johannesburg (SOWETO), Alexandra and other associated townships in the north eastern sections contributing to the higher LST values over those areas (Figure 3).

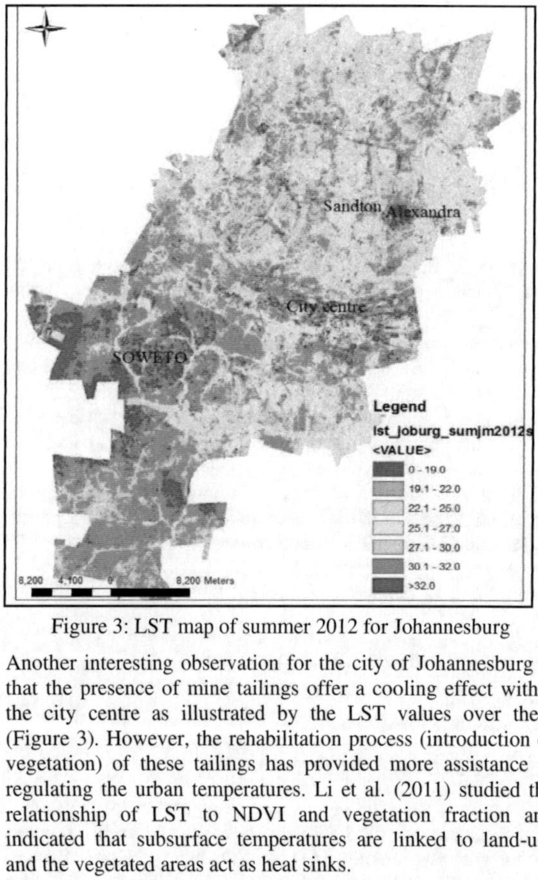

Figure 3: LST map of summer 2012 for Johannesburg

Another interesting observation for the city of Johannesburg is that the presence of mine tailings offer a cooling effect within the city centre as illustrated by the LST values over them (Figure 3). However, the rehabilitation process (introduction of vegetation) of these tailings has provided more assistance in regulating the urban temperatures. Li et al. (2011) studied the relationship of LST to NDVI and vegetation fraction and indicated that subsurface temperatures are linked to land-use and the vegetated areas act as heat sinks.

Human settlements have replaced existing vegetation with paved surfaces, which results in the sun's energy heating the man-made surfaces rather than being absorbed by vegetation and used for evapotranspiration (Taha, 1997) is considered prime cause of the UHI effect over these cities. The hottest temperatures were generally found in areas with the least vegetation (NDVI) and the greatest urban development. There is a strong relationship between LST and impervious surfaces as revealed in this study. This contributes to a large extent of the heat island problem thereby suggesting that town planners need to structure built-up areas with diverse structures and interspersed vegetation as well as water bodies in varying percentages to serve as the heat sinks and regulating urban climates (Zhang *et al.,* 2008).

There also exist some situations of anomalies in this study where the city centre of Johannesburg seems to be cooler than its surrounding areas in winter but with a well-developed UHI in summer (Figure 3). This is however, not strange as it confirms some empirical generalizations offered by Oke (1982). These generalizations include the UHI intensity being most well developed in summer or in the warm half of the year and that the UHI may disappear by day or the city may be cooler than the rural (surrounding) environs. However, there will be need for further research to investigate the factors contributing to these situations.

5. CONCLUSION

The mono-window algorithm provided a simple and highly effective method for retrieving land surface temperatures for the analysis over these cities. This affirms other similar studies elsewhere in the world which have applied this algorithm (Ifatimehin & Adeyemi, 2008; Lu *et al.,* 2009; Liu & Zhang, 2011; Khaled *et al.,* 2015). The LST derived from the ETM+ thermal band depicted the spatial variability of land surface temperature of the cities of Johannesburg and Cape Town.

The spatial variability and intensity of the surface temperatures across the cities especially in Johannesburg illustrated the contribution from high urbanisation experience. There was a further distinction in the planned and unplanned or poorly planned expansions of urban centres. The intensities did not only vary by space or land use or land cover but also over seasons wherein for the Johannesburg the phenomenon is present in summer and absent in winter. There is a need to further look into the impact of geographical features around these cities such as mountains or oceans and also more validation stations to the high spatial variability which was supported by only two weather stations.

ACKNOWLEDGEMENTS

The author express appreciation to the anonymous reviewers.

REFERENCES

Abutaleb, K., A. Ngie, A. Darwish, M. Ahmed, S. Arafat and F. Ahmed, 2015: Assessment of Urban Heat Island using remotely sensed imagery over Greater Cairo, Egypt. *Advances in Remote Sensing 4,* 35-47

Arnfield, A.J., 2003: Two decades of urban climate research: a review of turbulence, exchanges of energy and water, and the urban heat island. *International Journal of Climatology 23*, 1-26.

Basar, U.G., S. Kaya and M. Karaca, 2008: Evaluation of urban heat island in Istanbul using remote sensing techniques. Commission VII, WG VII/5. *The International Archives of the Photogrammetry, Remote Sensing and Spatial Information Sciences Vol. XXXVII. Part B7.* Beijing.

Brooks, D. R., 2006: Monitoring Solar Radiation and Its Transmission through the Atmosphere. The GLOBE Program's Aerosols, Water Vapor, and UV-A Monitoring Projects V 2.0 http://www.pages.drexel.edu/~brooksdr/DRB/papers/UsingThe Sun/using.htm Accessed 20/11/2013.

Chikoto, T., 2009: Informal settlements in South Africa. Unpublished BSc (Hons) in Faculty of Engineering, Built Environment and Information technology, University of Pretoria, South Africa.

Ifatimehin, O.O. and S. Adeyemi, 2008: A satellite remote sensing based land surface temperature retrieval from Landsat TM data. *Ethiopian Journal of Environmental Studies and Management 1(3),* 63-70.

Jiménez-Muňoz, J.C. and J.A. Sobrino, 2003: A generalized single-channel method for retrieving land surface temperature from remote sensing data. *Journal of Geophysical Research 108*, 4688-4694.

Joshi, J.P. and B. Bhatt, 2012: Estimating temporal land surface temperature using remote sensing: A study of Vadodara urban area, Gujarat. *International Journal of Geology, Earth and Environmental Sciences ISSN: 2277-2081 2(1),* 123-130.

Landsberg, H.E., 1981: *The Urban Climate*. New York: Academic Press.

Li, J., C. Song, L. Cao, F. Zhu, X. Meng and J. Wu, 2011: Impacts of landscape structure on surface urban heat islands: A case study of Shanghai, China. *Remote Sensing of Environment 115,* 3249-3263.

Liu, L. and Y. Zhang, 2011: Urban Heat Island analysis using the Landsat TM data and ASTER Data: A case study in Hong Kong. *Remote Sensing 3*, 1535-1552.

Liu,W., C. Ji, J. Zhong, X. Jiang and Z. Zheng, 2007: Temporal Characteristics of the Beijing Urban Heat Island. *Theoretical and Applied Climatology 87*, 213-21.

Lowry, W. P., 1977: Empirical estimation of urban effects on climate: A problem analysis. *Journal of Applied Meteorology 16(2),* 129-135.

Lu, Y., P. Feng, C. Shen and J. Sun, 2009: Urban Heat Island in summer of Nanjing based on TM data. In Proceedings of 2009 Joint Urban Remote Sensing Event, Shanghai, China, 20-22 May; pp. 1-5.

Mirzaei, P.A. and F. Haghighat, 2010: Approaches to study Urban Heat Island – Abilities and limitations. *Building and Environment 45,* 2192-201.

Ndebele, T., 2013: South Africa goes with the urbanisation flow. Press release from the South African Institute of Race Relations (SAIRR). www.sairr.org.za Tel: 011-482-7221 E-mail: thuthukani@sairr.org.za.

Ngie, A., K. Abutaleb, F. Ahmed, A. Darwish and M. Ahmed, 2014: Assessment of Urban Heat Island using remotely sensed imagery: A Review. *South African Geographical Journal.* ID: 924864 DOI:10.1080/03736245.2014.924864

Ngie, A., K. Abutaleb, F. Ahmed, O.J. Taiwo, A. Darwish and M. Ahmed, 2016: An estimation of land surface temperature from Landsat ETM+ images for Durban, South Africa. *Rwanda Journal, Series D, Volume 1, 2016, Life and Natural Sciences: Special issue II.* http://dx.doi.org/10.4314/rj.v1i2S.2D

Oke, T.R., 1973: City size and the urban heat island. *Atmospheric Environment 7,* 769-779.

Oke, T.R., 1982: The energetic basis of the urban heat island. *Quarterly Journal of the Royal Meteorological Society 108,* 1-24.

Oke, T.R., 1987: *Boundary Layer Climates*. New York: Routledge.

Orimoloye, I.R., S.P. Mazinyo, W. Nel, and A.M. Kalumba, 2018: Spatiotemporal monitoring of land surface temperature and estimated radiation using remote sensing: human health implications for East London, South Africa. *Environmental Earth Sciences 77(3)*, 1 – 10.

Qin, Z., A. Karnieli and P. Berliner, 2001: A mono-window algorithm for retrieving land surface temperature from Landsat TM data and its application to the Israel-Egypt border region. *International Journal of Remote Sensing 18,* 3719-3746.

Sharma, V.R. and K. Bisht, 2019: Estimation of Land Surface Temperature using LANDSAT DATA: A case study of Agra city, India. *International Journal of Advanced Research and Innovation of Ideas in Technology 5(2),* 1850-1857.

Sobrino, J.A., J.C. Jimenez-Munoz and L. Paolini, 2004: Land surface temperature retrieval from LANDSAT TM 5. *Remote Sensing of Environment 90,* 434-440.

Sobrino, J.A., Z. L. Li, M.P. Stoll and F. Becker, 1996: Multi-channel and multi-angle algorithms for estimating sea and land surface temperature with ASTER data. *International Journal of Remote Sensing 17,* 2089-2114.

Statistics South Africa (Stats SA), 2012: Census 2011 Statistical release- P0301.4/Statistics South Africa: Pretoria 78pp. www.statssa.gov.za

Sun, Q., J. Tan and Y. Xu, 2010: An ERDAS image processing method for retrieving LST and describing urban heat evolution: A case study in the Pearl River Delta Region in South China. *Environment Earth Science 59,* 1047-1055.

Taha, H., 1997: Urban climates and heat islands: Albedo, evapotranspiration and anthropogenic heat. *Energy and Buildings 25,* 99-103.

Voogt, J. A., 2004: Urban Heat Islands: Hotter Cities. American Institute of Biological Sciences. http://www.actionbioscience.org/environment/voogt.html Accessed 30/10/2013.

Voogt, J.A. and T.R. Oke, 2003: Thermal remote sensing of urban climates. *Remote Sensing of Environment 86,* 370-84.

Wagner, W. and A. Pruß, 2002: The IAPWS formulation 1995 for the thermodynamic properties of ordinary water substance for general and scientific use. *Journal of Physical and Chemical Reference Data 31(2),* pp. 387535.

Weng, Q., 2001: A remote Sensing-GIS evaluation of urban expansion and its impacts on surface temperature in the Zhujiang, Delta, China; *International Journal of Remote Sensing 22(10),* 1999-2014.

Zhang, Y., 2006: Land surface temperature retrieval from CBERS-02 IRMSS thermal infrared data and its applications in quantitative analysis of urban heat island effect. *Journal of Remote Sensing 10,* 789-797.

Zhang, Z., M. Ji, J. Shu, Z. Deng and Y. Wu, 2008: Surface urban heat island in Shanghai, China: Examining the relationship between land surface temperature and impervious surface fractions derived from Landsat ETM+ imagery. *The International Archives of the Photogrammetry, Remote Sensing and Spatial Information Sciences. Vol. XXXVII.* Part B8. Beijing.

ESTIMATION OF MAIZE NITRATE CONCENTRATIONS USING EO-1 DATA AND A NON-LINEAR REGRESSION MODEL

Adeline Ngie

AQCCI: Unit for Env. Sc. and Management, School of Geo-and Spatial Sciences, Faculty of Natural and Agricultural Sciences, North – West University, South Africa: adeline.ngie@nwu.ac.za

KEY WORDS: maize, waveband selection, image spectroscopy, cforest, spectral ratios, nitrate assessments, random forest

ABSTRACT:

Nitrogen compounds such as nitrates are considered the most important limiting factor for crop productivity. Monitoring the status of this element in crops has moved from destructive to non-destructive approaches. Remote sensing with ever evolving technologies has taken the lead on different crops across the world. This study assessed the potential of EO-1 data (Hyperion) to estimate nitrate concentrations in maize (*Zea mays*) leaves. The image was captured over the study area after the 11th week of planting. The random forest algorithm was useful for band selection to reduce data redundancy in the imagery, and regression analysis for nitrate predictions. Maize nitrate concentrations were detectable with key contributing wavebands as 752, 1043, 681, 851, 1820, 762, 862, 640, 1850, 609, 589, 569 and 650nm. From this list, a subset corresponding to previously identified bands was used to develop vegetation spectral ratios. There was improvement in accuracy of predictions from using: all selected wavebands, all developed ratios, and selected ratios as independent variables for the model with 752-681 contributing the most to an $R^2 = 0.90$; and RMSEP = 0.15. Therefore, selected bands of Hyperion to develop ratios could be used to monitor spatial variation of nitrate concentrations in maize from canopy level.

1. INTRODUCTION

Crop productivity is challenged among other factors by sustainable management regimes which are to be guided by timely monitoring of the nutrient status of the plants at critical growing stages (Fageria, 2009). Plant nutrient assessment has traditionally been done through numerous manual field visits through to tedious laboratory work. The process is laboured intensive and destructive as the many visits translate to collection of plant tissue samples (roots, stems and/or leaves) for analysis. Over the years, non-destructive methods have been developed and being applied to monitor plant nutrient status such as nitrogen. Some of the methods included the use of chlorophyll metres to measure the concentration of the pigment in plants and relating it to their nutrient status (Blackmer & Schepers, 1994). The concentration of the chlorophyll pigment is largely nitrogen (N) and usually enhanced through fertilisation that is required to be applied not just at the right time but also the right amounts required by the different crops (Fageria, 2009). The right application is important both for economic and environmental reasons. The nutrient variability spatially through soil quality is a confounding challenge to the right application of fertilizers as it requires a thorough assessment of N in the soil and/or crops, for which such assessments could be costlier over large areas. The application of remote sensing provides potential to rapidly and less costly to assess or monitor such nutrient status in not just soils but crops alike (Tilling et al., 2007).

The relationship between N and chlorophyll has been established (Lee et al., 1999) and has been proven as a linear relationship (Houlès et al., 2007; Ziadi et al., 2008; Haboudane et al., 2008). The use of remote sensing in assessing N through the chlorophyll content at canopy level has been attempted in the light of growth stages and other agronomic factors (Strachan et al., 2002; Nguyen & Lee, 2006; Nguyen et al., 2006; Chen et al., 2010). In a critical review Patane and Vibhute (2014) concluded on the concept of this relationship though with variations results not just from different techniques applied but also different crop biochemical constituents.

Hyperspectral data complexity has seriously challenged the management of hyperspectral data and usually overcome by selection of most relevant spectral wavelengths or bands (Clevers & Jongschaap, 2003; Thenkabail et al., 2004). The selectivity is on the basis that sunlight energy reaching vegetation surfaces is either absorbed or reflected back with the latter (reflectance) measured by a remote sensor at different wavelengths (Figure 1). The reflectance is more in the visible region (VIS) being influenced by the presence of leaf tissue pigments such as chlorophyll that should relate to leaf N status (Haboudane et al., 2002; Rodriguez et al., 2005). Chlorophyll *a* and *b* have been found to relate to the two absorption regions on the VIS being the blue (400-500nm) and red regions (600-670nm). Chlorophyll pigments are harboured in the chloroplast

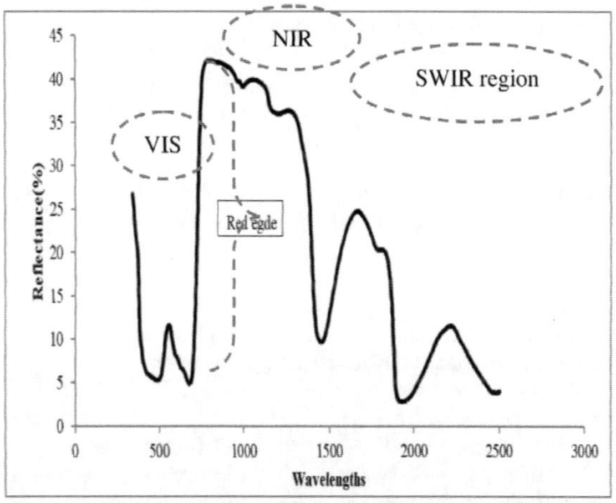

Figure 1: Hyperspectral curve for a healthy crop leaf (Adapted from Ngie et al., 2016)

which contains about 75% of total plant N (Lawlor, 2001). These absorption regions can serve as a measure of the N or its related compounds such as nitrate content in plants where the higher the absorption, the higher the chlorophyll content and so too will be the N content. In relation to this, hyperspectral data has been used over canopies of agricultural fields to characterise the variation of the chlorophyll pigment at an accuracy of 92% (Gitelson et al., 2005).

In transition from the VIS to the near-infrared (NIR) region is the red edge (Figure 1). This is also a region of interest for crop biochemical studies. This region lies between the 680-760nm and the reflectance have correlated positively to both leaf and canopy N status (Barnes et al., 2000; Cho & Skidmore, 2006). The next portion on the electromagnetic spectrum is the NIR region stretching from 700 to 1300nm and contributes to crop nutrient status by the leaf internal structure which reflects high energy levels as opposed to the absorptions in the VIS. The region lying from the 1300-2500nm is known as the shortwave infrared (SWIR) (Figure 1) which has more absorption regions as a result of water content in the leaves (Thenkabail et al., 2004). However, some wavelengths found in the region reflect energy accounted for by the protein and starch content in the leaves (Murray & Williams, 1987). The visible and near-infrared region (VNIR) has been described as the chlorophyll absorption feature (Kumar et al., 2003). This region shows through its reflectance a strong relationship with nitrogen content in plants (Mutanga & Skidmore, 2003; Zhao et al., 2005). However, a contrary situation has been illustrated of a weak relationship between the nitrogen content and canopy reflectance of crops such as rice under controlled conditions (Stroppiana et al., 2006) which have not deterred other recent researchers from obtaining success with this region in rice nitrogen contents through different sensors (Wang et al., 2013). There have also been other studies across other agricultural crops with varying but acceptable levels of accuracies (Ngie et al., 2014).

The wavelength range characteristics has been investigated as and another alternative approach for more efficient and accurate modelling or predicting N status in plants. The spectra measured from the visible and NIR regions (400-900nm) calibrated for plant N status resulted in an R^2 of 0.71 with error of prediction as 0.38% (Hansen & Schjoerring, 2003). In a similar study using spectra measured at the 530-1100 nm the results were better having R^2 of 0.81 with error of prediction as 0.27% (Alchanatis et al., 2005) and a subsequent increase wavelength range from 400-2500nm yielded an R^2 of 0.89 with prediction error of 0.64% (Morón et al., 2007). However, it should be noted that some of the differences could have accrued from the different analytical methods applied on the data sets or sensors used to measure the reflectance.

While some studies applied ratios of the reflectance value measured at the red and NIR region to detect plant nutrients (Jackson et al., 1981), other utilised single wavebands (Stroppiana et al., 2006). Most of the above cited studies have made use of reflectance values from single wavebands or optimising wavebands that could predict plant nitrogen content. It should be noted that one advantage in the use of vegetation indices has been the reduction in variations resulting from canopy geometry, irradiance and shading. They have also assisted in minimizing soil background effect on canopy reflectance (Jackson & Huete, 1991). Hence, the selection of wavebands for the creation of vegetation indices or ratios as input variables into predictive models using space-borne hyperspectral studies on biochemical content on field crops under irrigated or controlled conditions is relevant than using just single wavebands (Jain et al., 2007); (Zhu et al., 2008). Also most of these studies have utilised handheld spectroscopic devices to measure reflectance to assess N status except for a few that have utilised airborne and satellite-borne sensors (Huang et al., 2004; Oppelt & Mauser, 2004; Vigneau et al., 2011).

Hence, this study used EO-1 data through a non-linear regression models in assessing nitrate concentrations in maize growing under field conditions from canopy level. This is done in a subtropical area where moving through the fields can be challenged by its terrain. This is done through firstly, a selection of wavebands of importance in relation to maize nitrate content. Secondly, to use these selected wavebands to develop spectral vegetation ratios as input (with varied numbers) into a non-linear regression.

2. MATERIALS AND METHODS

2.1 Study area

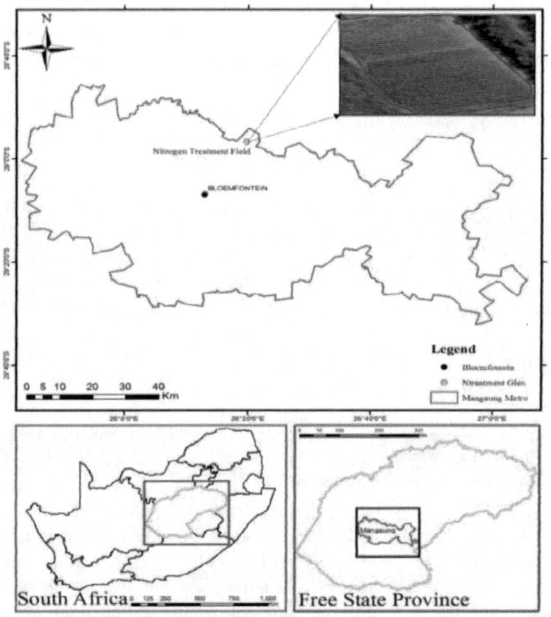

Figure 2: The experimental field in Bloemfontein within the Mangaung-metropolitan municipality, Free State South Africa (Insert image from Google Earth 4/10/2014)

The Glen Agricultural College in Bloemfontein in the Free State province of South Africa hosted the experimental field (Figure 2) of about 1.5ha in size and situated around 28°56‘41.37"S, 26°19‘46.51"E. The area which is about 20km north of the city of Bloemfontein receives long mean annual precipitation of about 600mm, mean annual maximum and minimum temperatures of 25° C and 8° C respectively (Botha et al., 2007). Its summers are hot and dry with scarce episodes of rainfall while the winters are frosty and cold. The local geology is primarily sedimentary rocks that form the basis on superficial deposits of rich agricultural soils mostly of sandy clay loamy texture (Botha et al., 2007). The cultivation of field crops in summer such as maize is done once in a farming season between December and July. The cultivation is usually supported through flooding-type irrigation schemes.

2.2 Field and laboratory analyses

The field was planted with maize and variable fertilize applications into three categories of low, medium and high or normal nitrate content. After the 11th week of planting with almost fully ground coverage, thirty leaf samples (10 across each nitrate category) were randomly collected (noting the coordinates of sample point) through excision on the third fully expanded leaf on the maize plants. The leaf samples were packaged in sampling bags and taken to the laboratory for chemical analyses. The samples were oven dried to eliminate moisture and the midrib region of the leaves excluded before crushing in a pulveriser. A sample solution was prepared from the crushed leaves as collected from the field.

The solutions were used to quantify the amount of nitrate in the maize leaves through ion chromatography (IC). The IC is a technique to analyze solutions containing complex mixtures of ions. In addition, it was considered as a rapid and sensitive technique for separation of anions. Plant tissue extraction with water was preferred since it reduces the challenges that are linked with safety, disposal, or masking of ions that might occur in extractions with acids as proven with eluent of corn leaf sap (Masson et al., 1996). The Dionex™ Potassium Hydroxide Eluent Generator Cartridge (EGC-KOH) system was used with a flow rate of 0.25mL per minute to analyse the anions and results recorded.

2.3 Image acquisition and pre-processing

Hyperion images were acquired over the study area after tasking on the United States Geological Society (USGS) website during the period of field visits. A cloud free image was obtained on the 2nd of April 2014 on the 171/080 path/row scene. There is usually a challenge of illumination to some of the Hyperion bands while others suffer from overlaps between the two spectrometers and end up without values or reflectance values set to zero during the Level 1B pre-processing (Datt et al., 2003). In reviewing the available bands of the acquired Hyperion images, over 44 bands out of the 242 were without reflectance values. These consisted of bands 1-7, 56-78 and 225- 242 with the remaining 196 bands radiometrically corrected and calibrated to at-sensor radiance (Beck, 2003; Green et al., 2003).

The images were geometrically referenced to a Landsat ETM$^+$ image (19 January 2012) which was already georeferenced (Universal Transverse Mercator (UTM), zone 35 South). In minimising the effects of systematic noise in the image, a de-streaking algorithm proposed by Datt et al. (2003) was applied to reduce the stripping effect. After which the radiance images were then corrected atmospherically and transformed to reflectance at canopy level using the MOD-TRAN based Fast Line-of-sight Atmospheric Analysis of Spectral Hypercube (FLAASH) algorithm that is built within the Environment for Visualising Images (ENVI version 5.0) software package. Through derivation of atmospheric properties including water vapour, surface albedo and others, FLAASH provides a well-adjusted input for the atmospheric correction (Thenkabail et al., 2013). The images were resampled to their initial spatial resolution of 30m, the nearest neighbour algorithm was performed wherein for an indicator of a good geometric correction, the root mean square error (RMSE) considered was at less than a pixel (Ferencz et al., 2004). Through the use of the collected ground truth points during sample collection, the maize canopy reflectance spectra were extracted from single pixels for statistical analysis.

2.4 Statistical analysis

The relationship between biochemical concentrations in plants and their measured reflectance maybe indeed nonlinear (Miglani et al., 2008). The nonlinear regression models predicted nitrogen concentrations in vegetation with air-borne hyperspectral data set at canopy level with higher R^2 values than the linear models. In predicting sugar cane leaf nitrogen concentrations using reflectance values from a space-borne sensor, a non-linear model again outperformed the multiple linear regression models with higher R^2 values (Abdel-Rahman et al., 2013). Hence, the choice of applying a non-linear model in this study was established and being the random forest (RF) ensemble.

The RF is a machine learning nonlinear algorithm that operates through two major parameters which include the *ntree* and *mtry*. The *ntree* is the number of trees that are to be used in an ensemble while the *mtry* is the number of variables that is being chosen randomly at each split. Recursive partitioning is used to divide data into regression trees and average the results of all trees. The operations are guided by the fact that every regression tree is developed to maximum size independently without pruning through bootstrapping samples from the training (2/3 of total samples) and testing (1/3 of total samples) data sets. Regression trees can at each node be combined and choosing randomly a subset of input variables (*mtry*) from which to calculate the split (Breiman, 2001).

The out-of-bag (OOB) error estimation derived through data predictions that are being considered in each tree is a means to evaluate performance of the RF ensemble. For each variable in the regression tree, the OOB error is calculated as the difference in mean square error of the data used to develop regression trees and that of the OOB. The comparison of the OOB errors to the previous or original ones yields values that indicate the importance of the variable wherein it shows how the error varies when a variable is permuted and all the other variables left unchanged (Prasad et al., 2006). However, for the predictive model, the accuracy for each run was by measuring the root mean squared error of prediction (RMSEP) which is considered more stable than the OOB (Abdel-Rahman et al., 2013).

The ensemble has the capability of performing variable of importance selection which reduces the high dimensionality in hyperspectral data sets and the predictive modelling. The process is to select a subset of relevant variables for use in model construction. The process is guided by the fact that the data contains many redundant wavelengths as identified by its low spectral ranges (Thenkabail et al., 2004; Kokaly, 2001). The original RF ensemble by Breiman (2001) can perform this task of selecting variables but has been criticised for bias selection (Strobl et al., 2007). The *cforest* function was introduced with the RF ensemble that performs the selection but with minimal bias (Strobl et al., 2009).

3. RESULTS AND DISCUSSION

3.1 Maize leaf nitrate concentrations

The maize nitrate concentrations measured in the laboratory from leaves collected during field visits that was about 13 days (12/03/2014) to the passing of the EO-1 sensor, recorded minimum and maximum amounts as 0.2% and 1.4% of total anions respectively with standard deviation of 0.3% for the three different fertilization categories. There was no scenario of 0% nitrate measured for any of the leaves collected which was affirmation to the application of N fertilizers during planting across the entire field to enable germination of all seeds. The categories of fertilization were created after germination through top dressing with regulated nitrate amounts. The amount of nitrates in the maize leaves measured in the laboratory corresponded with the different samples from the various categories.

3.2 Hyperion wavelength selection

The most important 26 wavelengths of the Hyperion image to maize nitrate content were identified for this study (Figure 3). The wavelengths were from the red region, NIR and the early mid-infrared regions of the electromagnetic spectrum. There were six wavelengths amongst the 26 selected that appeared with high OOB errors and corresponded to those identified in previous studies (Thenkabail et al., 2004; (Miglani et al., 2008). These included 609, 640, 650, 681, 752 and 1043 that corresponded to bands 26, 29, 30, 33, 40 and 90 according. It should be noted that the order of these wavelengths here was not derived from the OOB error listing as will be seen below (Figure 3) but rather sorted according to the values (highest to the lowest of the 26).

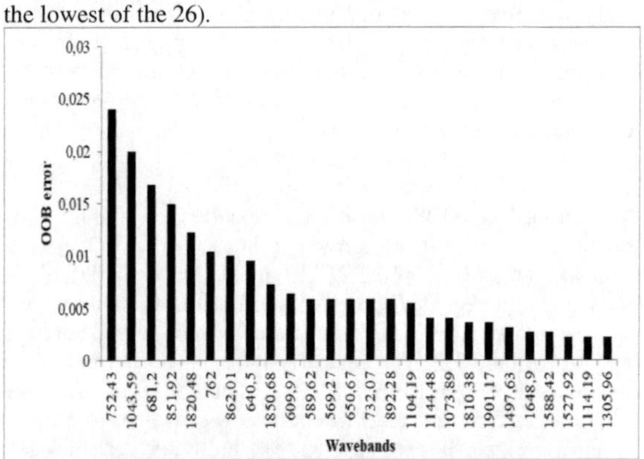

Figure 3: 26 wavelengths selected for maize canopy nitrate content prediction from Hyperion image

According to Thenkabail et al. (2004) the six identified bands in this study belonging to the region 599-650 are relatively sensitive to biomass; 671-681 are chlorophyll absorption region; 742-752 which is the red edge region is sensitive to vegetation stress and/or dynamics; and the 1003-1053 is related to plant moisture status, biomass and even the leaf area index (LAI) which are all functions of the biochemical status of the plants. The first three regions have jointly been labelled chlorophyll absorption bands and are closely related to nitrogen concentration in plants (Huang et al., 2004; Curran, 1989). In this study, some of the waveband regions in the most relevant (bigger OOB error values) 26, there was the selection of some around the protein absorption region at 1648 nm, which is in proximity to 1645 identified by Murray and Williams, (1987). The 752 nm was also identified by Vigneau et al. (2011) in addition to the other studies stated above as important in estimating N concentration in wheat plants.

3.3 Development of vegetation spectral ratios and nitrate assessment

The identified six wavelengths were used to develop all possible combinations of spectral ratios. This was done on spectra using the formula for the normalised difference vegetation index (NDVI) (Rouse et al., 1974). Hence, they were NDVI-based ratios. All the developed NDVI-based ratios or vegetation spectral ratios used as independent variable input into the predictive RF model after which there was another selection of variables based on their contribution in the permutation (Figure 4). Once again the results presented are the top most 26 ratios ranked by their OOB errors. Those ranked most important were created from the red edge bands including the 640, 650, 681 and the 752nm.

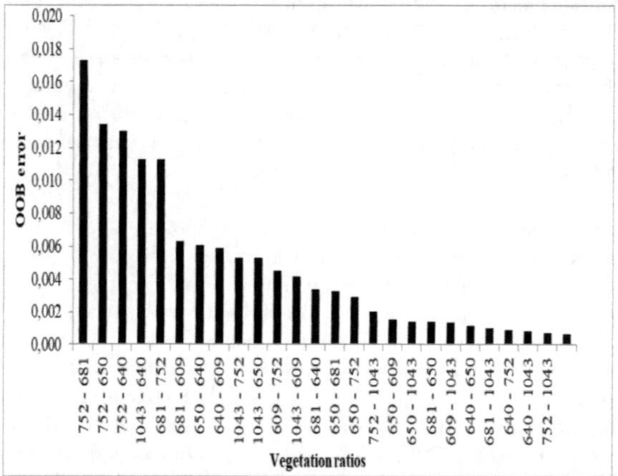

Figure 4: Selected vegetation-based ratios from 6 X 6 identified bands

The red edge-based ratio (752-681) and others are selected as the most important variables in the RF ensemble for nitrate content assessment in maize plants. The bands involved are related to those in the modified chlorophyll absorption in reflectance index (MCARI) that was designed to reduce the influence of soil background on photosynthetic active radiance (Daughtry et al., 2000). The index has been refined to improve sensitivity to the non-photosynthetic materials below crop canopy and established the transformed chlorophyll absorption in reflectance combined with the optimised soil-adjusted index (Haboudane et al., 2002). All these indices were made up of bands within the red edge region and have potential in the estimation of biochemical content of crops (Miglani et al., 2008; Bannari et al., 2008; Wang et al., 2017). The red edge position has been proven to be influenced by N concentration in the leaves and other critical crop parameters such as water status (Schlemmer et al., 2005); (Schlemmer et al., 2013). However, the selected variables in this study were nowhere around the known water absorption regions (970, 1450 and 1950nm). This means there was no influence from water absorption in the performance of the models (Figure 4). It would be of interest to perform the experiments and vary not just N concentrations but other critical parameters like water status since both scenarios could occur in a field.

The 752-681 vegetationbased ratio identified in this study, though with related wavebands in previous studies did not also match some of the ratios or indices reported in the past. In assessing the potential of EO-1 in estimating the chlorophyll content of wheat through a wide range of chlorophyll indices, also realised the most performing index was the normalised difference pigment index (NDPI) that is developed from reflectance values centred around wavelengths 430 and 680nm which belonged to the VIS and red edge region (Bannari et al., 2008). The NDPI was also identified as best input to the model testing potential of the VIS/NIR spectroscopy in predicting nitrogen concentrations in pear orchards (Wang et al., 2017). The common interest of the above studies and the present one is the fact that the wavebands comprising both ratios are from a similar region of the electromagnetic spectrum and the difference in specific bands could be attributed to difference in leaf structure or conditions under which spectra was collected. However, the studies all point in the direction of interest to the relationship between nitrogen or chlorophyll concentration in plants to spectral performance.

3.4 Maize leaf nitrate prediction

The prediction of maize leaf nitrate content using the imaging spectroscopy from canopy level was done through the RF ensemble as a non-linear regression model. The accuracies of the predictive models varied not just with the type of independent variable added but also the number of these variables being added. For instance in comparing the performance of the selected spectral bands and the developed vegetation spectral ratios, the latter recorded a higher R^2 of 0.82 with an RMSEP of 0.29 whereas the former had an R^2 of 0.76 and an RMSEP of 0.17 (
Figure 5(a.) & (b.)). The implication was that with the selected spectral bands of importance in predicting maize leaf nitrate content, the error margin was smaller than the unselected spectral ratios.

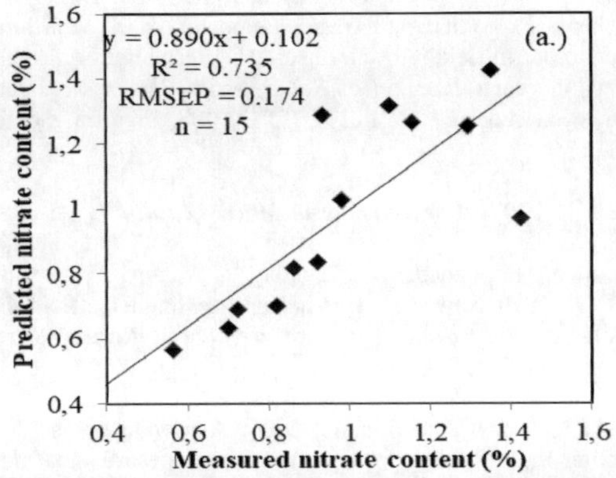

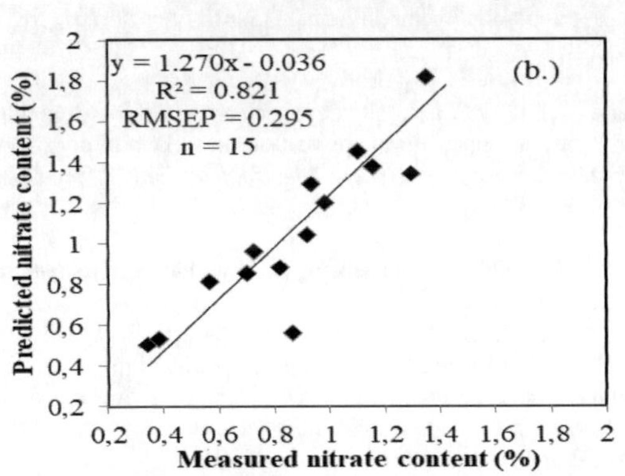

Figure 5: The relationship between measured and the predicted maize nitrate concentration through the RF model using: (a.) the selected waveband and (b.) all developed ratios – 6 X 6 wavebands

The selected vegetation spectral ratios also used as independent variables for the predictive model illustrated results of better accuracy with R^2 = 0.86 and RMSEP = 0.12

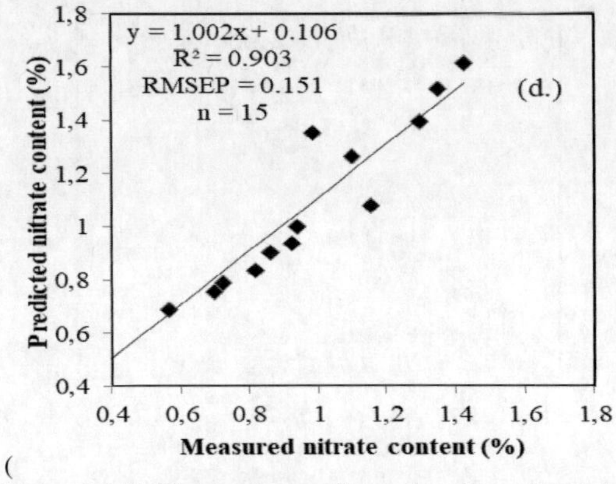

(
Figure 6 (c.)). The results were better than those obtained from all the possible vegetation spectral ratios. To further test on variable performances, the *cforest* ranking of the vegetation spectral ratios identified a subset of the top 10 most important ratios.

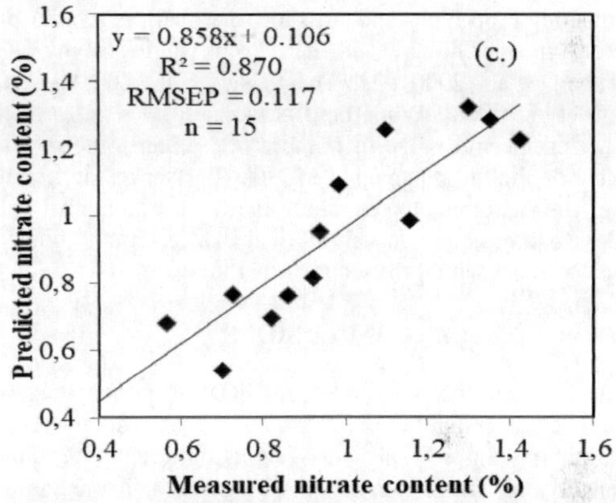

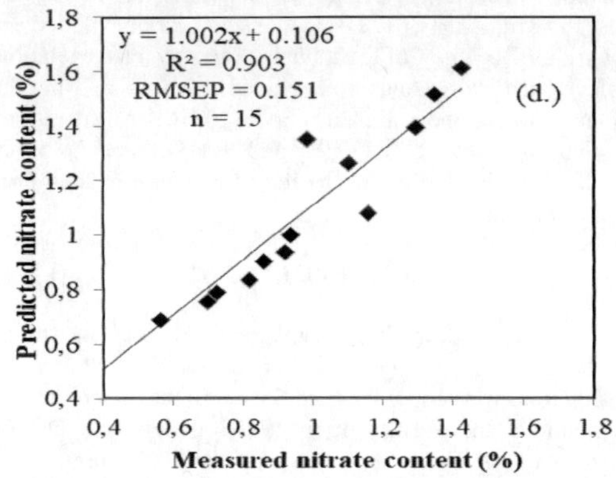

Figure 6: The relationship between measured and the predicted maize nitrate concentration from the RF model using: (c.) Selected 25 vegetation spectral ratios and (d.) selected top 10 ratios

These top 10 ratios were mostly developed from the red edge bands with the exception of 1043nm. These 10 vegetation spectral ratios outperformed all the other variable sets implemented in this study in predicting maize leaf nitrate content at canopy level with an R^2 of 0.905 but again with an

RMSEP of 0.151

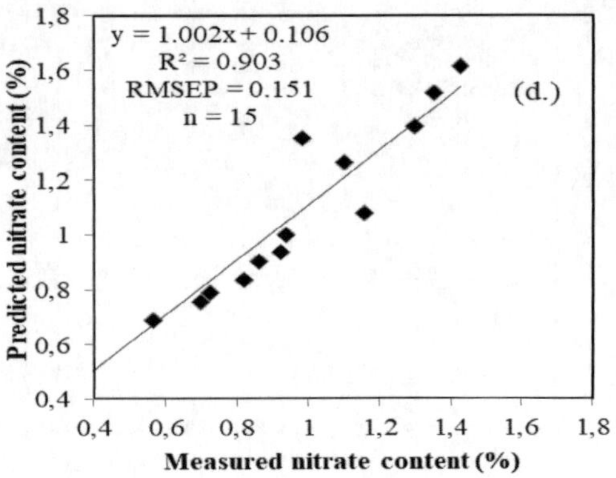

(

Figure 6 (d.)). The implication of this RMSEP value is that the predicted values of maize leaf nitrate content using the 10 ratios was further away from the measured amount than with the 26. This was quite interesting to realise that while narrowing down the variables to selected ones of importance could only get better results to an extent. There have been studies showing the optimal wavelength or band numbers for studies though on discriminating crop types and did show that there is a point of asymptote in the number of these wavelengths being 12 (Thenkabail et al., 2000), 22 (Thenkabail et al., 2004) or 26 (Miglani et al., 2008). While the first scenario was performed on a shorter spectral range (maximum of 1100nm), the recent ones were on the full range up to 2500nm. Further research will however be required to establish the optimal number of variables for predictive models of maize leaf nitrate content since that was not within the scope of this study.

4. CONCLUSIONS

The imaging spectroscopic data from EO-1 Hyperion images predicted maize leaf nitrate concentrations from canopy level under field condition. This was possible through the dual functionality of the random forest algorithm in reducing redundancy by selecting variables of importance and regression models for predictions. The selections of variable aided in improving the accuracy of the models. The spectral vegetation ratio developed with bands 752 and 681nm was relevant in contributing to the most accurate model for maize leaf nitrate concentrations with R^2 of 0.90 and RMSEP of 0.15. It consolidates the importance of the red edge region in assessing plant biochemical status.

ACKNOWLEDGEMENTS

Acknowledges funding from University of Johannesburg-Commonwealth scholarship for her degree and the field assistance from workers of the Free State provincial department of Agriculture and former departmental colleagues (Dr K. Abutaleb and Ms. R. Ashimwe) during data collection. Also appreciates the inputs from anonymous reviewers to this paper.

REFERENCES

Abdel-Rahman, E.M., F.B. Ahmed and R. Ismail, 2013: Random forest regression and spectral band selection for estimating sugarcane leaf nitrogen concentration using EO-1 Hyperion hyperspectral data. *Int. J. Remote Sens. 34(2),* 712-728.

Alchanatis, V., Z. Schmilovitch and M. Meron, 2005: In-Field assessment of single leaf nitrogen status by spectral reflectance measurements. *Precis. Agric. 6,* 25-39.

Bannari A., K.S. Khurshid, K. Staenz and J. Schwarz, 2008: Potential of hyperion EO-1 hyperspectral data for wheat crop chlorophyll content estimation. *Canadian J. remote sens. 34(1),* S139-S157.

Barnes, E.M., T.R. Clarke and S.E. Richards, 2000: Coincident detection of crop water stress, nitrogen status and canopy density using ground based Multispectral data, Proceedings of the 5th International Conference on Precision Agriculture. Bloomington, MN, USA.

Beck, R., 2003: *EO-1 User Guide-Version 2.3. Satellite Systems Branch*; USGS Earth Resources Observation Systems Data Center (EDC): Sioux Falls, SD, USA.

Blackmer, T.M. and J.S. Schepers, 1994: Techniques for monitoring crop nitrogen status in corn. *Communications in Soil Science and Plant Analysis Special Issue: Soil testing and plant Analysis: Precision nutrient management part II* 25, 1791-1800.

Botha, J.J., J.J. Anderson, D.C. Groenewald, N.N. Nhlabatsi, T.B. Zere, N. Mdibe and M.N., Baiphethi, 2007: On-farm application of in-field rainwater harvesting techniques on small plots in the central region of South Africa. Water Research Commission report, Vol. 1, No. TT 313/07, Pretoria, South Africa.

Breiman, L., 2001: Random Forests. *Mach. Learn. 45(1),* 5-32.

Chen, P., D. Haboudane, N. Tremblay, J. Wang, P. Vigneault and B. Li, 2010: New spectral indicator assessing the efficiency of crop nitrogen treatment in corn and wheat. *Remote Sens. Environ. 114(9),* 1987-1997.

Cho, M.A. and A.K. Skidmore, 2006: A new technique for extracting the red edge position from hyperspectral data: the linear extrapolation method. *Remote Sens. Environ.101,* 181-193.

Clevers, J.P.G.W. and R. Jongschaap, 2003: Imaging spectroscopy for agricultural application. In F. D. van der Meer and S. M. de Jong (eds), *Image Spectrometry, vol. 3,* pp. 157-199. London: Kluwer.

Curran, P. J., 1989: Remote sensing of foliar chemistry. *Remote Sens. Environ. 30,* 271-278.

Datt, B., T.R. McVicar, T.G. Van Niel, D.L. Jupp and J.S. Pearlman, 2003: Preprocessing EO-1 Hyperion hyperspectral data to support the application of agricultural indexes. *IEEE Trans. Geosci. Remote Sens. 41,* 1246-1259.

Daughtry, C. S. T., C.L. Walthall, M.S. Kim, E. Brown de Colstoun and J.E. Mcmurtrey, 2000: Estimating corn leaf chlorophyll concentration from leaf and canopy reflectance. *Remote Sens. Environ. 74,* 229-239.

Fageria, N.K., 2009: *The Use of Nutrients in Crop Plants.* New York: Taylor and Francis 430 pages.

Ferencz, Cs., P. Bognár, J. Lichtenberger, D. Hamar, Gy. Tarcsai, G. Timár, G. Molnár, S.Z. Pásztor, P. Steinbach, B. Székely, O. E. Ferencz and I. Ferencz-Árkos, 2004: Crop yield estimation by satellite remote sensing. *Int. J. Remote Sens. 25(20),* 4113-4149, DOI: 10.1080/01431160410001698870

Gitelson, A. A., A. Viña, V. Ciganda, D.C. Rundquist, and T.J. Arkebauer, 2005: Remote estimation of canopy chlorophyll content in crops. *Geophys. Res. Lett. 32,* L08403

Green, R.O., B.E. Pavri and T.G. Chrien, 2003: On orbit radiometric and spectral calibration characteristics of EO-1 Hyperion derived with an under flight of AVIRIS and *in situ* measurements at Salar de Arizaro, Argentina. *Remote Sens. Environ. 41,* 1194-1203.

Haboudane, D., J. R. Miller, N. Tremblay, P. J. Zarco-Tejada and L. Dextraze, 2002: Integrated narrow-band vegetation indices for prediction of crop chlorophyll content for application to precision agriculture. *Remote Sens. Environ. 81,* 416-426.

Haboudane, D., N. Tremblay, J. R. Miller and P. Vigneault, 2008: Remote Estimation of Crop Chlorophyll Content using spectral Indices derived from hyperspectral data. *IEEE Trans. Geosci. Remote Sens. 46(2),* 423-437.

Hansen, P.M. and J.K. Schjoerring, 2003: Reflectance measurement of canopy biomass and nitrogen status in wheat crops using normalized difference vegetation indices and partial least squares regression. *Remote Sens. Environ. 86,* 542-553.

Houlès, V., M. Guérif and B. Mary, 2007: Elaboration of a nitrogen nutrition indicator for winter wheat based on leaf area index and chlorophyll content for making nitrogen recommendations. *Eur. J. Agron. 27(1),* 1-11.

Huang, Z., B. J. Turner, S. J. Dury, I. R. Wallis and W. J. Foley, 2004: Estimating foliage nitrogen concentration from HYMAP data using continuum removal analysis. *Remote Sens. Environ. 93,* 18-29.

Jackson, R.D. and A. Huete, 1991: Interpreting vegetation indices. *Prev. Vet. Med. 11,* 185-200.

Jackson, R.D., C.A. Jones, G. Uehara and L.T. Santo, 1981: Remote detection of nutrient and water deficiencies in sugarcane under variable cloudiness. *Remote Sens. Environ. 11,* 327-337.

Jain, N., Ray, S. S., Sinph, J. P. and Panigrahy, S., 2007: Use of hyperspectral data to assess the effects of different nitrogen applications on a potato crop. *Precis. Agric. 8,* 225-239.

Kokaly, R.F., 2001: Investigating a physical basis for spectroscopic estimates of leaf nitrogen concentration. *Remote Sens. Environ. 75,* 153-161.

Kumar, L., K. Schmidt, S. Dury and A. Skidmore, 2003: Imaging spectrometry and vegetation science. In F. D. van der Meer and S. M. de Jong (eds), *Image Spectrometry 3,* 111-156 London: Kluwer.

Lawlor, D.W., 2001: *Photosynthesis* - Third Edition. Oxford: BIOS.

Lee, W., Searcy, S., Kataoka, T., 1999: Assessing nitrogen stress in corn varieties of varying color. In: 1999 ASAE Annual International Meeting. No. Paper No 99-3034 in ASAE Meeting Presentation. ASAE, 2950 Niles Rd., St. Joseph, MI 49085-9659 USA, Toronto, Ontario Canada.

Masson, P., G. Hilbert and D. Plenet, 1996: Ion chromatography methods for the simultaneous determination of mineral anions in plant sap. *J.Chromatography 752,* 298-303.
Miglani, A. S.S. Ray, R. Pandey and J.S. Parihar, 2008: Evaluation of EO-1 Hyperion Data for Agricultural Applications. *J. of the Indian Soc. Remote Sens. 36,* 255-266.

Morón, A., A. Garcìa, J. Sawchik and D. Cozzolino, 2007: Preliminary study on the use of near-infrared reflectance spectroscopy to assess nitrogen content of undried wheat plants. *J. Sci. Food Agric. 87(1),* 147-152.

Murray, J. and P.C. Williams, 1987: Chemical principles of near-infrared technology. In P. C. Williams, & K. Norris (eds.), *Near infrared technology in the agricultural and food industries,* pp. 17-37. St Paul, MN: American Association of Cereal Chemists.

Mutanga, O. and A.K. Skidmore, 2004: Narrow band vegetation indices overcome the saturation problem in biomass estimation. *Int. J.Remote Sens. 25,* 1-16.

Ngie, A., F. Ahmed and K. Abutaleb, (2014): Remote sensing potential for investigation of maize production: review of literature. *South Africa Journal of Geomatics 3(2),* 163-184.

Ngie, A., F. Ahmed and K. Abutaleb, (2016): Assessing maize foliar water stress levels under field conditions using in-situ spectroscopy. *Rwanda Journal, Series D, Volume 1, 2016, Life and Natural Sciences: Special issue II.* http://dx.doi.org/10.4314/rj.v1i2S.1D

Nguyen, H.T. and B. Lee, 2006: Assessment of rice leaf growth and nitrogen status by hyperspectral canopy reflectance and partial least square regression. *Eur. J. Agron. 24,* 349-356.

Nguyen, H.T., J.H. Kim, A.T. Nguyen, L.T. Nguyen, J.C. Shin and B-W. Lee, 2006: Using canopy reflectance and partial least squares regression to calculate within-field statistical variation in crop growth and nitrogen status of rice. *Precis. Agric. 7,* 249-264.

Oppelt, N. and W. Mauser, 2004: Hyperspectral monitoring of physiological parameters of wheat during a vegetation period using AVIS data. *Int. J Remote Sens. 25,* 145-159.

Patane, P. and A. Vibhute, 2014: Chlorophyll and nitrogen estimation techniques: A Review. *Int. J. Engineering Res. Reviews 2(4),* 33-41. ISSN 2348-697X.

Prasad, A. M., L.R. Iverson and A. Liaw, 2006: Newer classification and regression tree techniques: bagging and random forests for ecological prediction. *Ecosystems 9,* 181-199.

Rodriguez, D., G.J. Fitzgerald, R. Belford and L. Christensen, 2005: Detection of nitrogen deficiency in wheat from spectral reflectance indices and basic crop eco-biophysiological concepts. *Aust. J. Agric. Res. 57,* 781-789.

Rouse, J.W., R.H. Haas, J.A. Schell and D.W. Deering, 1974: Monitoring vegetation systems in the Great Plains with ERTS. In *Proceedings of Third ERTS Symposium*, Greenbelt, MD, December 1974; NASA SP-351-1, pp. 309-317.

Schlemmer, M., A. Gitelson, J. Schepers, R. Ferguson, Y. Peng and J. Shanahan, 2013: Remote estimation of nitrogen and chlorophyll contents in maize at leaf and canopy levels. *Int. J..Appl. Earth OBS 25,* 47-54.

Schlemmer, M.R., D.D. Francis, J.F. Shanahan and J.S. Schepers, 2005: Remotely measuring chlorophyll content in corn leaves with differing nitrogen levels and relative water content. *Agron. J. 97,* 106-112.

Strachan, I. B., Pattey, E. and J.B. Boisvert, 2002: Impact of nitrogen and environmental conditions on corn as detected by hyperspectral reflectance. *Remote Sens. Environ. 80,* 213-224.

Strobl, C. A.-L. Boulesteix, A. Zeileis and T. Hothorn, 2007: Bias in random forest variable importance measures: Illustrations, sources and a solution. *BMC Bioinformatics 8(25),* 1-21. doi:10.1186/1471-2105-8-25.

Strobl, C., T. Hothorn and A. Zeileis, 2009: Party on! A New, Conditional Variable-Importance measure for random forests available in the party package. *The R Journal 1/2,* 14-17. http://journal.r-project.org/archive/2009-2/RJournal_2009-2_Strobl~et~al.pdf Accessed 15/04/2015.

Stroppiana, D., M. Boschetti, P.A. Brivio and S. Bocchi, 2006: Remotely sensed estimation of rice nitrogen concentration for forcing crop growth models. *Italian Journal of Agrometeorology 3,* 50-57.

Thenkabail, P.S., E.A. Enclona, M.S. Ashton, B. Van Der Meer, 2004: Accuracy assessments of hyperspectral waveband performance for vegetation analysis applications. *Remote Sens. Environ. 91,* 354-376.

Thenkabail, P.S., I. Mariotto, M.K. Gumma, E.M. Middleton, D.R. Landis, F.K. Huemmrich, 2013: Selection of hyperspectral narrowbands (HNBs) and composition of hyperspectral two band vegetation indices (HVIs) for biophysical characterization and discrimination of crop types using field reflectance and Hyperion/EO-1 data. *IEEE J. Sel. Topics Appl. Earth Observ. Remote Sens. 6,* 427-439.

Thenkabail, P.S., R.B. Smith and E. De Pauw, 2000: Hyperspectral vegetation indices and their relationships with agricultural crop characteristics. *Remote Sens. Environ. 57,* 158-182.

Tilling, A.K., G.J. O'Leary, J.G. Ferwerda, S.D. Jones, G.J. Fitzgerald, D. Rodriguez and R. Belford, 2007: Remote sensing of nitrogen and water stress in wheat. *Field Crops Research 104,* 77-85.

Vigneau, N., M. Ecarnot, G. Rabatel and P. Roumet, 2011: Potential of field hyperspectral imaging as a non-destructive method to assess leaf nitrogen content in Wheat. *Field Crops Research 122(1),* 25-31.

Wang, J., C. Shen, N. Liu, X. Jin, X. Fan, C. Dong and Y. Xu: Non-destructive evaluation of the leaf nitrogen concentration by in-field visible/near-infrared spectroscopy in pear orchards. *Sensors* 2017, 17, 538; doi: 10.3390/s17030538 pg 1-15.

Wang, J., L. Sun, C. Shi and Q. Tian, 2013: Estimation of rice canopy nitrogen concentration by hyperspectral remote sensing. IEEE Proceedings of the 2nd International Conference on Agro-Geoinformatics, 52-55.

Zhao, D., Reddy, K. R., Kakani, V. G. and Reddy, V. R., 2005: Nitrogen deficiency effects on plant growth, leaf photosynthesis, and hyperspectral reflectance properties of sorghum. *Eur. J. Agron. 22,* 391-403.

Zhu, Y., Yao, X., Tian, Y., Liu, X. and W. Cao, 2008: Analysis of common canopy vegetation indices for indicating leaf nitrogen accumulations in wheat and rice. *Int. J Appl. Earth OBS 10,* 1-10.

Ziadi, N., M. Brassard, G. Bélanger, A. Claessens, N. Tremblay, A. Cambouris, M. Nolin, L.-E. Parent, 2008: Chlorophyll measurements and nitrogen nutrition index for the evaluation of corn nitrogen status. *Agron. J. 100(5),* 1264-1273.

UTILIZING LANDSAT AND SENTINEL-2 TO REMOTELY MONITOR AND EVALUATE THE PERFORMANCE OF WINTER COVER CROPS THROUGHOUT MARYLAND

Julio Peredo[1,*], Callum Wayman[1], Benjamin Whong[1], Alison Thieme[1], Logan R. Kline[1], Sunita Yadav[1], Bryan Eder[1], Victor Lenske[1], Diane Portillo[1], Sean McCartney[1], John Fitz[1], Perry Oddo[1], Jason Keppler[2], Dean Hively[3], John Bolten[4], Greg McCarty[5], Adam Lyon[2]

[1] Science Systems and Applications Inc. at NASA Goddard Space Flight Center, 8800 Greenbelt Road, Greenbelt, MD 20771 - juliop1@umbc.edu, (callum.wayman, benjaminwhong, alisonthieme, logan.kline1997, dr.biogeo)@gmail.com, beder@terpmail.umd.edu, (victor.lenske, dianeportillo26)@gmail.com, sean.mccartney@nasa.gov, johnfitz058@gmail.com, perry.oddo@nasa.gov
[2] Maryland Department of Agriculture, Annapolis, MD 21401 - (jason.keppler, adam.lyon)@maryland.gov
[3] USGS Lower Mississippi Gulf Science Center, Beltsville, MD 20705 - whively@usgs.gov
[4] NASA Goddard Space Flight Center, 8800 Greenbelt Road, Greenbelt, MD 20771 - john.bolten@nasa.gov
[5] USDA-ARS Hydrology & Remote Sensing Laboratory, Beltsville Agricultural Research Center, Beltsville MD 20705 - greg.mccarty@ars.usda.gov

KEY WORDS: Remote Sensing, Graphical User Interface, Google Earth Engine, Normalized Difference Vegetation Index, Biomass, Percent Ground Cover, Time Series

ABSTRACT:

Winter cover crops have been shown to limit erosion and nutrient runoff from agricultural land. To promote their usage, the Maryland Department of Agriculture (MDA) subsidizes farmers who plant cover crops. Conventional verification of cover crop planting and analysis of subsequent crop performance requires on-the-ground fieldwork, which is costly and labor intensive. In partnership with the MDA, NASA's DEVELOP program utilized imagery from Landsat 5, Landsat 8, and the European Space Agency's Sentinel-2 to create a decision support tool for satellite-based monitoring of cover crop performance throughout Maryland. Our teams created CCROP, an interactive graphical user interface, in Google Earth Engine which analyzes satellite imagery to calculate the normalized difference vegetation index (NDVI) of fields across the state. Linear regression models were applied to convert NDVI to estimates of crop biomass and percent green ground cover, with measure of fit (R^2) values ranging from 0.4 to 0.7. These crop metrics were implemented into an interactive filtering tool within CCROP which allows users to examine cover crop performance based on a variety of growing parameters. CCROP also includes a time series analysis routine for examining the progression of NDVI throughout the spring to help determine farmer-induced termination dates of cover crops. With this decision support tool, the MDA can analyze the effectiveness of cover crops throughout the state with reduced need to manually spot-check enrolled production fields, and can identify variables influencing overall cover crop performance to optimize implementation of their winter cover crop program via adaptive management approaches.

1. INTRODUCTION

1.1 Background Information

The Chesapeake Bay is the largest estuary in the United States, fed by a watershed spanning six states and encompassing 166,000 km^2 of diverse terrestrial and marine ecosystems (Phillips, McGee, 2016). Over 3,000 species of bivalves, birds, fish, and other organisms inhabit the watershed, and nearby coastal wetlands provide nurseries, food, and nesting sites (Chesapeake Bay Foundation, n.d.). The Chesapeake Bay also generates substantial commercial benefits, hosting a multi-billion dollar boating industry and fisheries capable of generating $2 billion per year (Maryland Department of Natural Resources, 2012). Therefore, the health of the Bay and neighboring watersheds is of great ecological and economic importance to the Mid-Atlantic region. However, soil erosion and nutrient leaching from agricultural lands pose a major threat to the estuary and its local waterways (Dauer et al., 2000). Sediment loading from farms can bury benthic organisms, prevent sunlight from reaching submerged aquatic vegetation, and transport nutrients such as nitrogen into vulnerable habitats (Gellis et al., 2004). Excess nitrogen can induce eutrophication and subsequent hypoxia in coastal waters, threatening the prolific fisheries of the Chesapeake Bay (Boesch et al., 2001; Malone et al., 1993).

"Cover crop" is a term used to refer to crops, typically large grains, which are planted in between growing seasons for the sake of nutrient retention and soil preservation (Dabney et al., 2001). Cover crops are a valuable component in land-use strategies for agroecosystems and provide a diverse set of benefits including soil enrichment, erosion reduction, and protection against nutrient loss (Hively et al., 2009). Their role in nitrogen and phosphorus uptake, as well as runoff prevention, improves water quality (Boesch et al., 2001). To promote cover crop use for Chesapeake Bay conservation, the Maryland Department of Agriculture (MDA) oversees the Maryland Agricultural Water Quality Cost-Share (MACS) Program, which offers subsidies to Maryland farmers who plant winter cover crops.

The effectiveness of cover crops depends on management practices and agronomic factors such as planting date, method, and crop species (Hively et al., 2009). Routine, accurate monitoring of cover crops is difficult using traditional methods. However, biomass, the total mass of vegetation in a given area, and normalized difference vegetation index (NDVI), a

*Corresponding author

commonly used indicator of green vegetation estimated by remote sensing techniques, are metrics which can reliably be used to evaluate cover crop performance (Hively et al., 2009; Prabahakara et al., 2015). To this end, satellite-based remote sensing can be used as a method to efficiently assess regional cover crop performance on a routine basis. Furthermore, the employment of remote sensing techniques can result in the accurate calculation and attainment of both biomass measurements and NDVI values.

The NASA DEVELOP program partnered with the MDA, the United States Geological Survey (USGS), and the United States Department of Agriculture (USDA) to create cover crop analysis procedures in Maryland. This two-year partnership began in the spring of 2017, when statistical analysis was performed to establish correlations between remotely sensed NDVI values and two more directly applicable performance metrics: biomass and percent ground cover (Thieme et al., submitted manuscript, 2019). They also automated the acquisition and analysis of satellite imagery via code in Google Earth Engine (GEE) and R (R Core Team, 2017) for the purpose of examining cover crop performance in four Maryland counties: Talbot, Somerset, Queen Anne's and Washington. The next stage of this partnership, which began in the fall of 2018, expanded the study area to the entire state of Maryland and created a graphical user interface (GUI) called CCROP in GEE to optimize end user analysis of cover crop data. This tool provided the Maryland Department of Agriculture (MDA) with a means to remotely verify cover crop growth and better visualize cover crop performance metrics via quantitative summary tables. These tabular outputs displayed cover crop data such as biomass, percent ground cover, and NDVI values of selected counties based on chosen GUI filtering methods. The final stage, beginning in the spring of 2019, concluded the partnership and focused on springtime analysis of cover crops and finalizing CCROP functionality. The team used the existing GEE GUI framework to incorporate an NDVI-based time series for the purpose of monitoring cover crops throughout their lifecycle. This added feature provides a user-friendly method to visualize crucial cover crop performance data such as per-pixel NDVI values, above ground biomass measurements, and planting and termination dates across the state of Maryland.

This project's study area included fields enrolled in the MDA's cover crop program across the state of Maryland (Figure 1). The final DEVELOP team focused on data from the winter (December 1 – January 31) and spring (March 1 – May 31) seasons from December 2006 to March 2019.

1.2 Project Partners & Objectives

Project development was conducted in collaboration with the MDA Office of Resource Conservation (ORC), the USGS Eastern Geographic Science Center (EGSC), the USDA Agricultural Research Service (ARS), and the Environmental Protection Agency (EPA) Chesapeake Bay Program (CBP).

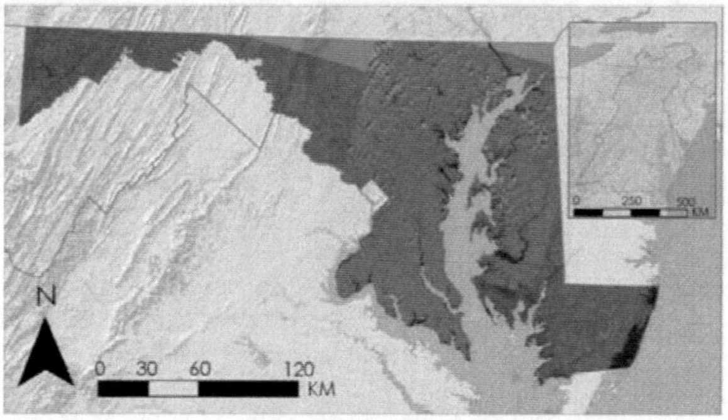

Figure 1. Study area, Maryland, as a mosaic of Landsat 8 OLI images. The Chesapeake Bay watershed is highlighted in the inset. Source: Landsat 8 OLI (NASA), Maryland County Boundaries (Maryland GIS Data Catalog), World Shaded Relief (ESRI), State Boundaries (ESRI).

2. METHODOLOGY

2.1 Data Acquisition

This project relies primarily on NASA Earth observations for its remote sensing analysis. The project teams used NASA Landsat archival imagery (atmospherically corrected to determine surface reflectance) and European Space Agency (ESA) Sentinel-2 imagery. These data were imported from the GEE repository for various winter cover crop planting seasons lasting from August to May. Data were analyzed for winter (December 15–January 31) and spring (March 1–May 31) from December 2014 to March 2019. See Table 1 for a complete description of sensors used.

Partners at USGS and USDA-ARS provided field-measured crop samples, which were pre-processed and used to develop calibration equations for converting NDVI into biomass and percent ground cover estimates, per Thieme et al. (submitted 2019).

The MDA provided ArcGIS-generated shapefiles (ESRI, Redlands, CA) containing management practice information and field boundaries for each field enrolled in the cover crop program. Data were received for 23 Maryland counties from 2017 to 2019. Prior to 2017, shapefiles of enrolled fields were provided for four counties: Talbot, Somerset, Queen Anne's and Washington.

Platform & Sensor	Level	Google Earth Engine Image Collection IDs	Available Years
Landsat 5 TM	2	LANDSAT/LT05/C01/T1_SR	2006–2012
Landsat 8 OLI	2	LANDSAT/LC08/C01/T1_SR	2013–2019
Sentinel-2 MSI	1C	COPERNICUS/S2	2015–2019

Table 1. Remote sensing products utilized in the development of CCROP.

2.2 Data Processing

Field shapefiles provided by the MDA and projected via the 1984 World Geodetic System (WKID 4326) were uploaded to GEE as tabular data. A 15-meter inner buffer was applied to diminish edge effects and refine field boundaries. This resulted in loss of fields whose geometry was contained entirely within this buffered region. An additional Maryland state boundary shapefile was imported, and Landsat and Sentinel-2 image collections were spatially filtered via these boundaries. The image collection was temporally filtered to include data acquired during pre-defined winter and spring intervals (December 15–January 31 and March 1–May 31, respectively).

The image collection for each season was then passed into our graphical user interface, CCROP, where it underwent three stages of processing. The CFmask algorithm, a multi-pass algorithm that uses decision trees to classify scenes pixel by pixel (United States Geological Survey, n.d.), was applied to remove cloud-covered pixels in Landsat and Sentinel-2 images. CCROP then inserted a property that denotes the date of image collection. Finally, NDVI was calculated on a pixel level using the standard NDVI equation (Equation 1):

$$\mathrm{NDVI} = \frac{(\mathrm{NIR} - \mathrm{Red})}{(\mathrm{NIR} + \mathrm{Red})} \quad (1)$$

These per-pixel NDVI values are displayed in the user interface for selected counties and dates.

2.3 Data Analysis

The maximum average NDVI values for each field in a selected season were used to calculate biomass and percent ground cover using linear regression equations calculated with the partner-provided calibration data (Table 2).

Performance Variable	Season	Equation	R^2
Biomass	Winter	ln(biomass) = 3.2022 + 5.3740*NDVI	0.562
Biomass	Spring	ln(biomass) = 4.7794 + 3.7453*NDVI	0.403
Percent ground cover	Winter	PGC = -21.904 + 116.305*NDVI	0.685
Percent ground cover	Spring	PGC = -10.783 + 107.566*NDVI	0.624

Table 2. Calibration models to predict biomass and percent vegetative ground cover from satellite-derived NDVI (adapted from Thieme et al., submitted 2019).

Biomass is then calculated by exponentiating these results, as per Equation 2 (Thieme et al., submitted 2019):

$$Biomass = e^{\ln(Biomass) + \frac{MSE}{2}} \quad (2)$$

where MSE represents the mean squared error.

The approximate amount paid to each farmer for cover crop yield per kilogram of nitrogen is calculated using the subsidy rates provided by the MDA. This calculation is made according to Equation 3:

$$CostField = \frac{Rate}{(\frac{Biomass}{2.47105 * 50})} \quad (3)$$

where Rate denotes the amount of money in USD paid to each farmer per acre and Biomass denotes the biomass of each field in kg of nitrogen per hectare. This biomass number is converted to kg of nitrogen per acre and divided by 50 based on the assumption that nitrogen uptake can be approximated as 2% of total biomass (Hively et al., 2009).

After data is analyzed for a season, CCROP was used to display NDVI values at the pixel scale for all enrolled fields on a given date. Days that can be displayed are dependent on the temporal resolution of remote sensing datasets. Sentinel-2's temporal resolution (5 days) is significantly higher than that of Landsat 5 and Landsat 8 (16 days). Imagery from the two satellites are not comparable as Sentinel-2 surface reflectance has yet to be fully incorporated in GEE; Sentinel-2 top of atmosphere data was used instead. In addition to NDVI, CCROP also can visualize biomass and percent ground cover values that are derived from the aforementioned linear regression equations.

3. RESULTS AND DISCUSSION

3.1 Derivation and Visualization of Filtered Agricultural Parameters

Thieme et al. (submitted 2019) established a reasonably strong (R^2 = 0.554-0.698) correlation between NDVI and agronomic parameters; these results are the groundwork by which CCROP conducts its analysis. These correlations varied depending on season and agronomic factor (Table 2).

CCROP provides a number of options for filtering a dataset to examine cover crop performance based on various factors. Experimentation with various combinations of these can yield insight into how to maximize cover crop effectiveness; for this study, patterns were identified in cover crop type and planting date. Figures 2 and 3 display the distribution of max NDVI for wheat fields in Queen Anne's County in spring 2018 versus max NDVI for triticale fields in the same county and time period.

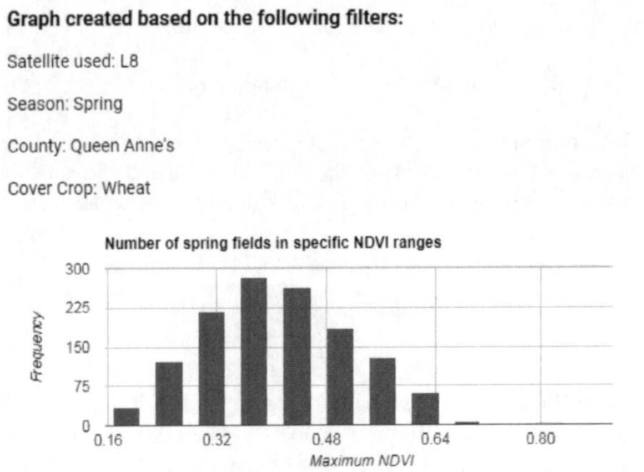

Figure 2. CCROP output histogram displaying number of fields versus NDVI for wheat fields in Queen Anne's County.

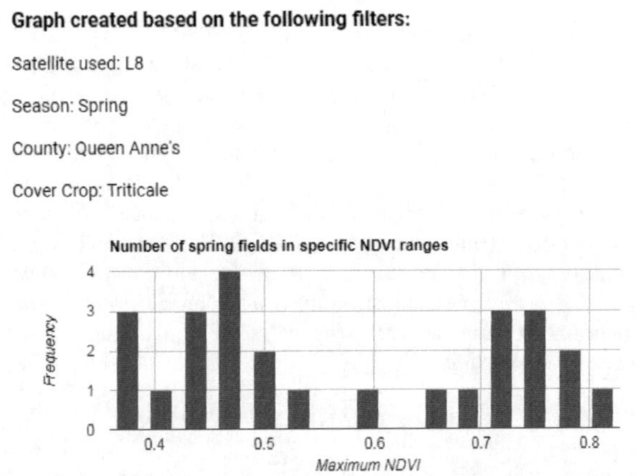

Figure 3. CCROP output histogram displaying number of fields versus NDVI for triticale fields in Queen Anne's County.

In these figures, triticale displays an average NDVI value of 0.56 while wheat displays an average of 0.38, which equates to a 47% difference in NDVI between the two crops. Although the sample size for triticale is comparatively smaller than that for wheat, these results suggest that triticale cover crops may perform better than wheat in Queen Anne's County. The user may then repeat this analysis in another county, or over the region as a whole, to examine this pattern in further detail.

CCROP can also be used to test scientific hypotheses on cover crop effectiveness. Prior studies have shown that cover crops are more effective if planted in early fall (Hively et al., 2009). To verify this, different planting date ranges can be selected in CCROP to examine the performance of cover crops planted in September 2017 (Figure 4) compared to those planted in November 2017 (Figure 5).

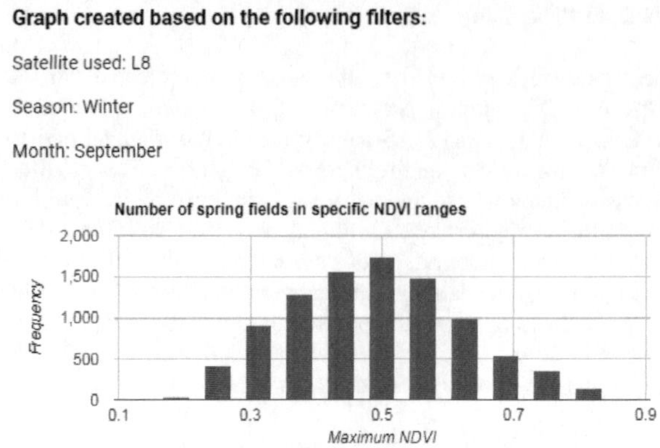

Figure 4. CCROP output histogram showing number of fields versus maximum NDVI for fields planted in September.

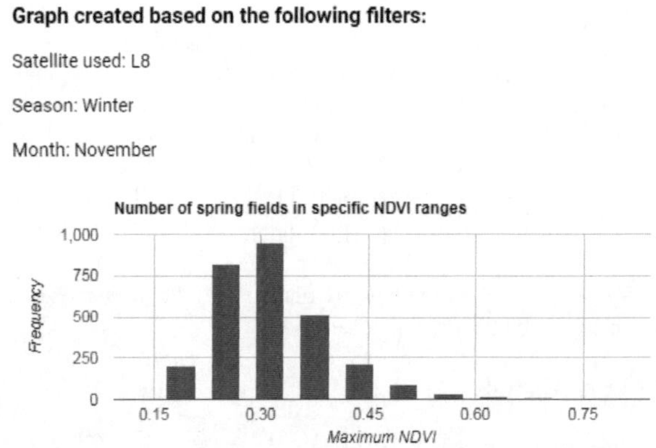

Figure 5. CCROP output histogram showing number of fields versus maximum NDVI for fields planted in November.

These figures support the hypothesis that early planting, before October 1, yields better cover crop performance: the NDVI peak of the November curve is nearly 0.2 lower than the September curve, corresponding to a difference of 20% ground cover (per the conversion equation in Table 2).

3.2 Time Series Analysis of Cover Crop Life Cycles

A time series component was added to CCROP to employ operational remote sensing technology for the purposes of assessing cover crop performance over time and identifying crop termination dates. This new functionality harnesses CCROP's ability to analyze imagery from both Landsat 8 and Sentinel-2 satellites, an integration which improves temporal coverage within the boundaries of our study area. CCROP produces histograms and imagery displaying the distribution of NDVI values on a county level for each date with available data. Comparing outputs from various days in the season reveals long-term trends in cover crop performance. Additionally, by examining GUI output from dates near the termination deadlines, possible crop kill dates can be identified based on changes in NDVI. Figures 6 and 7 display NDVI distributions for Kent County, MD, for December 21, 2017 and March 18, 2018, respectively.

The frequency distributions in Figures 6 and 7 demonstrate the change in amount of vegetation present in the enrolled cover crop fields in Kent County using two specific days as points of comparison: December 21, 2017 and March 18, 2018. These dates were selected for comparison due to the fact that they fell within the cover crop growing season, had limited cloud cover interference, and allowed a significant time margin for analysis. After the histograms were output by CCROP's time series tool, the data were sorted to determine the percentage of fields which fell below the MDA-provided NDVI threshold of 0.3, a cut-off value that denotes either low performing or terminated fields. On December 21st, approximately 8% of enrolled cover crop fields in Kent County fell below this threshold. On March 18th, approximately 22% of cover crop fields in Kent County were below 0.3 NDVI, marking a 157% relative increase over this time period. This percentage increase in low NDVI fields is a significant finding as it suggests that a significant portion of Kent County cover crops were killed off within the earliest termination window set by the MDA, March 1st to March 15th.

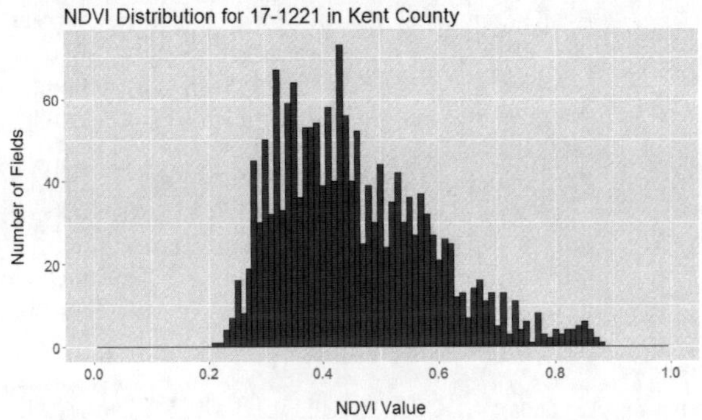

Figure 6. NDVI histogram of enrolled fields in Kent County, Landsat imagery on December 21, 2017. Date is displayed as YY-MMdd.

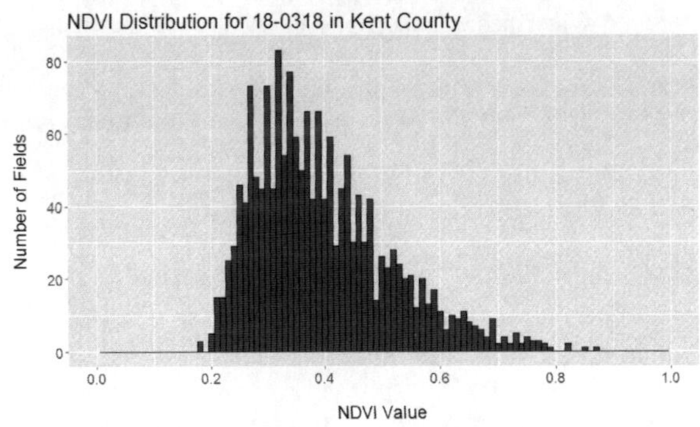

Figure 7. NDVI histogram of enrolled fields in Kent County, Landsat imagery on March 18, 2018. Date is displayed as YY-MMdd.

CCROP can be used to perform similar analysis for all Maryland counties, provided cloud-free data is available over the entire region for the dates of interest. Table 3 displays percentages of low NDVI fields for various Maryland counties in winter and spring.

County	**Low perfor-ming December fields (%)**	**Low perfor-ming March fields (%)**	**Relative Difference (%)**
Caroline	5	3	-40
Cecil	10	21	+110
Kent	8	22	+175
Oueen	9	11	+22
Somerset	2	1*	-50
Talbot	7	11	+57
Washington	11	15*	+36

Table 3. Percentages of low performing (below 0.3 NDVI) fields for various Maryland counties in winter and spring. *Denotes use of April imagery when no high-quality March imagery was available

Table 3 supports the results found in Kent and Talbot counties; of the Maryland counties examined, five out of seven had a higher ratio of low NDVI in the spring, and the two counties that display the opposite trend show relatively small disparity between winter and spring values. Cover crop effectiveness is positively correlated with the duration of time that the crops remain alive in the spring (Lyon, personal communication, 2019). Therefore, it is significant that such a large proportion of cover crops in multiple counties are eliminated in these early weeks.

While differences in crop termination dates can explain the above findings, it is important to note that numerous factors, such as cover crop type and planting dates of traditional harvest crops, can affect the results. The latter may mask observed termination dates when farmers plant new crops soon after cover crop termination given the GUI's current inability to conduct land/crop classification. NDVI values alone indicate only vegetation presence, not type. An uncertain amount of observed springtime greenness is caused by fields that may contain non-cover crop vegetation. However, in this case, these false positives could increase the average NDVI for fields in which they were present, thus reducing the ratio of low NDVI fields calculated with CCROP.

4. CONCLUSIONS

Developed in partnership with the Maryland Department of Agriculture, CCROP has potential to bolster the MDA's current decision making process regarding cover crop management practices. CCROP provides both scientific and practical benefits to this end. Scientifically, its filtering ability allows users to examine cover crop effectiveness with various permutations of management practices to determine the conditions that yield the best performance. For instance, CCROP's outputs suggest that wheat, the most commonly planted cover crop, has a lower-skewing maximum NDVI distribution than its less common counterpart, triticale.

Practically, CCROP's time series feature enables users to detect trends in NDVI over time, potentially including remote identification of springtime crop termination. This would reduce the MDA's need to manually verify termination on fields individually, freeing time and resources for other conservation efforts. This project demonstrates the potential to apply NASA Earth observations and other remote sensing data to agriculture and conservation purposes on a statewide scale. As this technology improves, so will CCROP's capabilities; integration of Sentinel-2 surface reflectance, for instance, would enable unification of Landsat and Sentinel data to create a more temporally contiguous time series.

5. ACKNOWLEDGMENTS

This material contains modified Copernicus Sentinel data (2017-2019), processed by ESA.

Any opinions, findings, and conclusions or recommendations expressed in this material are those of the author(s) and do not necessarily reflect the views of the National Aeronautics and Space Administration.

This material is based upon work supported by NASA through contract NNL16AA05C.

6. REFERENCES

Boesch, D.F., Brinsfield, R.B., Magnien, R.E., 2001. Chesapeake bay eutrophication. *Journal of Environmental Quality* 30, 303–320.

Chesapeake Bay Foundation, n.d.. Chesapeake wildlife. URL: http://www.cbf.org/about-the-bay/more-than-just-the-bay/chesapeake-wildlife/

Dabney, S.M., Delgado, J.A., Reeves, D.W., 2001. Using winter cover crops to improve soil and water quality. Communications in Soil Science and Plant Analysis 32, 1221–1250.

Dauer, D.M., Ranasinghe, J.A., Weisberg, S.B., 2000. Relationships between benthic community condition, water quality, sediment quality, nutrient loads, and land use patterns in Chesapeake Bay. Estuaries 23, 80–96.

Gellis, A.C., Banks, W.S., Langland, M.J., Martucci, S.K., 2004. Summary of suspended-sediment data for streams draining the Chesapeake Bay watershed, water years 1952–2002. US Geological Survey Scientific Investigations Report 5056, 1–59.

Hively, W.D., Duiker, S., McCarty, G., Prabhakara, K., 2015. Remote sensing to monitor cover crop adoption in southeastern Pennsylvania. *Journal of Soil and Water Conservation* 70, 340–352. https://doi.org/10.2489/jswc.70.6.340

Hively, W.D., Lang, M., McCarty, G.W., Keppler, J., Sadeghi, A., McConnell, L.L., 2009. Using satellite remote sensing to estimate winter cover crop nutrient uptake efficiency. *Journal of Soil and Water Conservation* 64, 303–313. https://doi.org/10.2489/jswc.64.5.303

Malone, T.C., Boynton, W., Horton, T., Stevenson, C., 1993. Nutrient loadings to surface waters: Chesapeake Bay case study. Keeping pace with science and engineering: Case studies in environmental regulation 8–38. Retrieved from: https://www.nap.edu/read/2127/chapter/3

Maryland Department of Natural Resources, 2012. Accounting for Maryland's Ecosystem Services: Integrating the value of nature into decision making. Maryland Department of Natural Resources, Annapolis, MD.

Phillips, S., McGee, B., 2016. Ecosystem service benefits of a cleaner Chesapeake Bay. *Coastal Management* 44, 241–258. https://doi.org/10.1080/08920753.2016.1160205

Prabhakara, K., Hively, W.D., McCarty, G.W., 2015. Evaluating the relationship between biomass, percent groundcover and remote sensing indices across six winter cover crop fields in Maryland, United States. *International Journal of Applied Earth Observation and Geoinformation* 39, 88–102. https://doi.org/10.1016/j.jag.2015.03.002

R Core Team (2017). R: A language and environment for statistical computing. R Foundation for Statistical Computing, Vienna, Austria. http://www.R-project.org/

Thieme, A., Yadav, S., Oddo, P. C., Fitz, J. M., McCartney, S., King, L., Keppler, J., McCarty, G., Hively, W. D., 2019. Using NASA Earth Observations and Google Earth Engine to Map Winter Cover Crop Conservation Performance in the Chesapeake Bay Watershed. Manuscript submitted for publication.

United States Geological Survey, n.d.. What is the C Function of Mask (CFMask) Algorithm? Retrieved from: https://landsat.usgs.gov/what-cfmask

OBJECT-BASED IMAGE ANALYSIS APPROACH TO DETERMINE THE FALLOW PERIODS FOR SHIFTING CULTIVATION IN INDIGENOUS COMMUNITIES IN GUYANA

Haimwant Persaud[1], Rehana Thomas[1], Pradeepa Bholanauth[2], Towana Smartt[2] Pete Watt[3]

[1] Ministry of Natural Resources, Geospatial Information Management Unit, Georgetown, Guyana – hv.persaud@gmail.com, rehanathomas95@yahoo.com

[2] Guyana Forestry Commission, Georgetown, Guyana – project.coordinator@forestry.gov.gy, towanasmartt@yahoo.com

[3] Indufor Asia Pacific, Auckland, New Zealand - pete.watt@indufor-ap.com

KEYWORDS: Guyana, Shifting cultivation, Object-based image analysis, Time series satellite data, Landcover change

ABSTRACT:

Shifting cultivation is an agricultural practice that is the basis of subsistence for the Indigenous population in Guyana and has impacted on a total forest area of 13,922ha to varying degrees of impact on forest carbon. Generally, within these communities, there are two types of shifting cultivation: pioneer and rotational. Pioneer shifting cultivation involves the cutting of primary forest and subsequent cropping and then abandonment. Rotational shifting cultivation involves revisiting areas on a rotational cycle. In Guyana, shifting cultivation is not included in the sustainable land use system since no work has been done to understand the rotational cycles. This study utilized an Object-based image analysis (OBIA) of time-series satellite data (Landsat TM5 and OLI) for the period 2004 to 2017 to determine the dynamics of land cover, time-series changes, and prevailing shifting cultivation cycle in the indigenous communities of Jawalla and Phillipai in the western section of Guyana. OBIA proved to be an efficient method for shifting cultivation and sustainable forest management analyses in Guyana. The findings of this study indicate that short fallows are associated with shifting cultivation in Guyana and the size of the patches cleared each year has been increasing. These trends have potential ecological and livelihood implications that can impact the flow of ecosystem services and the sustainability of livelihoods.

1. INTRODUCTION

1.1 Background

Shifting cultivation is an integral land use activity for the Indigenous population of Guyana. Malmer *et al.* (2005) define shifting cultivation as a form of "non-sedentary" agriculture that involves the clearing and burning of patches of forest for the purpose of cultivation. It is a type of agricultural system carried out extensively in the tropical regions under other terms such as swidden farming or slash and burn farming (Gleave, 1996).

The practice of shifting cultivation is characterized by a subsequent return to the once-cultivated lands to restart the process (Gleave, 1996). The period between cultivations is referred to as the fallow period, the duration of which determines soil and vegetation (forest) replenishment in the area (Erni, 2015). There are many variations of this form of agriculture across the world (Erni, 2015), and even more opinions on the level of sustainability it entails (Erni, 2015). While it is argued that shifting cultivation is an ideal form of agriculture for the tropics, it is important to note that this is dependent on tolerable human population densities and long fallow periods (Erni, 2015).

A range of factors, inclusive of population growth, influences the length of fallow (Gleave, 1996). Studies (Erni, 2016; Gleave, 1996) show that fallow periods are increasingly becoming shorter in tropical regions, resulting in grim degradation and deforestation issues. In Guyana, shifting cultivation is categorized as either Pioneer Shifting Cultivation or Rotational Shifting Cultivation (Brown *et al.*, 2014). Pioneer shifting cultivation refers to the burning of primary forests for the purpose of cultivation, while rotational shifting cultivation refers to the cultivation of areas that have already been burnt and cultivated before (Brown *et al.*, 2014). Over the years, Guyana has seen an increase in the total forest area degraded as a result of shifting cultivation. Further, community surveys conducted in Indigenous communities in Guyana indicate short fallow periods and long cultivation periods (Brown *et al.*, 2014).

Against this backdrop, this study is conducted to determine the fallow periods for shifting cultivation in Guyana and to assess the ecological and livelihood implications of such, using remote sensing technologies.

The study is confined to two Titled Indigenous areas in Guyana; i.e. the communities of Jawalla and Phillipai in Administrative Region seven of Guyana.

1.2 Remote Sensing for Land Use and Land Cover Analysis

Remote sensing methods are tremendously useful for a range of land use and land cover analysis. Analyzing and monitoring land cover change using multi-temporal satellite imagery and advanced remote sensing techniques presents an efficient means of understanding the landscape dynamics of the earth. Guyana, for the past decade, has employed the use of this dynamic technology to map land use and land cover change under its REDD+ Monitoring Reporting and Verification System (MRVS). Shifting cultivation, in particular, has been mapped using modern satellite imagery such as Rapid Eye, Landsat, and Sentinel (GFC, 2018).

1.3 Object-based Image Analysis

This study adopts the object-based approach to image classification to determine the length of the fallow period of shifting cultivation in Indigenous communities of Guyana. This brand of image analysis has become popular for its ability to classify imagery beyond pixels (Blaschke, 2010). Blaschke *et al.* (2004) and Hossain & Chen (2019) explain the main shortcomings of pixel-based image analysis, highlighting its inability to consider the shape and contextual characteristics of the imagery.

The difference with object-based analysis is that it focuses on image-objects instead of pixels. Image-objects, according to Blaschke *et al.* (2004) are groups of pixels that translate to features in the real world. Moreover, as discussed by Hossain & Chen (2019), pixel by pixel analysis fails to consider the spatial variation of land covers in high-resolution imagery.

Object-based image analysis is built on image segmentation and involves the development of automated methods to generate meaningful objects from imagery (Hay & Castilla, 2008). Image segmentation refers to the partitioning of an image into sections (objects) that vary based on properties such as colour, size, shape, etc. (Hossain & Chen, 2019).

2. STUDY AREA

This study was conducted in Jawalla and Phillipai, two Titled Indigenous villages in Western Guyana, geographically, the area is located between 5.231 to 5.753°N latitude and 60.117° to 60.675°W longitude (Figure 1). Together the villages occupy an area of 120,595.92 hectares (1,205.96 Km^2). The villages are populated with the Akawaios indigenous sub-tribe who are thought to be the linguistic descendants of the Karinya and are mainly found in the Upper Mazaruni area, Barama, Upper Pomeroon, Demerara, Wenamu and Upper Cuyuni Rivers.

The adjoining villages, both located in remote locations of the Pakaraima Mountains, Region number 7, Cuyuni – Mazaruni. The main economic activity in this area is alluvial mining with a combined population of approximately 2,370 individuals. The area experiences year-round heavy rainfall. The mean precipitation within the last decade is 22.86 mm/pentad and ranges between 1.71 mm/pentad (low) and 137 mm/pentad (high). The daily surface temperature is around 17.3°C and 28.7°C with an average within the last decade of 22.73°C. The topography is extremely irregular and comprises of mountains, highlands, small basins, and valleys. The vegetation in this area is characterized as a moist, evergreen forest of mixed species types.

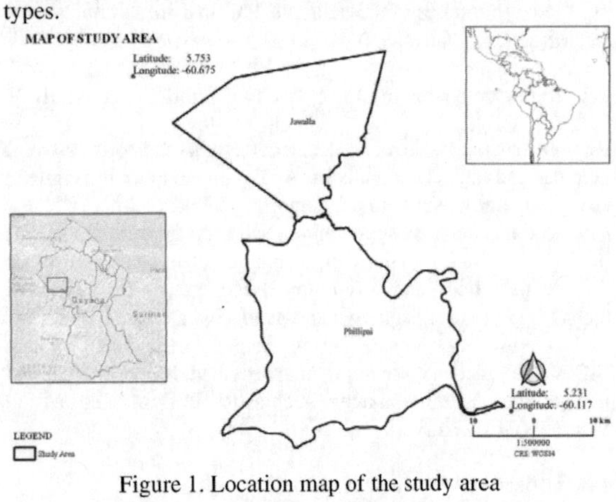

Figure 1. Location map of the study area

3. MATERIALS AND METHODS

3.1 Satellite data acquisition

The present study involves the collection and analysis of 14-years Landsat satellite images (TM and OLI). The required satellite imageries were acquired and processed utilizing the services of Google Earth Engine (GEE) for the development of the land cover change maps.

The processing sequence included the pre-processing of the satellite data after which the cloud/shadow mask was applied, subsequently, the sub-setting of satellite images was performed for extracting study area from all images, taking the geo-referenced outline boundary of Jawalla and Phillipai as AOI (Area of Interest).

Vegetation indices in the form of Normalize Difference Vegetation Index (NDVI) and Enhance Vegetation Index (EVI) were calculated and subsequent band stacking was performed. The stacked images were then downloaded after which Geometric corrections and the segmentation process were performed.

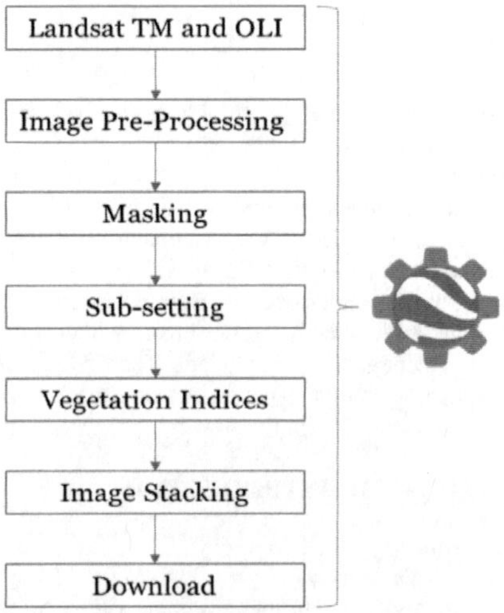

Figure 2: Diagram of the image processing steps

3.2 Object-based image analysis

3.2.1 Image Segmentation: Object-based image analysis requires the creation of objects or separated regions in an image. The pre-processed images were subjected to image segmentation in ENVI. This involved the breaking up of each image into small segments, i.e. objects, based on the unique image properties such as texture, shape, and colour so that the characteristics of the image objects within each region have high homogeneity and strong statistical correlations.

3.2.2 Image classification: Subsequent to image segmentation, image classes were created and matched with the objects created in the previous step, i.e. creating examples for image classification (see figure 2). These image classes were used as examples for the classification of the entire images. Seven Land Use Land Cover (LULC) classes were considered for this purpose (figure 2). Considering the power of object-based methods, an attempt has been made to separate each class, e.g., new cultivation, under cultivation, regrowth, high and low forest, etc., in the classification schemes.

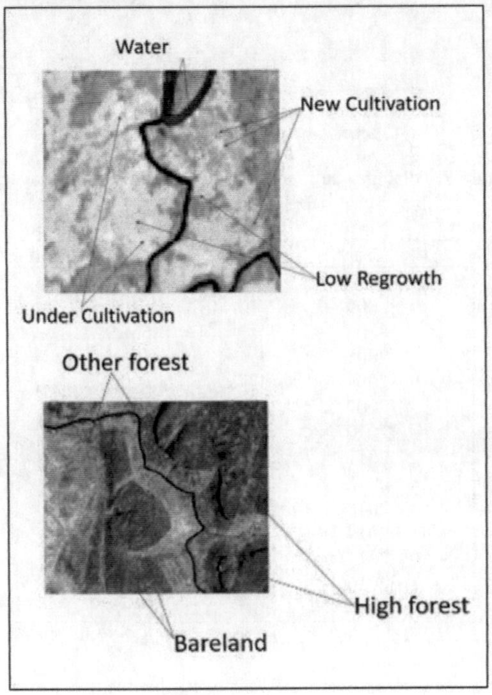

Figure 3. Classes delineated for classification

3.3 Post-classification sieving

The results from the classification were further screened to remove isolated pixels based on a size (number of pixels) threshold and generalize classification of the images through a sieving process. Normally classified images often manifest a salt-and-pepper appearance due to the inherent spectral variability encountered by a classifier. In such situations, it is often desirable to smooth the classified images to show only the dominant presumably correct classification. Thus, post-classification processes were applied over a classified image to eliminate isolated pixels and to generate an apparently less noisy image. In this study, the post-classification process applied was sieving. Once this process was completed quality checks were done to verify the integrity of the image classification. This step was performed in QGIS 3.8 software.

4. RESULTS AND DISCUSSION

The results obtained through the analysis of multi-temporal satellite imageries are diagrammatically illustrated in Figure 3. The seven land cover classes identified previously were analyzed, and a year by year account of the annual area under shifting cultivation within the study area was identified continuously for the entire study period. This analysis gives a fair idea about the prevailing condition of the practice within the communities. Figure 3 shows the spatial distribution of burnt areas for each year under analysis. These results demonstrate that the shifting cultivation plots are concentrated (more or less) in a common area, i.e. following the path of a river that passes through the study area.

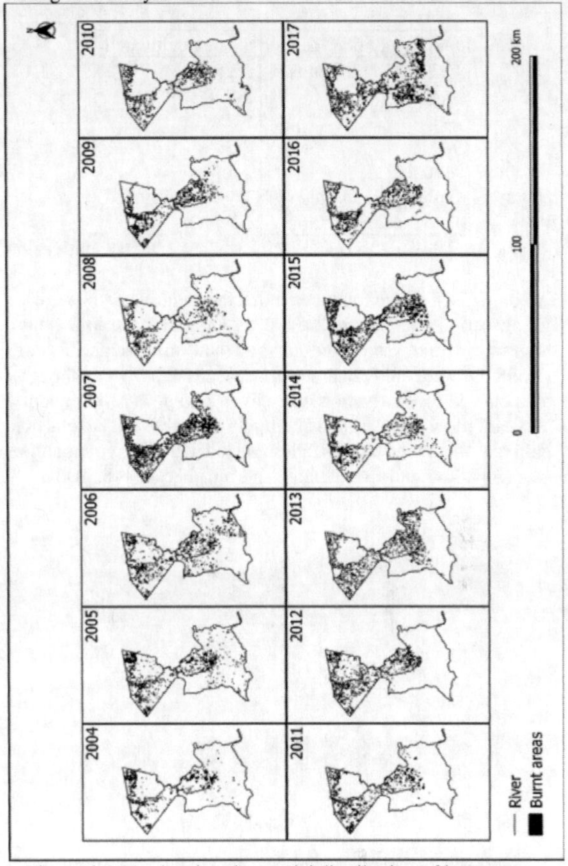

Figure 4. Map showing the spatial distribution of burnt areas each year 2004-2017

The percent of the total cultivated area under varying fallow cycles is observed in figure 4. The results of the classification were simplified and grouped into five principal groups, areas that were identified as single period cultivation, and areas with fallow years of 1-3 years, 4-6 years, 7-9 years and ≥ 10 years.

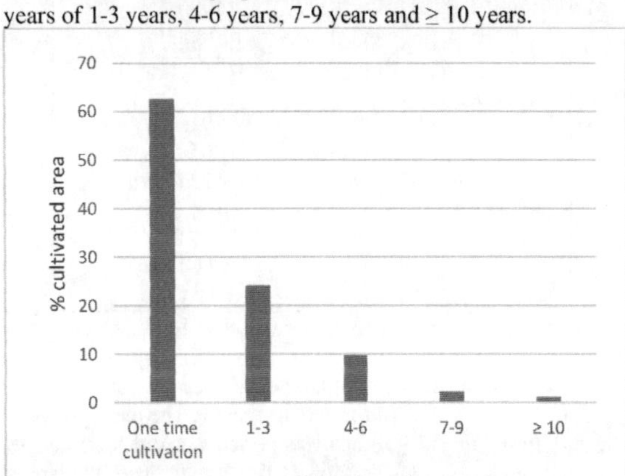

Figure 5. Chart showing the percent of area cultivated under various fallow periods

As illustrated in figure 4, most of the cultivated area consists of a one-time instance of burning (62.6%) while the remaining 37.4% was distributed among the longer fallow periods. The fallow periods and their respective values are represented in table 1.

Fallow period (years)	% area cultivated
One time cultivation	62.59
1-3	24.20
4-6	9.78
7-9	2.30
≥ 10	1.14
Fallow period (years)	% area cultivated
One time cultivation	62.59

Table 1. Fallow periods and their respective % area cultivated

A comparison of the total area in hectares cultivated under each fallow period (1-13 years) 2004 -2017 showed that the maximum area of two thousand three hundred and ninety-eight hectares (2,398 ha) were cultivated under a 2 year fallow, while the least area cleared for cultivation was eleven hectares (11 ha) under a 13 year fallow as illustrated in figure 5 and table 2. The total area cultivated over the 13 years fallow period within the communities was eight thousand five hundred and nine hectares (8,509 ha).

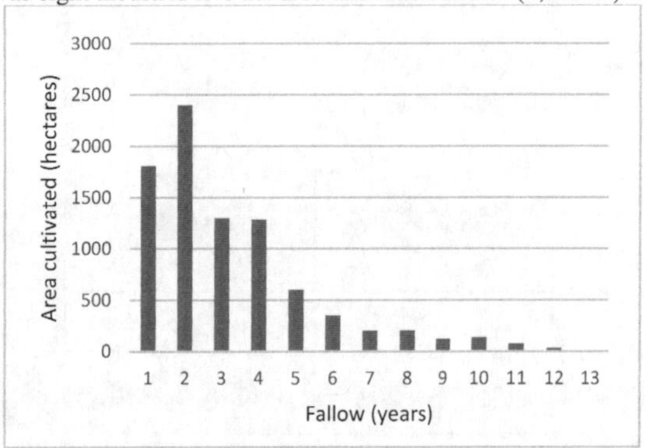

Figure 6: Chart showing the total area in hectares cultivated under various fallow periods

Fallow period (years)	Area cultivated (hectares)
1	1808.74
2	2398.31
3	1296.2
4	1280.36
5	597.92
6	346.35
7	194.77
8	205.43
9	122.08
10	136.28
11	77.7
12	33.45
13	11.17

Table 2. Fallow periods and the respective total area cleared

An analysis of the total area of land cleared each year for shifting agriculture each year is illustrated in figure 6. The mean land area cleared during the 13 year analysis period was one thousand six hundred and thirty-three hectares (1,633 ha). In 2017, the largest area was cleared amounting to two thousand six hundred hectares (2,600 ha). The smallest observable areas cleared was seen in 2008, the total area cleared in that year was five hundred and forty-four hectares (544 ha) (table 3).

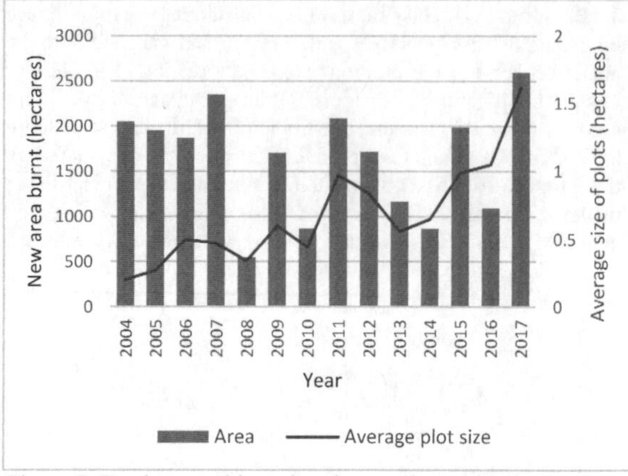

Figure 7. Total area cultivated each year along with the average size of plots burnt

Year	Area cultivated (hectares)
2004	2051.86
2005	1954.59
2006	1866.29
2007	2349.97
2008	544.02
2009	1707.92
2010	871.45
2011	2086.54
2012	1717.97
2013	1161.5
2014	865.78
2015	1991.16
2016	1092.36
2017	2599.35

Table 3. Total area cultivated each year

The areas cleared over time fluctuates, the average size of the plots cleared per year was 0.68 hectares. While there is no observable trend in the total area cleared, the results indicate that there has been a gradual increase in the size of plots burnt each year.

4.1 Fallow periods for shifting cultivation in Guyana

The length of time that an area previously burnt and cultivated is left to replenish is referred to as the fallow in shifting cultivation practices (Erni, 2015). In many tropical countries, where this type of agriculture is readily practiced, fallow periods are becoming significantly shorter owing to a variety of reasons (Dalle & Blois, 2006). Shorter fallows are concerning because this type of agriculture is sustainable only if the soil and natural vegetation are left to recover; the alternative is permanent cropland (Yadav *et al.*, 2012).

The results of this study show that Guyana's case is similar to many other tropical countries, i.e. short fallow periods are associated with shifting cultivation. The results depicted in figures 1 and 2 show a significant disparity between the total area cultivated under fallow cycles less than 5 and that over 5 years. This can be attributed to an increasing population in these communities, thereby an increased demand for food; proximity to farmlands; changes in land tenures (Dalle & Blois, 2006).

In addition to short fallows, the results indicate significant instances of one-time cultivation. This implies that, in many cases, burnt croplands are not returned to; rather new areas are being burnt and cultivated. Figure 1 indicates the high percent of burnt land which has been cultivated just once within the 13- year period of this study. Further, the results depicted in figure 6 show the large areas which are newly burnt every year (pioneer shifting cultivation). This indicates that old burnt areas are not often returned to for subsequent cultivations. This trend is indicative of arbitrary agricultural practice which can have substantial ecological consequences. Also concerning, is the increasing size of patches cleared for agriculture. Figure 6 shows that the average sizes of the patches cleared have been steadily increasing since 2013. This practice can potentially change the ecological conditions at the level of the farmlands, as well as, at the landscape level (Dalle & Blois, 2006).

4.2 Ecological and livelihood implications of shifting cultivation practices in Guyana

Based on the results of this study, the practice of shifting cultivation in Phillipai and Jawalla is associated with short fallows, large areas being newly burnt each year and larger sizes of plots being burnt. These results are concerning given the potential impacts on the ecology and livelihood sustainability of people.

Short fallows can potentially reduce species diversity and can alter the vegetative composition of forests. Yadav *et al.* (2012) find that frequent cycles of burning and cultivation have directly affected tree species diversity and have led to the degradation of forest land in parts of India. Further, studies (Dalle & Blois, 2006) show that shorten fallows are closely associated with an increase of pioneer species, thus altering the natural vegetative composition of forest landscapes (Delang *et al.*, 2016).

In addition to vegetation impacts, short fallows do not permit complete recovery of soil nutrients and can result in poor soil properties. Further, the clearing of large patches of primary forest creates fragmented landscapes; modifies ecosystems, thereby affecting the flow of ecosystem services and displaces wildlife (Dalle & Blois, 2006).

While these issues associated with shifting cultivation are as a result of increasing population pressures, the very sustainability of people's livelihoods can be affected by short fallows and arbitrary burning. Forest degradation and soil depletion at current rates would create a landscape that can no longer support livelihood activities such as shifting cultivation (Delang *et al.*, 2016).

The view that shortened fallows in shifting cultivation equate to yield declines has been challenged by Mertz (2002) on the premise that this might not be the case for many locations as yields vary based on site-specific conditions, such as soil type, vegetation type, and climate. Nonetheless, while the connection between yields and length of fallow is not straightforward, the ecological implications of short fallows are unquestionable.

4.3 Object-based image analysis for shifting cultivation analysis in Guyana

This study employed the use of a contemporary remote sensing technique, i.e. OBIA. This technique proved efficient in achieving the objectives of this research and provided accurate results due to the flexibility and thoroughness associated with image segmentation and classification. Further, due to the automation aspect associated with the use of the ENVI software, all burnt patches were detected as opposed to only considering those areas that are of a certain size. This is helpful as it gives a holistic view of the shifting cultivation scenario in the interior regions of Guyana. Moreover, this study is evidence that new and efficient remote sensing techniques can be applied to a range of land and forest management issues in Guyana to guide decision making and contribute to sustainable forest management.

5. CONCLUSION

Shifting cultivation is an important aspect of the livelihoods of indigenous communities in Guyana. For many years, it has been practiced in the interior regions of Guyana with minimal negative impacts. However, with increasing population pressures, this agricultural practice has changed. This paper analyzed the trends associated with fallow periods of shifting cultivation in the communities of Phillipai and Jawalla. It was found that most of the area under shifting cultivation experiences short fallows and each year, large areas are being burnt for agriculture. Further, the size of the patches burn each year has been on the rise for the past years. These practices have both ecological and livelihood implications.

REFERENCES

Blaschke, T., Burnett, C., Pekkarinen, A., 2004: Image Segmentation Methods for Object-based Analysis and Classification. In: de Jong, S.M., van der Meer, F.D. (Eds.) *Remote Sensing Image Analysis: Including the Spatial Domain*. Netherlands: Kluwer Academic Publishers, 211-236.

Blaschke, T., 2010: Object-based image analysis for remote sensing. *ISPRS Journal of Photogrammetry and Remote Sensing*. 65 (1), 2-16.

Brown, S, Goslee, K., Casarim, F., Harris, N., Petrova, S., 2014: *Sampling design and implementation plan for Guyana's REDD+ FCMA: version 2*. Submitted by Winrock International to the Guyana Forestry Commission.

Dalle, S.P., Blois, S., 2006: Shorter fallow cycles affect the availability of noncrop plant resources in a shifting cultivation system. *Ecology and Society*. 11(2).

Delang, C.O., Weiyi, X., Brooke, B., Chun, K.P., 2016: The effect of fallow period length on the abundance and diversity of useable plant assemblages in shifting cultivation system (swidden agriculture) in Northern Laos. *Polish Journal of Ecology*. 64 (3), 350-356.

Erni, C., 2015: *Shifting Cultivation, livelihood and food security: new and old challenges for Indigenous peoples in Asia*. Bangkok: FAO, IWGIA & AIPP.

GFC, 2018: *Guyana REDD+ Monitoring and Verification System (MRVS): Year 7 Summary Report*. Georgetown: GFC.

Gleave, M.B. (1996). The length of the fallow period in tropical fallow farming systems: a discussion with evidence from Sierra Leone. *The Geographical Journal*. 162 (1), 14-24.

Hay, G.J., Castilla, G., 2008: Geographic Object-Based Image Analysis (GEOBIA): A new name for a new discipline. In: Blaschke, T., Lang, S. & Hay, G. (Eds.) *Object-Based Image Analysis*. Heidelberg, Berlin, New York: Springer. Pp. 93-112.

Hossain, M. & Chen, D. (2019). Segmentation for Object-based image analysis (OBIA): A review of algorithms and challenges from a remote sensing perspective. *ISPRS Journal of Photogrammetry and Remote Sensing*. 150, 115-134.

Malmer, A., van Noordwijk, M., Bruijnzeel, L. A., 2005: Effects of shifting cultivation and forest fire. In: Bonell, M., Bruijnzeel, L. A. (Eds). *Forests, water and people in the humid tropics: past, present and future hydrological research for integrated land and water management.* Cambridge: Cambridge University Press, 533-560.

Mertz, O., 2002: The relationship between the length of fallow and crop yields in shifting cultivation: a rethinking. *Agroforestry Systems.* 55, 149-159.

Yadav, P.K., Kapoor, M., Sarma, K., 2012: Impact of slash-and-burn agriculture on forest ecosystem in Garo Hills landscape of Meghalaya, North-East India. *Journal of Biodiversity Management and Forestry.* 1(1), 1-6.

ANALYSIS OF LANDSLIDE REACTIVATION USING SATELLITE DATA: A CASE STUDY OF KOTRUPI LANDSLIDE, MANDI, HIMACHAL PRADESH, INDIA

N. Singh[1], S. K. Gupta[1,*], D. P. Shukla[1]

[1] School of Engineering, Indian Institute of Technology Mandi, India (*sharadgupta27@gmail.com)

KEY WORDS: Temporal Change detection, Reactivated Landslide, Kotrupi (HP), Google Earth, Geomorphological Analysis, Remote Sensing

ABSTRACT:

Landslide is a global natural hazard that occurs frequently in the areas of incompetent weak rocks, undulating topography, steep slopes and incessant rainfall. In the night of 12 August 2017, a massive landslide took place at Kotrupi, Mandi district, Himachal Pradesh, India. The slide was so huge that it eroded more than 300-meter stretch of NH-154 killing over 50 people with more than 40 missing. Local residents report that this area has always been unstable where small landslides had occurred in the past. The landslide scar could be seen on the past satellite images from December 2001 to March 2017 on Google Earth. A huge landslide occurred at this location on 13 August 1977. After two decades on 13 August 1997, the landslide reactivated and some part of the slope failed, which can be seen on satellite images of the year 2001. The landslide reactivated again on 13 August 2007, but not much attention was given to it, as it was a small event and did not affect much. Again, after a decade, in the night of 12 August 2017 this landslide was reactivated. There is the possibility of reoccurrence of slope instability from upper reaches of the crown area of the main slide complex as well as the debris, which have been already accumulated on hill and valley side. Based on the geological, geotechnical and geophysical investigations the site stability can be done but its monitoring from satellite provides the information for its future preventive measures.

1. INTRODUCTION

The Himalayan mountain range runs from West-Northwest to East-Southeast in an arc of 2400 km length having rich geo-diversity. The increasing population along the range and at the foothills have increased anthropogenic activities (Pande, Uniyal, 2007, Ray et al., 2007). Increased human activities are the main drivers for changing climate, extreme events and geomorphological changes in the Himalayan region (Petley et al., 2007, Bennett et al., 2016, Froude, Petley, 2018). Climate change can affect the timing, size, and frequency of natural hazards in the Himalaya in an uncontrolled manner with changing topography (Haritashya et al., 2006). The steep slopes with thin soil cover without any vegetation, over impervious bedrock and high orographic precipitation are the favorable conditions for landslides and floods in the Himalayan region (Haritashya et al., 2006). Being the youngest mountain range in the world and structurally unstable, Himalayan region is highly prone to topographically controlled natural hazards.

Globally, many lives are lost, huge amount of properties and revenue is lost each year due to landslide (Lee, Pradhan, 2007, Tiwari, Ajmera, 2017). Global landslide database shows that distribution of landslide is heterogeneous, have caused largest causalities in Asia especially in the Himalayan region (Froude, Petley, 2018). Landslides are more prevalent in the Himalayan region mainly due to varying lithology, changing weather, extreme precipitation, density of drainage, tectonic activities, orography, topography, and ongoing seismicity (Ray et al., 2007, Pareek et al., 2010, Tiwari, Ajmera, 2017). Precipitation coupled with weak lithology and unattended slope mitigation leads to pervasive landslides along road corridors, which impose high risk on urban settlements (Lee, Pradhan, 2007, Kundu et al., 2013, Sarkar et al., 2015, Bennett et al., 2016, Froude, Petley, 2018).

The present landslide was caused due to reactivation of already three times failed slope, which have occurred at about 20-year regular interval since 1977. Locals recall that scarp had already started developing due to slope failure on 13 August 1977 (GSI 2017). After two decades, on 13 August 1997, the landslide reactivated, and scarp widened in dimensions. Again, on 13 August 2007, slide reactivated, but the slope failure was not large and did not affect people, hence not much attention was given. In 2018 as well, the slide reactivated but since the slope failure was not that prominent, again not much attention was given. However, the landslide event in August 2017 was most devastating. The people residing in the area of landslide zone (near toe and run off zone) were concerned about the reactivation of the landslide in future as it had happened in past. If that happens then their livelihood will be at stake.

The repetitive nature of the landslide is the major concern in the study area. Here, we have carried out detailed analysis of geological, seismicity, climatic and soil moisture data to understand the probable cause of this deadly landslide. We visited the landslide site many times i.e. November 2017, August 2018, February 2019, April 2019. We carried out detailed geological field investigations. Further, we analyzed meteorological and seismic data to understand the possible triggering and reactivation of the landslide. We have used past satellite images for analysis of landslide reactivation.

2. STUDY AREA

The Kotrupi landslide is located on Mandi - Joginder Nagar - Pathankot National Highway (NH)-154 with center coordinates of 31°54'46.29"N and 76°53'31.90"E (as shown in Figure. 1). The landslide falls in seismic zone V, which represents highly vulnerable seismic area. The landslide lies in the Mandi district which is geographically located within 31°13'-32°05'N latitudes and 76°37'-77°25'E longitudes. The area of the slide is measured to be around 133,674 m2 and width is 190 m, while the run-out length is 1155 m as seen on high-resolution satellite image. The elevation of the district varies between 550 m to 3960 m a.m.s.l. Mandi district has varying climate from sub-tropical climate in the valley area to temperate at hilltops. In high altitude regions, the weather remains colder throughout the year with snowfall above 2000 m a.m.s.l in winter season. The temperature varies between 30°-40° C in summers in low-lying areas. The district receives good amount of rainfall in the monsoon season from July to

*Corresponding author

September along with sporadic rainfall in post winters season i.e. February to March due to western disturbances. The annual average rainfall of Mandi district is 1331.50 mm, which varies in between 700-2000 mm from place to place (CGWB report 2013).

2.1 Landslide Event on 13 August 2017

A massive landslide occurred near Kotrupi village in Mandi district during early hours of 13 August 2017 (01:00-02:00 AM local time). The debris from landslide killed more than 50 people when two HRTC (Himachal Road Transport Corporation) buses (i.e. Manali-Katra and Chamba-Manali) along with other private vehicles were swept away and buried under the debris (Roy et al., 2018, Sharma et al., 2018, Pradhan et al., 2019) (GSI 2017). As per the reports, approximately six bodies were recovered from first b us, w hereas n othing c ould b e recovered from second bus (Pradhan et al., 2019). News reports suggests that approximately 300 m of road stretch of national highway (NH-154) has been completely buried under debris, thus disrupting communication on an important route. Immense quantity of debris released from the landslide blocked the drainage up to 1 km in the downstream. The area was subsiding due to presence of nallas, which made the region swampier. Due to this catastrophe, the new alignment of the NH was carried out by Public works department (PWD). As the Kotrupi landslide witnessed minor scarp failure in July 2018, the road was washed off and the authorities changed the road alignment again. The present stretch of NH-154 follows the same alignment, which existed before the incident of 13th August 2017. The Kotrupi landslide and the road alignments are shown in Figure-1.

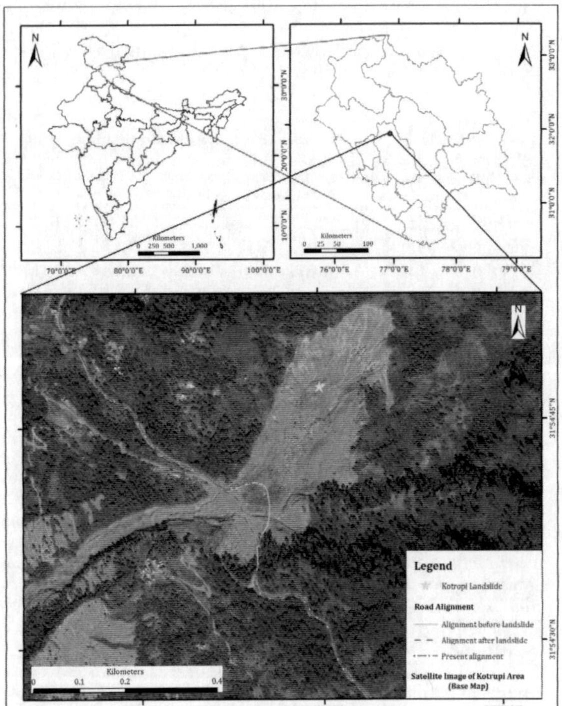

Figure 1. Kotrupi landslide location as seen on satellite images. The present road alignment, before landslide and change in alignment soon after the landslide are shown with red, green and blue lines

2.2 Geology of the Landslide Site

The primary rock types found in and around the Kotrupi landslide are red shale, dolomite of Shali formation and sandstone, shale of Siwalik group (Roy et al., 2018, Sharma et al., 2018). The red shale rock present in the outcrop is loose, highly crushed and almost weathered into the soil. The shale beds are present with interspaces, which are filled w ith l oose s oil a nd c rushed r ock material. Highly crushed and thinly bedded weathered sandstone intercalated with shale beds is present at few places.

The study area (Kotrupi landslide) is sandwiched between Siwaliks and Shali formation along the thrusted contact of MBT. In the west of MBT, local faults like Drang, Jwala Mukhi and Ropar Manali faults (Thakur et al., 2019) passes as shown in Figure-2. These structural settings in the vicinity of the landslide area create favorable conditions for high vulnerability of the slope failures. Due to its weak and fragile topography and lithology, the area in between Ghatasani and Padhar region encounters many landslides. However, the Kotrupi landslide is one of its kind due to its large size and losses occurred.

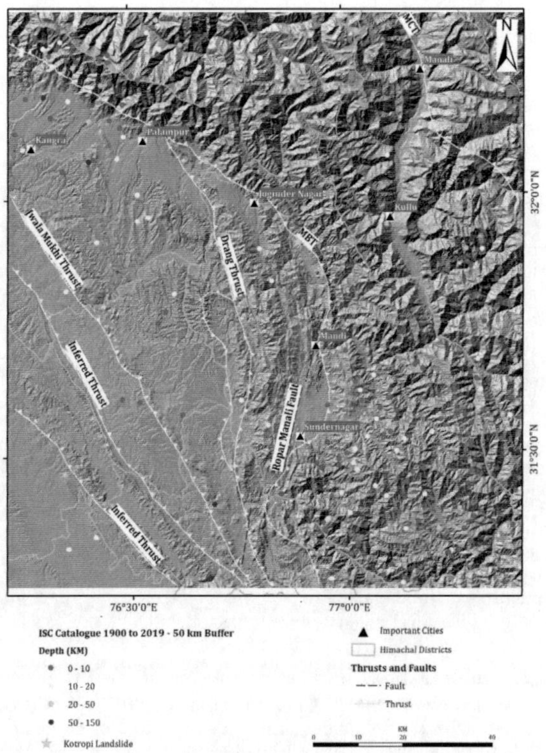

Figure 2. Ropar-Manali fault that passes through the Kotrupi landslide location. The earthquake epicenters within 50 km buffer z one o f s tudy a rea h as b een marked. The main boundary thrust (MBT) and Main central Thrust (MCT) is marked in the map (Modified after (Thakur et al., 2019))

3. DATA AND METHODOLOGY

To understand the landslide reactivation, it is important to evaluate both internal and external factors that may have caused the slope failure. We have studied the probable cause of the Kotrupi landslide using geological data, past earthquakes, precipitation and soil moisture data. The historic records of past earthquake events can unfold the future aspects of seismic activities in the Himalayan region (Bilham, 2019). Hence, we have analyzed the frequency and magnitude of the earthquakes within 50 km radius around the Kotrupi landslide that have occurred during 1970-2018. In addition, landslide inventory was also prepared to understand the past activities and relate it to mass movements in the region. Earthquake data for the study area was collected from International Seismic Centre (ISC) for the period 1970-2016. While for the years 2017 and 2018, United States Geological Survey (USGS) earthquake data inventory was utilized.

The daily rainfall data with the spatial resolution of 0.25°X 0.25°, was collected from TRMM (Tropical Rainfall Measuring Mission) satellite for the months of July and August 2010 - 2018. The daily data was used for studying the effect of unprecedented rainfall prior to the failure of the slope. The 3-hourly data of soil moisture with the spatial resolution of 0.25°X0.25° at depth levels 1-10, 10-40, 40-100 and 100-200 cm, was collected from GLDAS (Global Land Data Assimilation System). The 3 hourly data were converted into daily data for correlating with the daily rainfall data. However, we have found that the microwave satellite data is not sensitive to soil moisture at higher depth more than 20 cm in Himalaya terrain due to presence of thin soil cover and dense vegetation.

4. RESULTS AND DISCUSSIONS

The Kotrupi landslide is one of the biggest of its type that took place in Himachal Pradesh. The initial report suggested that it occurred due to excessive rainfall. However, landslide scar can be seen on the previous year's satellite images from December 2001 to March 2017 on Google Earth. From the satellite images, it can be inferred that this catastrophe is not a mere one-day event rather decades-long process. Cracks were developed well before the occurrence of the event; however, rainfall might have triggered the slide. Local residents report that this area has always been unstable where small landslides had occurred in the past. A huge landslide occurred at this location on 13 August 1977. After two decades on 3 August 1997, the landslide reactivated and some part of the slope failed, which can be seen on satellite images of the year 2001 (Figure3). The landslide reactivated again on 13 August 2007, but not much attention was given to it, as it was a small event and did not affect much. Again, after a decade, in the early hours of 13 August 2017 this landslide was reactivated. The Figure 3 shows various images of Kotrupi area prior to landslide and after landslide. The scar can be seen expanding from surficial c rack t o m assive l andside f rom 2 001 t o 2 018 in Figure-3(a)-3(h).

We have considered both external and internal factors such as rainfall, slope, soil moisture, lineament density, faults (local and regional), earthquakes and lithology, to understand the reactivation of Kotrupi landslide.

4.1 Magnitude and Frequency of Earthquakes

To understand the impacts of external factors such as earthquakes on the landslide and surrounding region, the earthquake event data was analyzed. The magnitude and frequency of earthquakes are shown in Figure-4(a) and 4(b). During 1970-1997, total 97 earthquakes (mostly shallow, depth 10-20 km & magnitude 4-5) were recorded, whereas 272 earthquakes (magnitude >5.0) were recorded during 1997 to 2018, which indicate an increase in the frequency of the earthquakes in recent years. The depth of an earthquake plays an important role in controlling the topography of any region. Shallow earthquakes with high magnitude in weak and fragile topography damage more as compared to a deep-seated earthquake. Some of the great examples of earthquake-induced landslides in the past are Kangra earthquake of 1905 in

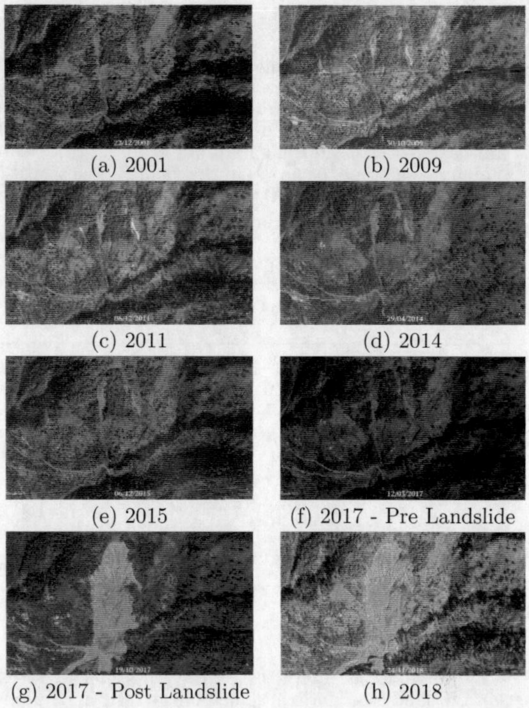

Figure 3. Landslide Reactivation – The existence of crack (Year 2001) can be clearly seen in (a). The development and enlargement of crack can be seen in subsequent images (b)-(f). Landslide occurrence can be seen in (g) and (h). Change in Road alignment is also visible

the Himachal Pradesh, Uttarkashi earthquake 1991 of Uttarakhand and Gorkha earthquake in Nepal of 2015 (Pradhan et al., 2006).

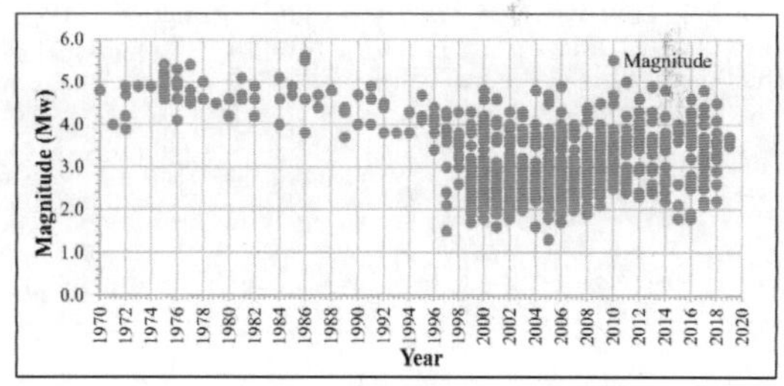

(a)

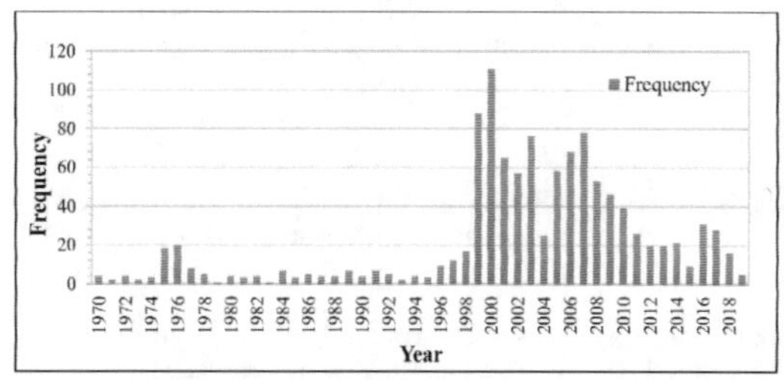

(b)

Figure 4. (a) Magnitude and (b) Frequency distribution of earthquakes for the periods 1970-2018 occurred within 50km radius of the Kotrupi landslide

Nearly 55% of the earthquakes which have occurred within 50 km radius of study area, are shallow earthquake with focal depth of 10-20 km as given in Table-1. Shallow earthquakes of 4-5 Mw magnitude tends to be very violent especially if the geology of the area is weak and fragile. Earthquakes have triggered landslides in and around epicenter of earthquakes specially if the earthquake is of shallow type (Pradhan et al., 2006). The depth of MBT is observed to be about 10 km while the depth of MCT is around 10-20 km near Kotrupi landslide (Thakur et al., 2019). From the earthquake events occurred in the vicinity of Kotrupi it can be established that MCT is still active around Kotrupi landslide site.

Table 1. Number of earthquake events occurred within the radius of 50 km of the landslide location at different depths.

S. No.	Depth of occurrence of the earthquake (km)	No of events of earthquakes
1	<10	371
2	10 - 20	252
3	20 - 35	237
4	35 - 50	180
5	>50	106

4.2 Rainfall as Triggering Factor for Landslide

The hydrologic characteristics of the slope, shear stress acting on the slope and the mechanical resistance along potential slip surface plays a prominent role in affecting the failure mechanism of the slope (Berti et al., 2012). Although the correlation between precipitation and landslides is well known, it's invariably a vital question to ask – what affects the occurrence of landslides more, antecedent rainfall (i.e. rainfall in the days immediately preceding a landslide event) or triggering rainfall (i.e. rainfalls at the time that the landslide occurs) (Rahardjo et al., 2001). We have tried to understand the variation in soil moisture with rainfall for the months of July and August from 2010-2018. During August 2017, total 293.88 mm rainfall was recorded, however, 194.81 mm rainfall occurred between 05-13 August and only 99.07 mm rainfall was recorded on other days which can be seen in Figure-5.

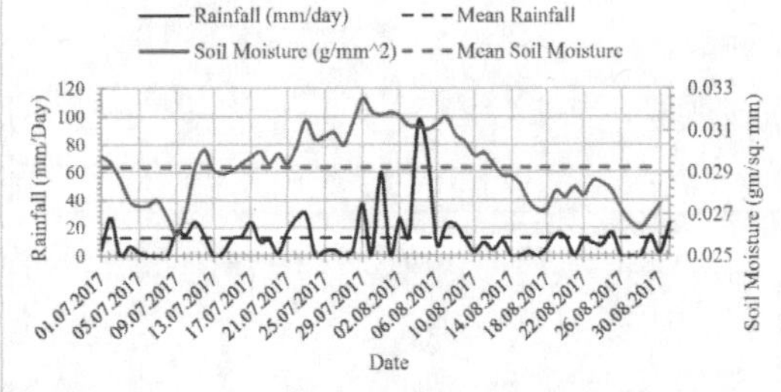

Figure 5. Variations of rainfall with soil moisture for the months of July-August 2017

This seven-day antecedent rainfall intensity prior to landslide was 1.19 mm/hr. As already discussed, the rocks present within the vicinity of the sliding area are highly crushed and weathered. The low rainfall intensity with continuous downpour may have allowed water to percolate within the rock formation/slope. The percolated water may have increased pore water pressure which further reduced the strength of the rock/soil lying between MBT and MCT thrusts. It is evident from Figure 5 that the failure of slope on 13 August 2017 could be a consequence of unprecedented rainfall occurred 5-7 days prior to landslide.

4.3 Effect of Rainfall on Soil Moisture

The study of antecedent soil moisture content is necessary to understand the hydrologic response of a hilly slope in controlling the landslide incidence (Tohari, 2018). Excessive precipitation can trigger landslide in two ways i.e. either by saturating the pore water pressure or by flooding the area which further leads to surface erosion and expose the slope profile p resent b eneath (Dortch e t al., 2009). In the Figure-6, we have shown the soil moisture (gm/mm2) variations at depth of 0-10 cm and rainfall (mm/day) within the landslide location for the month of July and August from 2010 to 2018. The rainfall and soil moisture vary from 30 mm/day and 0.031 g/mm2 on 25 July 2017 to 96 mm/day and 0.032 g/mm2 on 06 August 2017 respectively. Although the soil moisture and rainfall follow the same trend, there is a time lag of 2-3 days. Similarly, graphs plotted for 2010 to 2018 for rainfall and soil moisture content shows that both parameters follow the same trend for previous years as well (Figure6). It is evident that, the rainfall intensity and hence soil moisture starts rising by the end July month (between 25-28) which was observed during the Kotrupi landslide as well.

The landslide might have happened due to a combination of following geo-scientific c auses: d evelopment o f tension cracks, sheared/thrusted rock mass having poorly drained slope, antecedent rainfall, rise in soil moisture content as well as increased seismic activities, however the single cause of failure cannot be reached. There is the possibility of re-occurrence of slope instability from upper reaches of the crown area of the main slide complex as well as the debris, which have been already accumulated on hill and valley side.

5. CONCLUSIONS

In this study, we have analyzed various parameters such as seismic, hydro-meteorological, soil moisture, geology and lineaments which are likely responsible for triggering the Kotrupi landslide on National Highway (NH-154) on the night of 13 August 2017. This landslide is the biggest of its type that took place in Himachal Pradesh. The recurrent behavior of Kotrupi landslide has put a lot of pressure on not only the government but also the people living near that landslide. The preliminary reports prepared by various agencies suggested that it occurred due to excessive rainfall, however the effect of antecedent rainfall was not considered.

Moreover, the landslide scar could be seen on the satellite images from December 2001 to March 2017 on Google Earth. The earlier occurrences of small landslides already carved the way for bigger landslides which was waiting to happen. This was also confirmed b y local residents who have witnessed all the small incidents happening in the landslide regime. The freely available multi-temporal satellite images from Google Earth provides a good platform for comparative study and analyzing the continuous behavior of recurrent slides.

About 55% of the earthquakes occurred at shallow depth during 1970-2019. The frequency of earthquakes within 50 km radius of the landslide has increased from 1997. However, the magnitude varies in between 4-5. The area as mentioned above lies in between MCT and MBT and is crossed by several local and regional faults that makes the area highly prone to seismic activity. The highly crushed and weathered rock in the vicinity is the proof of stress accumulation within the region, which has repeatedly vented out in form of slope failure.

Daily soil moisture and rainfall data for the months of July and August was utilized for a duration of 2010 to 2018 to understand the impact of rainfall and soil moisture on the observed landslide in August 2017. The rainfall 15 days prior to the landslides may have allowed

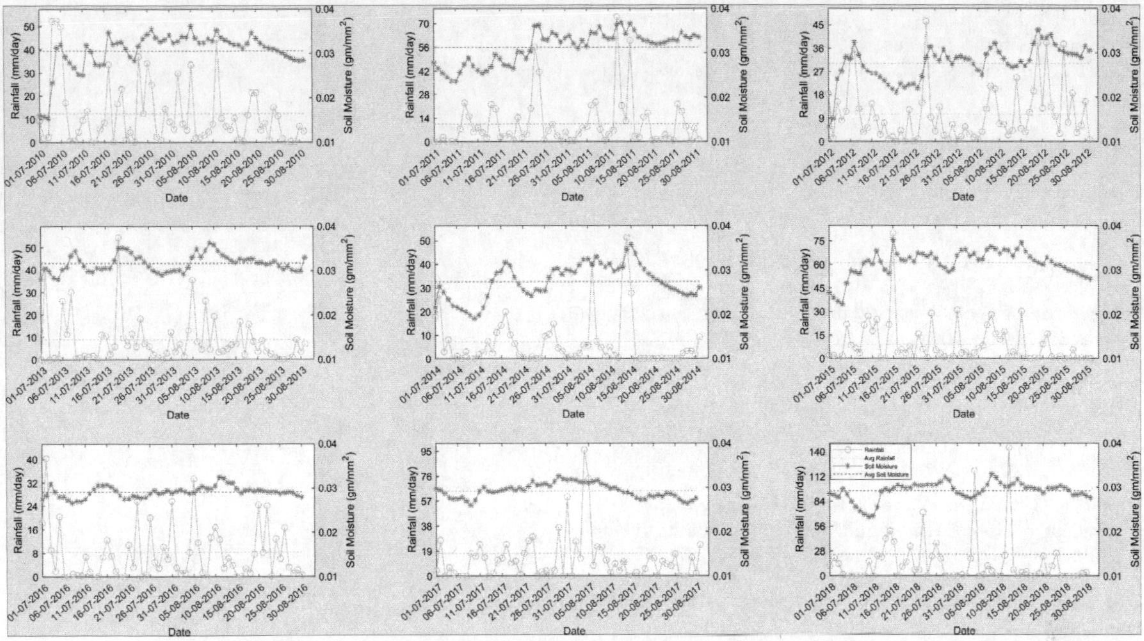

Figure 6. Soil moisture (cm) profile for July- August Months at depth of 0-10 cm with rainfall intensity for 2010, 2011, 2012, 2013, 2014, 2015, 2016, 2017, 2018 and 2019

slow infiltration/percolation o f w ater i nto t he soil/rock through cracks and may have likely to increase pore water pressure leading to decrease in strength of the slope. As predicted through various models, the rainfall shows increasing trend in the future, hence the increasing rainfall could be an alarming scenario as it can directly lead to slope instability and landslide activation.

Based on the analysis carried out in the present study, it can be concluded that the reactivation of Kotrupi landslide is a combination of various causes such as development of tension cracks, sheared/thrusted rock mass having poorly drained slope, antecedent rainfall, rise in soil moisture content as well as increased seismic activities, hence the single cause of failure cannot be reached. The possibility of re-occurrence of Kotrupi landslide cannot be ignored as the debris and loose material is accumulated on hill and valley side. The slope instability from the crown area of the main slide complex is also possible as the current slide has exposed the mountain surface to open environment.

ACKNOWLEDGEMENTS

The authors would like to thank Prof. Ramesh P. Singh for sparing his valuable time and visit the landslide site. We are grateful to NASA Giovanni team for making satellite data available for the present study. We would also like to thank Google Inc. for providing Google Earth Pro. We would also like to thank USGS for providing Landat 8 images free of cost.

REFERENCES

Bennett, G. L., Roering, J. J., Mackey, B. H., Handwerger, A. L., Schmidt, D. A., Guillod, B. P., 2016. Historic drought puts the brakes on earthflows i n N orthern California. Geophysical Research Letters, 43(11), 5725–5731.

Berti, M., Martina, M. L., Franceschini, S., Pignone, S., Simoni, A., Pizziolo, M., 2012. Probabilistic rainfall thresholds for landslide occurrence using a Bayesian approach. Journal of Geophysical Research: Earth Surface, 117(4), 1–20.

Bilham, R., 2019. Himalayan earthquakes: a review of historical seismicity and early 21st century slip potential. Geological Society, London, Special Publications, SP483.16.

Dortch, J. M., Owen, L. A., Haneberg, W. C., Caffee, M. W., Dietsch, C., Kamp, U., 2009. Nature and timing of large landslides in the Himalaya and Transhimalaya of northern India. Quaternary Science Reviews, 28(11-12), 1037–1054.

Froude, M. J., Petley, D., 2018. Global fatal landslide occurrence from 2004 to 2016. Natural Hazards and Earth System Sciences, 18, 2161–2181.

Haritashya, U. K., Singh, P., Kumar, N., Singh, Y., 2006. Hydrological importance of an unusual hazard in a mountainous basin: flood and landslide. Hydrological Processes: An International Journal, 20(14), 3147–3154.

Kundu, S., Saha, A., Sharma, D., Pant, C., 2013. Remote sensing and GIS based landslide susceptibility assessment using binary logistic regression model: a case study in the Ganeshganga Watershed, Himalayas. Journal of the Indian Society of Remote Sensing, 41(3), 697–709.

Lee, S., Pradhan, B., 2007. Landslide hazard mapping at Selangor, Malaysia using frequency ratio and logistic regression models. Landslides, 4(1), 33–41.

Pande, R. K., Uniyal, A., 2007. The fury of nature in Uttaranchal: Uttarkashi landslide of the year 2003. Disaster Prevention and Management: An International Journal, 16(4), 562–575.

Pareek, N., Sharma, M. L., Arora, M. K., 2010. Impact of seismic factors on landslide susceptibility zonation: a case study in part of Indian Himalayas. Landslides, 7(2), 191–201.

Petley, D. N., Hearn, G. J., Hart, A., Rosser, N. J., Dunning, S. A., Oven, K., Mitchell, W. A., 2007. Trends in landslide occurrence in Nepal. Natural hazards, 43(1), 23–44.

Pradhan, B., Singh, R., Buchroithner, M., 2006. Estimation of stress and its use in evaluation of landslide prone regions using remote sensing data. Advances in Space Research, 37(4), 698–709.

Pradhan, S., Panda, S., Roul, A., Thakur, M., 2019. Insights into the recent Kotropi landslide of August 2017, India: a geological investigation and slope stability analysis. Landslides, 16(8), 1529–1537.

Rahardjo, H., Li, X. W., Toll, D. G., Leong, E. C., 2001. The effect of antecedent rainfall on slope stability. Geotechnical and Geological Engineering, 19(3-4), 371–399.

Ray, P. K., Dimri, S., Lakhera, R. C., Sati, S., 2007. Fuzzy-based method for landslide hazard assessment in active seismic zone of Himalaya. Landslides, 4(2), 101–111.

Roy, P., Martha, T. R., Jain, N., Kumar, V. K., 2018. Reactivation of minor scars to major landslides – a satellite-based analysis of Kotropi landslide (13 August 2017) in Himachal Pradesh, India. Current science, 115(3).

Sarkar, S., Kanungo, D. P., Sharma, S., 2015. Landslide hazard assessment in the upper Alaknanda valley of Indian Himalayas. Geomatics, Natural Hazards and Risk, 6(4), 308–325. http://dx.doi.org/10.1080/19475705.2013.847501.

Sharma, P., Rawat, S., Gupta, A. K., 2018. Study and Remedy of Kotropi Landslide in Himachal Pradesh, India. Indian Geotechnical Journal, 1–17. https://doi.org/10.1007/s40098-018-0343-1.

Thakur, V., Jayangondaperumal, R., Joevivek, V., 2019. Seismotectonics of central and NW Himalaya: plate boundary–wedge thrust earthquakes in thin-and thick-skinned tectonic framework. Geological Society, London, Special Publications, 481(1), 41–63.

Tiwari, B., Ajmera, B., 2017. Landslides triggered by earthquakes from 1920 to 2015. Workshop on World Landslide Forum, Springer, 5–15.

Tohari, A., 2018. Study of rainfall-induced landslide: a review. IOP Conference Series: Earth and Environmental Science, 118, IOP Publishing.

NO-COST LANDSAT DATA REDEFINES STUDENT RESEARCH PROJECTS IN APPLIED REMOTE SENSING CLASSES AT UW

Ramesh Sivanpillai

Wyoming GIS Center, University of Wyoming, Laramie, WY 82071, USA - sivan@uwyo.edu

KEY WORDS: Undergraduate Curriculum, Workforce Development, Research Projects, WyomingView

ABSTRACT:

Students enrolled in an applied remote sensing class at University of Wyoming are required to complete a term research project. They have to apply the concepts learned throughout the semester to address a real-world problem in a natural resource management topic that uses remotely sensed imagery data. Prior to the availability of no-cost Landsat data, students in this class had to restrict the scope of their projects to images available in the smaller archives maintained by academic and research institutions. Since the US Geological Survey (USGS) made the entire Landsat archive available at no-cost, opportunities have increased for students to use those data in class projects. Now, students can download as many Landsat images as is needed to address the scope of their project. Using examples from previous years, this paper describes how no-cost Landsat imagery has allowed students enrolled in an applied remote sensing class to add depth and breadth to their projects.

1. INTRODUCTION

Research projects that are a part of most college remote sensing courses require students to apply the concepts they have learned in lectures and labs to address real world problems. If these projects are conducted in areas that are familiar to the students, they will be able to better relate the image information to conditions on the ground and enhance their understanding (Sivanpillai and Driese, 2008). In order to complete such projects, students need to have access to remotely sensed data acquired at the appropriate time, i.e., year, season, or month.

Historically, the cost associated with obtaining imagery for classroom projects was the primary limiting factor. In many US institutions, instructors would provide sample images obtained from data or software vendors. At certain institutions faculty used imagery purchased through other research grants in their classrooms. These images were acquired for some random location in the world i.e., a lake and its adjoining area in the south-eastern US or a random landscape in Africa. Students enrolled in remote sensing courses in different years or semesters would process the same set of images. In research active institutions, faculty would hire undergraduate students to complete one or more research tasks, and those students would gain additional research experience. However, other students, especially those enrolled in teaching institutions were not able to gain this research experience during their academic careers.

The Landsat satellite program (http://landsat.usgs.gov) has been collecting Earth Observation (EO) data since 1972. Imagery data acquired by Landsat satellites 1 – 5, 7 and 8 constitute the longest and continuous collection of moderate resolution EO data (Loomis et al. 2015). The spatial and spectral resolutions of Thematic Mapper (TM), Enhanced Thematic Mapper (ETM+), and Operational Land Imaginer (OLI) sensors on board Landsat 5, 7 and 8 are well suited for monitoring changes caused by natural and anthropogenic factors on the surface of the Earth. The cost associated with purchasing a Landsat image or scene (~180 x 180 km) was prohibitive for classroom teaching and term project research activities.

Under those conditions, students had to define the scope of their term projects to match the characteristics of the imagery data they could access. For example, a student interested in studying the effect of drought on vegetation growth might not have access to images acquired during those years. Similarly, another student interested in mapping land cover changes in a region, might not have access to images acquired before and after certain policies were implemented. Hence the scope of the projects had to be modified to match the already available data.

The AmericaView (AV) program established in the late 1990s allowed users to access free Landsat imagery from the archives of its member states (Landenberger et al. 2011). Also, users could also purchase Landsat 5 and 7 images or scenes for US$540 and US$600 respectively that were no in the archive. However this program was focused primarily on users in the United States. The Global Land Cover Facility (GLCF) operated by University of Maryland (USA) since 1997 also provided several Landsat images and other remotely sensed data acquired for different parts of the world (Lindsay et al., 1999). However, the GLCF archive did not include all Landsat scenes and all the newly acquired images were not uploaded to it. Nevertheless, remotely sensed data contained in the GLCF archive gave students and instructors and opportunity to obtain data from different areas of the world.

In 2008, the USGS opened the entire Landsat collection to users worldwide (Zhu, 2017). A user can download any number or volume of Landsat data for any location in the world (Wulder et al. 2012). This created an unprecedented opportunity for teachers and students enrolled in remote sensing courses. Teachers and students can download images acquired in certain years, seasons or months. As long as they are not covered by clouds and cloud shadows, these images could be used in teaching and term research projects. The 30 meter spatial resolution of Landsat 5, 7 and 8 data also might not be suitable for mapping land cover or tracking changes in highly fragmented landscapes. Nevertheless, studies have demonstrated the utility of these data for mapping urban growth and land conversion throughout the world. Studies have used the spectral information collected by Landsat satellites 5, 7 and 8 to derive secondary information, i.e. spectral indices, which are valuable to monitoring or tracking changes in crop growth (Rufin et al., 2019), forest (Hansen et al., 2013; Hughes et al., 2017), and water quantity and quality (Arvor et al., 2018).

This paper describes how no-cost Landsat data has transformed undergraduate student research projects in an applied remote sensing course at the University of Wyoming. Section 2 provides background information about this course, Section 3 describes the changes in the projects completed by these students with Landsat data, and Section 4 includes suggestions and recommendations for the future.

2. COURSE BACKGROUND

The University of Wyoming (UW) offers an applied remote sensing course every fall semester. This course is intended for upper level (3rd and 4th year) undergraduate students majoring in Rangeland Ecology & Watershed Management, Agroecology, and Plant Sciences. This course has been continuously taught since 2005 in a lecture and laboratory format and has attracted students from other degree programs such as Geography.

UW junior (3rd year) and senior (4th year) students from the College of Agriculture & Natural Resources as well as the College of Arts & Sciences have registered for this class. The number of students in this class has varied between 14 and 38 (average of 27). Majority of the students who have completed this course are from small, rural communities in Wyoming. Often students were from families with a farming or ranching background or have gained experience from working on farms or ranches. Some students have also worked for federal (i.e., national) and state agencies to fight wildfires or treat weed infested areas. The remaining students came from adjacent (Colorado, Montana, Idaho, and South Dakota) or more distant (Michigan, Wisconsin, Maryland, and Texas) states, with some or no background in farming and ranching.

3. STUDENT RESEARCH PROJECTS

As part of the course requirement, students are required to complete a term research project that takes approximately six weeks to complete. Projects have to focus on the applications of remotely sensed data to address natural resource or management issues. Often students select study areas that are familiar to them, such as family property, work sites, or public lands such as forests, and lakes and reservoirs.

3.1 Pre No-cost Landsat Era

Students had access to 400+ Landsat images and 1,800 ASTER images that were part of the WyomingView's data archive (Sivanpillai, 2008). The Landsat images were mostly concentrated in certain parts of the state with intermittent temporal coverage. This prevented students from undertaking projects that required images that were unavailable for specific years, seasons, or months. For example, a student might have found two or more scenes from the spring season within the data archive, for their area of interest but not those that were acquired in late summer or early autumn. Often students would find images that only partially covered their study area in a scene. The adjacent scene would also not be present in the state archive. Such limitations forced students to modify their original research objective to match the existing inventory in WyomingView's data archive.

Despite these limitations, several students found the images they needed and used them in their project (Sivanpillai, 2008). Most of these projects focused on mapping land cover or conditions on the ground using one to three Landsat images. Students utilized changes in the spectral reflectance values of healthy and stressed vegetation, clear and turbid water, vegetation in burnt and unburned areas, etc. and generated land cover maps. However, very few studies focused on using four or more Landsat images because of the temporal gaps or presence of clouds, snow and/or ice.

3.2 No-cost Landsat Era I (2009 – 2017)

Since the fall semester of 2011, students have been downloading as many Landsat images as they need for their projects. Images are useful as long as clouds and shadows are not over their study areas or the location is not covered by ice or snow. The number of projects that focused on tracking intra- and inter-annual changes increased many fold, in comparison to change analyses using two or at most 3 images in the Pre No-cost Landsat era. Almost half of the projects conducted by the students in this remote sensing course now use four or more Landsat images (Figure 1).

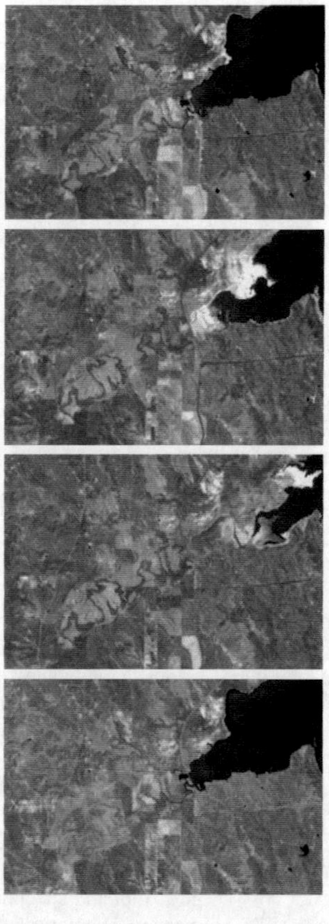

Figure 1. Landsat images acquired in 1999 (top), 2002, 2005, and 2011 (bottom) show vegetation growth or response in dry (2002 and 2005) and wet years (1999 and 2011) by Keyhole Reservoir near Moorcroft, Wyoming.

Landsat images are used for monitoring intra-annual changes in vegetation growth in Wyoming's rangelands, which are influenced by the amount of precipitation and other conditions. Landsat images acquired every 16 days captured changes at the start of the 2007 growing season (Figure 2).

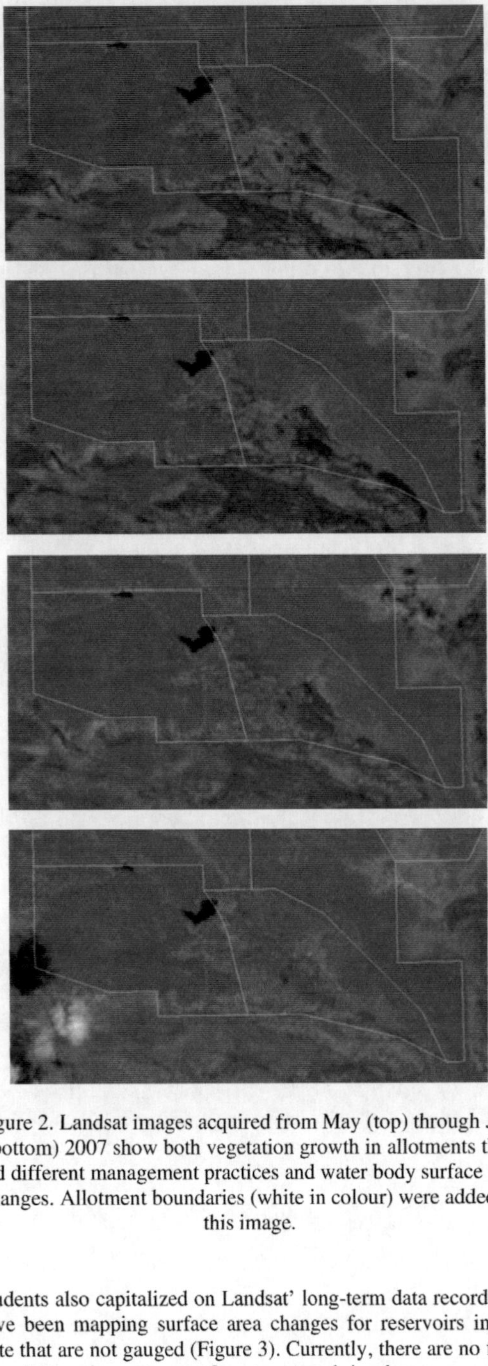

Figure 2. Landsat images acquired from May (top) through July (bottom) 2007 show both vegetation growth in allotments that had different management practices and water body surface area changes. Allotment boundaries (white in colour) were added to this image.

Students also capitalized on Landsat' long-term data record and have been mapping surface area changes for reservoirs in the state that are not gauged (Figure 3). Currently, there are no field records on the amount of water stored in these reservoirs. Students processed images acquired from 1984 and extracted their surface areas. Plotting these values and comparing them to precipitation data from nearby weather stations, students identified potential drivers of those changes.

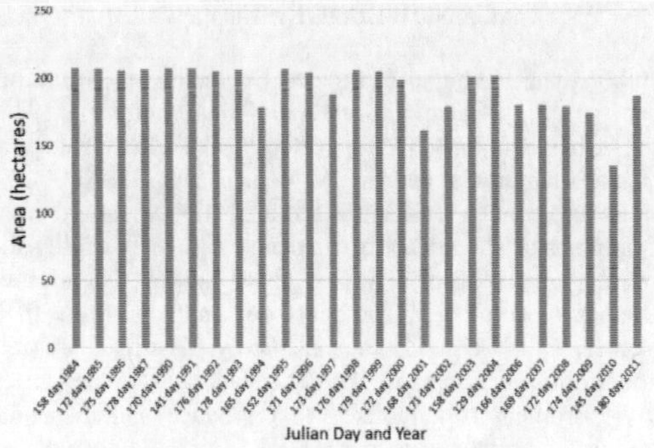

Figure 3. Surface area estimates for Ray Lake derived from Landsat 5 TM images acquired from 1984 through 2011 (Colleen Friday; Fall 2015).

However, prior to using multiple Landsat images in a single project, students have to check the associated metadata files for determining the positional accuracy of the image and make any necessary adjustments. This step requires spending extra time prior to extracting the information from the Landsat images.

3.3 No-cost Landsat Era II (2017 – Present)

In 2017 the USGS released the Landsat images in a Collection and Tier Level system/structure after reprocessing the entire archive (Micijevic et al., 2016). Images meeting the highest quality standards were placed in the first tier, and those that failed to meet those standards were placed in the second tier. Landsat images placed in the first tier are also suitable for pixel-by-pixel level change analyses.

Students, and any users, can now quickly identify and download high quality Landsat scenes for their area of interest. The Collection Level product also eliminates the time needed to align the images prior to information extraction and allows instructors to bring the classroom discussion to a higher level. Students can now focus on any of the image processing steps they propose to use in their project.

In early 2020, the USGS plans to reprocess and release Collection Level 2. Users can download Landsat and the European Space Agency's Sentinel 2 data and increase the temporal resolution in their projects. The new collection will create additional opportunities for monitoring more sites in Wyoming that experience relatively more cloud cover.

4. CONCLUSION

Availability of no-cost Landsat data has crreated unprecedented opportunities for students to complete their term research projects. Students and instructors can access Tier 1 (highest quality) imagery data dating back to 1972 for conducting change analyses without having to realign the images. By working with images acquired for an area or region that is familiar to them, students develop a better and deeper understanding of remote sensing concepts introduced in lectures and laboratory exercises. Finally, products generated by students can provide valuable feedback to the USGS and other space agencies that are developing the next generation of remote sensing sensors.

ACKNOWLEDGEMENTS

The author thanks Abigail Gettinger, University of Wyoming, for proof reading the earlier versions of this manuscript. This material is based upon work supported by the U.S. Geological Survey under Grant/Cooperative Agreement No. G18AP00077. The views and conclusions contained in this document are those of the author and should not be interpreted as representing the opinions or policies of the U.S. Geological Survey. Mention of trade names or commercial products does not constitute their endorsement by the U.S. Geological Survey. This manuscript is submitted for publication with the understanding that the United States Government is authorized to reproduce and distribute reprints for Governmental purposes.

REFERENCES

Arvor, D., Daher, F.R.G., Braind, D., Dufour, S., Rollet, A-J., Simones, M., Ferraz, R.P.D., 2018. Monitoring Thirty Years of Small Water Reservoirs Proliferation in the Southern Brazilian Amazon with Landsat time series. *ISPRS Journal of Photogrammetry and Remote Sensing*, 145: 225-237.

Hansen M.C., Potapov, P.V., Moore, R., Hancher, M., Turubanova, S.A, Tyukavina,, A, Thau, D., Stehman, S.V., Goetz, S.J., Loveland, T.R., Kommareddy, A., Egorov, A., Chini, L., Justice, C.O., Townshend, J.R.G., 2013. High-Resolution Global Maps of 21st-Century Forest Cover Change, *Science*, 342(6160):850–853.

Hughes, M.J., Kaylor, S.D., Hayes, D.J., 2017. Patch-Based Forest Change Detection from Landsat Time Series. *Forests* 8(5), 166. DOI: 10.3390/f8050166.

Landenberger, R., Patterson, M., Sivanpillai, R., Lawrence, R, 2011. AmericaView: Promoting Remote Sensing Science and Technology through Cooperative Partnerships. *Photogrammetric Engineering and Remote Sensing* 77(10), 969-974.

Lindsay, F. E., Townshend, J.R.G., Jaja, J., Humphries, J., Plaisant, C, Shneiderman, B., 1999. Developing the Next Generation of Earth Science Data Systems: the Global Land Cover Facility. *IEEE 1999 International Geoscience and Remote Sensing Symposium*. IGARSS'99 (Cat. No.99CH36293). DOI: 10.1109/IGARSS.1999.773583.

Loomis, J., Koontz, S., Miller, H., Richardson, L., 2015. Valuing geospatial information: Using the contingent valuation method to estimate the economic benefits of Landsat satellite imagery, *Photogrammetric Engineering & Remote Sensing*, 81(8):647–656.

Micijevic, E., Haque, M.O., Mishra, N., 2016. Radiometric Calibration Updates to the Landsat collection. Volume 9972, Earth Observing Systems XXI; 99720D (2016). DOI: 10.1117/12.2239426.

Rufin, P., Frantz, D., Ernst, S., Rabe, A., Griffithis, P., Ozdogan, M., Hostert, P., 2019. Mapping Cropping Practices on a National Scale Using Intra-Annual Landsat Time Series Binning. *Remote Sensing*, 11(3), 232. DOI: 10.3390/rs11030232.

Sivanpillai, R., 2008. Remote Sensing Education through Interactive Learning. *Proceeding of the Pecora 17 – The Future of Land Imaging ... Going Operational*, Denver CO, USA.

Sivanpillai, R., Driese, K.L., 2008. Promoting Remote Sensing Education and Research through WyomingView data resources. *Journal of Geography* 107(4-5): 154-160.

Wulder, M.A., J.G. Masek, W.B. Cohen, T.R., Loveland, and C.E. Woodcock, 2012. Opening the Archive: How Dree Data has Enabled the Science and Monitoring Promise of Landsat, *Remote Sensing of Environment*, 122: 02–10.

Zhu, Z., 2017. Change Detection Using Landsat Time Series: A Review of Frequencies, Preprocessing, Algorithms, and Applications. *International Journal of Photogrammetry and Remote Sensing,* 130: 370-384.

INTEROPERABILITY OF VARIOUS DATA STREAMS WITHIN GUYANA's MRV SYSTEM

Pete Watt[1] , Pradeepa Bholanath[2], Nasheta Dewnath[2] ,Towana Smartt[2] , Chaplin Chan[1], Daniel Donoghue[3]

[1] Indufor Asia Pacific Ltd, Auckland City 1147, New Zealand - pete.watt@indufor-ap.com, chaplin.chan@indufor-ap.com
[2] Guyana Forestry Commission, 1 Water Street, Georgetown, Guyana - project.coordinator@forestry.gov.gy, ndewnath@gmail.com, towanasmartt@yahoo.com
[3] Durham University, Stockton Road, Durham, DH1 3LE, UK- danny.donoghue@durham.ac.uk

KEY WORDS: Guyana, MRVS, Satellite Imagery, Forest Change Monitoring , REDD+, Data Interoperability, Data Agnostic Systems, Earth Observation

ABSTRACT:

In 2010 Guyana started work on the development of a national Monitoring Reporting and Verification System (MRVs) to quantify and measure the changes in the country's forest cover carbon and carbon emissions. A necessary part of this process involved the identification of reliable Earth Observation data of sufficient resolution to detect and quantify land use change in Guyana's forests. Over the past 10 years the MRVs has evaluated and integrated many data streams used to; map national-scale forest change, support the analysis, and importantly datasets suitable to determine the accuracy of the change area mapped. Guyana's approach has evolved over time, to accommodate new technologies, but at its core the MRVs recognises the importance of local management, existing datasets and linking these elements to appropriate EO data such as Landsat, RapidEye, Sentinel, Planet Scope and very high spatial resolution aerial imagery. From the outset the MRVs development was divided into phases. This approach recognises that not all MRVs reporting functions can be satisfied immediately. For Phase 1 (Years 2010 to 2014) of the MRVS, historical change analysis was conducted using Landsat 30 m resolution imagery. Being a persistently cloudy country alternative EO data sources were included, with Landsat and DMC imagery largely superseded by 5 m resolution RapidEye imagery. After five years of monitoring the forest change baselines, methods, reporting processes and standard operating procedures had been well established and able to provide the required performance-based indicators. The focus of MRVs phase 2 (Years 2015 to 2019) was to retain the reporting standards already achieved while also streamlining processes, improving functionality and reducing operational costs (i.e. the reliance on commercial image data) post-2019. Process improvements and operational research targeted two areas; the feasibility of using freely available Landsat and 10 metre resolution Sentinel data to map countrywide deforestation, and development of a sample-based approach to estimate degradation from aerial imagery and Planet Scope (3-5 metres resolution). Guyana's move to integrate multiple data streams has been driven by the need for higher temporal resolution, repeated monitoring, and the creation of a data agnostic system to supports multiple decision-making processes on forest management. While the originally the MRVs was intended to support REDD+ reporting, the flexibility of the system design has meant it is proving to be invaluable tool for natural resources management in Guyana.

1. INTRODUCTION

Guyana is a tropical country located on the Caribbean coast of South America between 2° to 8° N and 57° to 61° W. It shares common borders with Venezuela, Brazil and Suriname. The country has a land area of 21.1 million ha with a forest area of 18.5 million ha (87%). The Government of Guyana aims to protect and maintain its forests by attracting investment to foster growth and development that follows a low carbon emissions pathway. The Conference of the Parties (e.g. COP16 and COP21) of the United Nations Framework Convention on Climate Change (UNFCCC) requests countries to report reference greenhouse gas emission levels and/or forest change reference levels using certified, consistent and verifiable methodologies and common metrics so these can be assessed by the Intergovernmental Panel on Climate Change (IPCC). This involves individual countries developing monitoring, reporting and verification (MRV) systems able to generate detailed annual reports on national land use change (activity data) and related CO_2 emission factors (EF) in an IPCC format for audit by the UNFCCC Reduced Emission from Deforestation and Forest Degradation (UN-REDD) programme. The reporting requirements include initiatives on conservation, sustainable management of forests and enhancement of forest stocks (REDD+). The audit function presents considerable challenges for many developing countries that lack national forest inventories, access to frequently updated national land cover maps, technical expertise and resources.

Prior to 2009 Guyana reported approximate forest change values, collated and published via the FAO global Forest Resources Assessments (FRA). This value can be described as a best estimate of national forest change. In 2009 the Governments of Guyana and Norway agreed to cooperate on broader emission reduction goals under the umbrella of UNFCCC-REDD+. The objective to develop an MRV system (MRVs) allowing for a comprehensive, consistent, transparent and verifiable assessment of forest area change (GOFC-GOLD, 2016; GFOI, 2016). While committing up to US$250 million over the period of the Agreement (5 years), the level of support was to be determined by Guyana's delivery of results against two sets of indicators:

– *Indicators of enabling activities*: a set of policies and safeguards designed to ensure REDD+ contributes to the achievement of the goals set out in the Agreement. These indicators spoke to arrangements to ensure systematic and transparent multi-stakeholder consultations throughout the process; protection of the rights of indigenous peoples; ensuring environmental integrity and biodiversity protection; ensuring continuous improvements in forest governance; and providing transparent, accountable oversight and governance of the financial support received.

- *REDD-plus Performance Indicators*: a set of forest-based greenhouse gas emissions-related indicators. It was agreed that these would gradually be substituted as the MRVs became fully operational. They were developed based on conservative estimates while encouraging the development of a more accurate MRVs over time through building national capacities. These REDD+ Performance Indicators sought to contribute to the development of the national MRVs, based on internationally accepted methodologies and following the IPCC reporting principles of completeness, consistency, transparency, uncertainty, comparability, and encouragement of independent international review of results.

The agreement divided Guyana's MRVs programme into two stages: phase 1 running from 2010 to 2014 was designed to establish several interim measures, while phase 2 running from 2015 to 2020 was planned to consolidate and expand capacities for national REDD+ monitoring, thereby underpinning results-based REDD+ compensation in the long-term. To date, the GFC has completed eight national assessments. The results of each have been independently verified by third party verifiers.

With one year remaining to the end of Norway's funding commitment this paper intends to showcase Guyana's MRVS and the associated satellite imagery and GIS datasets that were used for the national assessments. It also provides a summary of Guyana's progress towards implementation of REDD+ and how interoperability of various data streams has helped to build and extend Guyana's MRVS to become a versatile platform that supports natural resource decision-making.

2. MRV SYSTEM DESIGN

Guyana's MRVS design and development started during a period when no comparable national-scale systems existed. While definitions, guidance material and good practice guidelines had been drafted (GOFC-GOLD, 2008), a template for fulfilling the reporting requirements under REDD+ was not provided. As early adopters this presented the opportunity to mould the design to expand beyond the core REDD+ monitoring and reporting functions.

With a blank canvas Guyana set about identifying key system requirements to enable annual reporting of deforestation and degradation. The Guyana Forestry Commission (GFC) was the responsible agency tasked with the monitoring and reporting function. Indufor Asia Pacific (IAP) was engaged to support ongoing system development.

Development began in earnest in 2009. Challenges that needed to be addressed were identified at the early stages of the system design: Guyana's forest definition, annual reporting timelines, high levels of cloud cover, and lack of familiarity of the GFC GIS staff with remote sensing software. A need for the system to be data agnostic and capable of incorporating multiple space-borne datasets became evident from the outset. As pointed out by Lehmann et al. (2015), multi-sensor frameworks will continue to generate significant interest for international forest monitoring initiatives because of the increasing availability of spaceborne data.

2.1 Guyana's Forest Definition and Establishing Benchmarks

The definition of what constitutes a 'forest' is the corner stone of any MRVs as it defines forest land and the monitoring extent. Guyana adopted the definition as outlined in the Marrakech Accords (UNFCCC, 2001); land is classified as forest if there is a minimum 30% tree canopy cover at a minimum tree height of 5 m covering a minimum area of 1 ha.

In accordance with the agreement with the Norwegian government the national forest cover as at 1990 was used as a starting point. From this baseline forest change is documented to determine the rate of forest change. Under the agreement this information is used to inform the progress of several Interim measures relative to benchmark values – set as at September 2009. Within this framework the steps and core inputs required to establish the MRVs are as follows:

- Determining the forest area as at 1990 using medium resolution satellite images (Landsat) by excluding non-forest areas (including existing infrastructure)
- Accounting for forest to non-forest land use changes between 1990 and 2009 using a temporal series of satellite data.
- Establishing a benchmark period (1990-2009) and a 30 September 2009 'benchmark map' as a reference point.
- Comparing annual change post-2009 against the 2009 benchmark.

In keeping with international best practice, the developed process uses a wall-to-wall approach that enables complete, consistent, and transparent monitoring of land use and land use changes over time. The approach allows for land cover change greater than one hectare in size to be tracked through time and attributed by its driver such as mining or shifting agriculture. This is in accordance with Approach 3 in the *Good Practice Guidance for Land Use, Land-Use Change and Forestry* IPCC report (Penman et al., 2003).

2.2 Capacity Building

The most important aspect of the Guyana/Norway agreement was the requirement to build national capacity to enable the reporting of annual national forest change. The design of the MRVs sought to capitalise on the professional skills and knowledge of the GFC's staff in the use of their GIS software which is used extensively in support of GFC's management functions. Over the years of development of Guyana's MRVs, significant capacity has been built within the local organisational structure of the Guyana Forestry Commission where the system is housed. The system started at a point with predominantly external consultants' support and has developed to a point where only new development features involve the inputs of consultants. GFC's proactive management and ownership of the MRVs reporting requirements have smoothed this transition, underpinned by the retention of core staff and the development of Standard Operating Procedures (SOP).

2.3 System Design

An overview of the processes, datasets and outputs of the MRVs is given in Figure 1. Access to the monitoring products of the MRVs by the National Agencies responsible for the management of the land resources is planned.

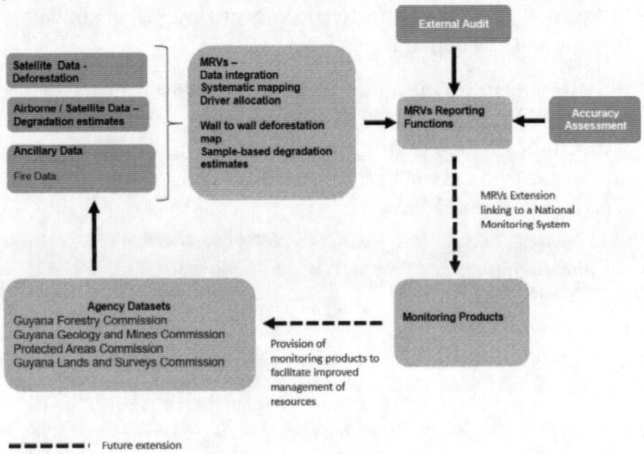

Figure 1. Overview of Guyana's MRVs

2.4 Software

The MRVs design focused on presenting the user with an intuitive visual GIS front-end using ArcGIS software thereby utilizing and building on the GFC staff familiarity with existing data input tools, data queries, spatial analyses, map manipulation and reporting. Customised extensions for systematic deforestation mapping and driver allocation were designed and implemented. Further improvements to the MRVs made post 2015 in line with the second phase of the MRV included implementation of a sample-based design to provide estimates of degradation.

Outside of the GIS framework, Guyana-specific image processing routines were developed using ENVI image processing software to normalise imagery, automate the extraction of deforestation boundaries, and identify areas of persistent cloud cover. At a later phase of the MRVs development the image processing routines were moved from commercial software to a cloud-based system supported by Google to take advantage of the online image datasets and the computation speed offered by the cloud. This strategy sits within the MRV Roadmap's mandate to continue to develop the MRVs along a low-cost pathway, where both image and software costs are minimised.

2.5 Data Structure

The spatial and attribute and related data are stored in a consolidated geodatabase backend for which data models and storage protocols were designed and implemented. The database holds all relevant spatial data, metadata about the dataset, its location on the network and anticipated update frequency.

Satellite image datasets downloaded from the image providers are stored as raster datasets and catalogued based on the analysis period. Metadata information includes the sensor, path and row (Tile ID), and the processing routines that have been applied to the images.

All results are subject to an external third-party annual audit which is intended to review the analysis processes and consistency and accuracy of the results. In line with the guiding principles for an MRV GFC's system strives to support the auditing function by documenting and archiving all information in the geodatabase.

2.6 Management Level GIS Datasets

Several Government agencies that are involved in the management and allocation of land resources in Guyana, hold spatial datasets. Since 2010 GFC has coordinated the storage of these datasets for the MRVS. These agencies fall under the responsibility of the Ministry of Natural Resources and the Ministry of the Presidency in Guyana and are responsible for forestry, mining, and land use planning and coordination. The main management data layers are summarised below, in Table 1 and are mainly used to provide context/background information that supports the annual mapping exercise.

	Agency	Role	Data Held
Ministry Natural Resources	Guyana Forestry Commission (GFC)	Management of forest resources	Resource management related datasets
	Guyana Geology and Mines Commission (GGMC)	Management of mining and mineral resources	Mining concessions, active mining areas
Department of Environment	Protected Areas Commission	Management of Protected Areas System in Guyana	Spatial representations of all protected areas
Ministry of the Presidency	Guyana Lands and Surveys Commission (GL&SC)	Management of land titling and surveying of land	Land tenure, settlement extents and country boundary

Table 1. Agency datasets provided

3. INTEGRATION of MULTIPLE SATELLITE DATA STREAMS

The datasets used for the change analysis have evolved over time – moving progressively from establishment of baselines required to monitor deforestation to the inclusion of datasets and development of methods required to estimate and report degradation.

Mapping of the 1990 forest extent was undertaken using Landsat imagery supported by 1991 JERS-1 radar data and historical 1960-era aerial photography. The latter datasets were used to verify the location of the forest/non-forest fringe and vegetation composition. Change analysis from 1990 to 2009 was conducted primarily with Landsat imagery, supported as required with IRS and CBERS images. As pointed out by Schingler (2015), the establishment of robust reference points would be challenging without science-grade legacy datasets such as Landsat.

After 2009 Guyana's MRVs moved to annual reporting of deforestation. Under the Norway / Guyana agreement the schedule of performance-based payments was referenced against the 2009 benchmark map that provided a snapshot of forest change as at 30 September 2009. The agreement imposed the constraint of including images acquired between August and December of each year. Given the persistent cloud cover and the temporal frequency of Landsat 5 (16 days) and image quality issues affecting Landsat 7, commercial DMC imagery was tasked. This ensured the acquisition of enough cloud-free images required to map and report deforestation by forest change driver.

From 2011 onwards these datasets were primarily superseded with high-resolution images from RapidEye. With a constellation of five satellites RapidEye allowed higher temporal and spatial resolution (5 m). The advantages were twofold, reducing risk of cloud obscuring change and improving the ability to assess degradation. Improvements post 2015 included a shift to the use of free Sentinel 2 imagery and testing and use of 3 m PlanetScope data.

4. NATIONAL FOREST MONITORING - PROGRESSIVE IMPROVEMENTS

The phased approach of the MRVs allowed for the progressive development of the system. Targeted improvements were included to keep pace with evolving reporting requirements and Guyana's continued desire to better understand the forest change drivers and their overall contribution to carbon emissions. Early on mining was identified as one of the main drivers of deforestation and degradation, contributing to around 90% of the forest loss in any one year.

Figure 2. Example of a typical ribbon pattern of alluvial mining practice.

The legacy of Guyana's MRVs offers a unique insight with nearly 30 years of continuous national-scale monitoring. A defining feature is the accuracy of the monitoring dataset which has routinely met and exceeded annual audit requirements. It has consequently been used as a reference point in several recent scientific papers, e.g. Bovolo & Donoghue (2017), Pickering *et al.*, (2019), Galiatsatos *et al.*, (submitted).

Data interoperability was an important consideration in the design and development of the MRVs given the multiple types of data being analysed, the need for data sharing, and updates to the system. This recognises that to improve accuracy, detect small scale changes, increase monitoring frequency and cost- effectiveness a process of continual review and adaptation is required.

Figure 3 identifies the timing of various improvements made to the MRVs, including the shift to annual reporting, national estimation of degradation, improvements in the accuracy assessment processes, and lastly deployment of a low-cost Guyana-managed MRVs.

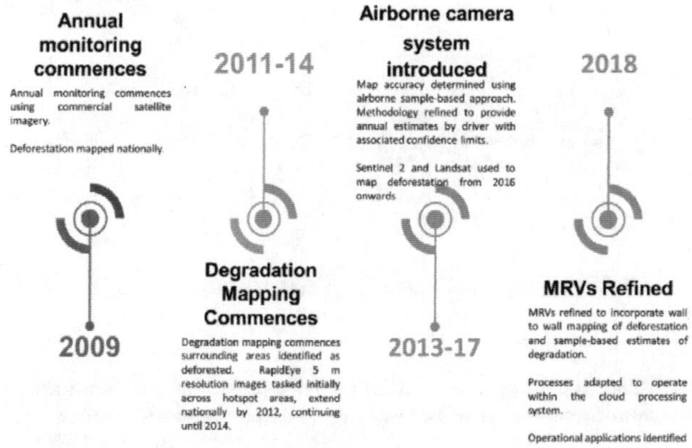

Figure 3. MRVs progressive improvement timeline

Some developments were intuitive, for example the move in 2009 to commercial imagery. This was prompted by the constraints placed on permissible image collection dates (August to December) and the risk of not attaining sufficient cloud-free imagery to report on the results-based indicators - a point amplified by the 2011 failure of Landsat 5 after 25+ years of imaging. Commercial DMC satellite imagery was introduced in 2010, and while suitable for deforestation events (>1 ha) it was not a viable alternative to Landsat, nor a solution suitable for reporting on forest degradation impacts.

Reporting degradation impact is an area of continual review and study within the remote sensing community (e.g. Herold et al., 2011, FAO, 2011, Mitchell et al., 2017). Under the Norwegian agreement Guyana was required to identify and map degradation surrounding deforestation events. While a country-specific criterion was used the underlying IPCC definition of degradation still applied (IPCC, 2016). As a first step RapidEye imagery was tasked over 54% of Guyana to capture the main areas of change. The improved resolution enabled better identification of change boundaries, drivers of change and areas of forest degradation. RapidEye continued to be used between 2011 to 2014 with annual coverages routinely captured between August and December each year. This was complimented with some processing of spaceborne SAR (ALOS Palsar) data. The resulting map products included national maps of deforestation and degradation.

While these outputs met the reporting requirements several challenges were identified, namely having enough cloud-free coverage with adequate spatial and temporal resolution to accurately capture degradation extent. Outside the technical domain the financial cost of RapidEye and increasing requirement to provide degradation estimates for drivers such as shifting cultivation, infrastructure and forest harvest were also relevant.

Since inception the quality of annual forest change estimates had been independently evaluated using recognised statistical methods (Herold *et al*., 2006; Powell *et al.*, 2004; Khorram, 1999). To mitigate the cloud risk, high-resolution imagery was captured using a Cessna-mounted aerial multispectral imaging system post 2013. The strategy employed uses the imaging system to capture high-quality image data at sites pre-determined by a two-stage stratified-random sample design that provides good coverage of the strata with high and medium risk of change.

The camera system (Aeroptic/ aka GeoVantage) is transferable to various models of a light aircraft able to capture data between 25 to 60 cm (varied by the altitude of the aircraft at the time of capture) - a resolution capable of identifying forest degradation with some certainty. For reference Figure 4. shows small-scale clearance of forest between two years.

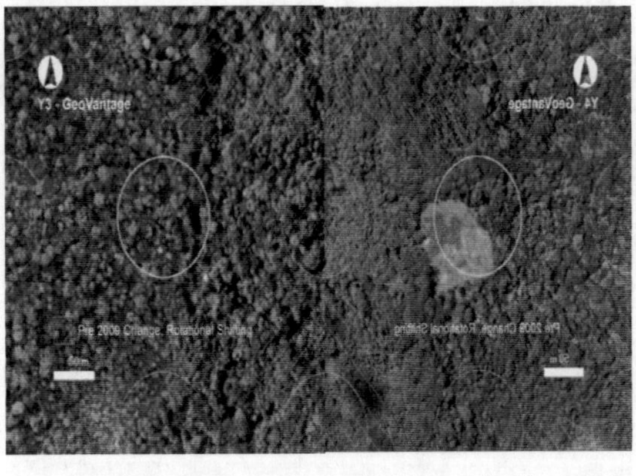

Figure 4. Aerial survey comparative time series data years 2012 (left) and 2013 (right)

Full sample coverage was achieved by including satellite imagery such as PlanetScope over stratum with a low risk of forest change and over areas where it is not possible to safely operate a small aircraft.

While the process was originally intended to provide area estimates of the deforestation and degradation as a comparison against the MRVs mapping it became evident that the sampling approach is very efficient, prompting the decision to shift from mapping degradation to obtaining a national estimate using the sampling approach.

The routine use of the MRVs data is now being integrated within Guyana's ministries, and specifically mandated uses are being explored for areas of resource management. The main constraint identified is data lag associated with annual reports, in that the reporting conducted by the GFC is retrospective, rather than providing frequently updated data in a near real time monitoring system. The GFC is currently advancing a process of developing a Continuous Resources Monitoring System building on the platform of the MRVS.

5. DISCUSSION

Guyana's experience has shown that there are shortcomings in the use of satellite imagery for national assessment of forest cover. Some of the factors that affect the practicality of using certain datasets include cost, revisit frequency, resolution, need for historic coverage, cloud cover and desire for continuous and real-time monitoring versus one-off monitoring. Guyana has attempted to strike a balance by evaluating various options, striving to achieve the most feasible and practical outcome.

From 1990 to 2009, there were limited land cover datasets at 20-30 m resolution. Landsat uniquely fulfils this role with the program referenced as a 'gold standard' by which other sensors have been compared due to its historical legacy, availability of data, global imaging, and superior calibration (Goward et al., 2017). Thus, it serves as the central reference point to which most moderate and high-resolution optical satellite systems are compared against even to the present day (Wulder et al., 2019). For Guyana Landsat was the only option that offered historic 1990 to 2009 coverage. However insufficient data prevented the production of annual forest change maps resulting in a decision to create forest change datasets covering three epochs: 1990-2000, 2000-2005, and 2006-2009.

Guyana's first year of annual assessment was 2010 for which annual deforestation estimates were the priority. Deforestation under the applied forest definition, focused on reporting all forest change greater than 1 hectare in size. Medium resolution sources like Landsat continued to be a practical option, however the risk of incomplete coverage meant alternative satellites such as DMC were considered. DMC was discontinued due to coverage and quality issues. Between 2011 and 2014 RapidEye progressively became the sensor of choice, out of recognition that smaller scale change events, such as new infrastructure and incremental change in mining footprints (greater than 1 hectare in size) prevailed over larger scale ones. Analysis showed that while Landsat was efficient for deforestation report it was not appropriate for forest degradation (at 0.5-1 hectare scale) reporting.

As Guyana embarked on the second phase of its MRV System (2015 to 2019), lower cost satellite options were explored in common with many other countries. Certainly, recent trends point to ever increasing availability of satellites designed to monitor land cover at greater spatial and temporal resolution, leading to greater visibility and monitoring of tropical deforestation (Reiche et al., 2018).

For deforestation mapping a combination of ESA Sentinel 2 and Landsat imagery provides a sound alternative to RapidEye, with close enough resolution and sufficient temporal coverage to generate an annual snapshot of forest change. This is the intended approach for the MRVs for 2019-2020. Sentinel's five-day revisit rate will be leveraged to develop a more real time, continuous monitoring system hosted within the MRVs.

A measure of the MRVs robustness, and consistency can be attained by comparing results of national estimates against those calculated from an independent sample-based assessment. Accuracy assessment reports on the accuracy of the forest change area and attaches confidence limits (i.e. +/- x ha). Guyana is fortunate in that the comparison has been a feature of each reporting period since 2010. The results show a close correspondence across all periods when compared to the area of deforestation generated from the wall to wall national map (Figure 5).

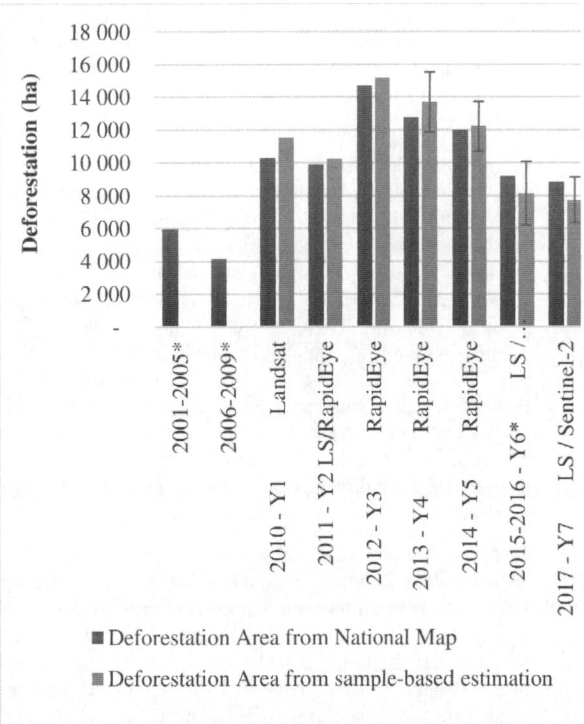

Figure 5. Area comparison between accuracy assessment and the national map

For example, in 2012 the two estimates were within 500 ha using RapidEye, the results of progressive assessments show that the differences also fall within the confidence limits of the sample-based estimate. Essentially this supports the notion that the transition from RapidEye to Sentinel 2 has had little impact on the ability of the MRVs to accurately capture deforestation.

5.1 Enhancing Resource Management

Following the commencement of the MRV System in 2009, the outputs generated have significantly enhanced the timeliness, quality and effectiveness of decision-making process as applied to Guyana's natural resources sector. The following summary (Table 5) captures the main decision-making and policy impacts derived from the system.

Data Product Generated	Natural Resources Application	Impact
Spatial Mapping of Protected Areas	Informing the Management and Monitoring Plan for Protected Areas in Guyana. Support decision making on the allocation of potential new Protected Areas.	Ensure zero-deforestation in Protected Areas and maintenance of Intact Forest Landscapes.
Informing Academic Studies and Research of long-term impacts on forest structure	Using a time series of spatial data from various sources integrated under the MRV System and inform trend analysis of forest utilization impacts on long term sustainability of forest resources in Guyana.	Sustainable management of forest resources informed by scientific research and spatial products, conducted locally in Guyana.
Informing a compliance monitoring programme for the extractives sector.	A compliance programme for proactive monitoring, leading to improved allocated for extractive purposes so to limit forest impacts.	Maintained low rate of deforestation over the last 5 years in Guyana.
Informing national policy and reporting on forest at the level of SDGs, International Commitments (Paris Agreement), and Bilateral Agreements.	Guyana's reporting on progress on its national Green State Development Strategy, Paris Agreement Commitments, areas under the Rio Conventions, and SDG Goal 13 and 15 have been strengthened as a result of an integrated system of monitoring forest using a range of satellite data streams.	A data agnostic system built by Guyana with strong local capacities to operate, that serves national and international monitoring requirements, and which inform decision making and policy setting for the natural resources sector in Guyana.

Table 5. Summary of natural resources applications and impacts

6. CONCLUSION

From inception the MRVs was designed to include data from multiple sources which can be combined in a consistent manner. The stepwise approach allowing time to bridge gaps in capacity allowing the integration of image processing and time series analysis routines. Today, the potential of the data generated through annual mapping of forest change extends well beyond the intended MRV function to include a range of functions relating to policies, decision-making, integration of compliance functions, and more effective management within the natural resources sector.

Guyana requires spatially explicit forest-based change assessment, as well as precise estimates of the annual area of change in forest cover and degradation. From Guyana's MRVs reporting, it is concluded that there are significant efficiencies in terms of processing time and cost that are achieved through the interoperability of data sets within the national system. It is still the reality that very high-resolution data (3m to 5m) is often difficult to obtain at an affordable cost from optical satellites. Yet, as countries like Guyana that have a more predominant change matrix of smaller scale change, monitoring less than 1-hectare pockets across the country is necessary to allow for comprehensive, accurate reporting and decision making on forests. There is therefore the clear need to find the balance between what is available and suited to the purpose for Guyana and what is cost effective and accessible. Guyana has adapted its methods over the last 8 years, whilst maintaining uniform standards on definitions, drivers and operating procedures. Over the years, specific methods required refining, to support a flexible, user friendly system. Guyana has been able to adapt these features of the MRV System whilst maintaining the fundamentals of this system.

Recent and evolving developments in the remote sensing environment has opened up many possibilities over the last 4 years, including with a broader range of high resolution optical satellites, higher revisit rates, and online platforms that support a faster and more efficient use of that data, all possible now at country level.

To embrace this dynamic, the MRVS has been built to be data agnostic and this has provided a versatile platform that grows, develops and allows improvements as these became necessary for Guyana. Currently, the MRVs which has been created from the interoperability of various satellite data streams has significantly strengthened decision-making processes. Decisions are more informed, responsive, and well-planned leading to better management of extractive activities, protected areas, and planning for development within forests.

Guyana has shown how the evolving needs of its system has been enabled from the interoperability of various data sets and how these have been integrated to provide the decision support and policy platform for forest management in Guyana.

ACKNOWLEDGEMENTS

The Guyana Forestry Commission would like to thank the Government of the Kingdom of Norway for its support of Guyana's work on the MRVs development from 2010 to present. The GFC also recognises the role of its close partners Indufor Asia Pacific and Winrock International in the development of the MRV System.

REFERENCES

Bovolo, C., Donoghue, D., 2017. Has regional forest loss been underestimated? Environmental Research Letters, 12:111003.

Food and Agriculture Organization (FAO), 2011. In: Assessing forest degradation. Towards the development of globally applicable guidelines. UN FAO, Forest Resources Assessment Working Paper 177. http://www.fao.org/docrep/015/i2479e/i2479e00.pdf

Galiatsatos N., Donoghue D.N.M., Watt P., Bholanath P., Pickering J., Hansen M.C., Mahmood, A.R.J., Submitted Sensors. An assessment of global forest change datasets for national forest monitoring.

GFOI, 2016. Integration of remote-sensing and ground-based observations for estimation of emissions and removals of greenhouse gases in forests: Methods and Guidance from the Global Forest Observations Initiative, Edition 2.0, Food and Agriculture Organization, Rome.

GOFC-GOLD, 2008. Reducing greenhouse gas emissions from deforestation and degradation in developing countries: a sourcebook of methods and procedures for monitoring and reporting. GOFC-GOLD Report version COP13-2. Global Observations of Forest Cover and Land Dynamics Project Office, Natural Resources Canada, Alberta, Canada. http://www.gofcgold.wur.nl/redd/index.php.

GOFC-GOLD, 2016. A sourcebook of methods and procedures for monitoring and reporting anthropogenic greenhouse gas emissions and removals associated with deforestation, gains and losses of carbon stocks in forests remaining forests, and forestation. GOFC-GOLD Report version COP22-1, (GOFC-GOLD Land Cover Project Office, Wageningen University, The Netherlands).

Goward, S.N., Williams, D.L., Arvidson, T., Rocchio, L.E., Irons, J.R., Russell, C.A., Johnston, S.S., 2017. Landsat's Enduring Legacy: Pioneering Global Land Observations from Space. American Society for Photogrammetry and Remote Sensing, Bethesda, Maryland.

Herold, M., DeFries, R., Achard, F., Skole, D., Townshend, J., 2006. Report of the workshop on monitoring tropical deforestation for compensated reductions GOFC-GOLD Symposium on Forest and Land Cover Observations, Jena, Germany, 21–22 March 2006

Herold M, Román-Cuesta R.M., Mollicone D., Hirata Y., van Laake P., Asner G.P., Souza C., Skutsch M., Avitabile V., MacDicken K., 2011. Options for monitoring and estimating historical carbon emissions from forest degradation in the context of REDD+. *Carbon Balance Manag.*6:13.

Khorram S., 1999. Accuracy Assessment of Remote Sensing-Derived Change Detection. Ed. S Khorram Bethesda, Maryland: American Society for Photogrammetry and Remote Sensing.

Lehmann E.A., Caccetta P., Lowell K., Mitchell A., Zhou Z-s., Held A., Milne T., and Tapley I., 2015. SAR and optical remote sensing: Assessment of complementarity and interoperability in the context of a large-scale operational forest monitoring system. *Remote Sensing of Environment*. Volume 156, Pages 335-348.

Penman J., Gytarski M., Hiraishi T., Krug T., Kruger D., Pipatti R., Buendia L., Miwa K., et al., 2003. Good Practice Guidance for Land Use, Land-Use Change and Forestry. Intergovernmental Panel on Climate Change, National Greenhouse Gas Inventories Programme (IPCC-NGGIP).

Pickering, J., Stehman, S.V., Tyukavina, A., Potapov, P., Watt, P., Jantz, S.M., Bholanath, P., Hansen, M.C., 2019. Quantifying the trade-off between cost and precision in estimating area of forest loss and degradation using probability sampling in Guyana. *Remote Sens. Environ.* 221, 122–135.

Powell, R.L., Matzke, N., de Souza Jr., C., Clarke, M., Numata, I., Hess, L.L., Roberts, D.A., 2004. Sources of error in accuracy assessment of thematic land-cover maps in the Brazilian Amazon, *Remote Sens. Environ,* 90:221-234.

Reiche J., Hamunyela E., Verbesselt J., Hoekman D., Herold M., 2018. Improving near-real time deforestation monitoring in tropical dry forests by combining dense Sentinel-1 time series with Landsat and Alos-2 PALSAR-2. Remote Sensing of Environment. Volume 204, January 2018, Pages 147-161.

Schingler R.H. Jr., 2015. Perspectives on Public - Private Partnerships in U.S. Earth Observing Programs. Joint Hearing of the Subcommittees for Space and Environment of the House Science Committee (2015)10 p. http://docs.house.gov/meetings/SY/SY16/20151117/104181/HHRG-114-SY16-Wstate-SchinglerR-20151117.pdf.

UNFCCC, 2001. The Marrakesh Accords & The Marrakesh Declaration, advance text, http://www.unfcccc.int/cop7/documents/accords_draft.pdf .

UNFCCC, 2015. Secretariat Report of the Conference of the Parties on its twenty-first session, held in Paris from 30 November to 11 December 2015. Addendum. Part two: Action taken by the Conference of the Parties at its twenty-first session. In Proceedings of the Report of the Conference of the Parties on its twenty-first session, held in Paris from 30 November to 13 December 2015. Part one: Proceedings.; Paris, 2016; p. 36.

Wulder, M.A., White, J.C., Goward, S.N., Masek, J.G., Irons, J.R., Herold, M., Cohen, W.B., Loveland, T.R., Woodcock, C.E., 2008. Landsat continuity: issues and opportunities for land cover monitoring. *Remote Sens. Environ.* 112 (3), 955–969.

Wulder M. A., Loveland T.R., Roy D. P. Crawford C.J., Masek J G., Woodcock C. E., Allen R. G., Anderson M. C., Belward A. S., Cohen W. B., Dwyer J., Arb A., Gao F., Griffiths P., Helder D., Hermosilla T., Hipple F. D., Hostert P., Hughes M. J., Huntington J., Johnson D. M., Kennedy R., Kilic A., Li Z., Lymburner L., McCorckel J., Pahlevan N., Scambos T. A., Schaaf C., Schott J. R., Sheng Y., Storey J., Vermote E., Vogelmann J., White J. C., Wynne R. H., Zhu Z., 2019. Current status of Landsat program, science, and application. *Remote Sensing of Environment*. 225; 127-147.

AUTOMATED LAND COVER CHANGE DETECTION THROUGH RAPID UAS UPDATES OF DIGITAL SURFACE MODELS

C. T. White[a], A. Petrasova[a], W. Reckling[b], H. Mitasova[ab]

[a] Center for Geospatial Analytics, North Carolina State University, United States - (ctwhite, akratoc, hmitaso)@ncsu.edu
[b] Department of Marine, Earth, and Atmospheric Sciences, North Carolina State University, United States - (wjreckli, hmitaso)@ncsu.edu

Commission VI, WG VI/4

KEY WORDS: Google Earth Engine, GRASS GIS, PlanetScope, random forest, data fusion *ISPRS Archives*, *ISPRS Annals*

ABSTRACT:

Up to date geospatial data provide the foundation for the development of smart and connected communities. While high-resolution 2D imagery is becoming widely available at less than monthly intervals and several infrastructure layers (e.g., roads, building footprints) are updated on a continuous basis, digital surface models (DSM) are generated less frequently and become quickly obsolete in rapidly developing regions. We present a methodology for continuous and efficient updates of DSM based on automated change detection from high-resolution satellite imagery that is used to develop UAS deployment plan, data acquisition, and DSM generation for targeted areas. The resulting UAS-derived DSM is then seamlessly fused with existing (usually lidar-based) DSM. We demonstrate our methodology in a rapidly developing watershed in the Triangle Region, North Carolina. The change detection maps were created using pixel-based classification methods on monthly composite data generated from PlanetScope satellites (3m resolution) as input for UAS flight planning, data acquisition, and processing. In future work a GRASS GIS script using a moving window resampling process will create flight areas to resample the change detection output into 10 acres flight areas for the UAS flight planning software, and a plugin for WebODM will be developed using GRASS GIS to enable seamless updates to centralized repositories of DSM.

1. INTRODUCTION

Rapid urbanization is changing the form and function of communities all over the world. It is predicted that 68% of the world's population will live in urban areas by 2050, and already 82% of North Americans live in urban areas in 2018 (2018 Revision of World Urbanization Prospects | Multimedia Library - United Nations Department of Economic and Social Affairs, n.d.). Accurate, up to data digital surface models (DSM) are essential for urban planning, especially in rapidly growing regions. DSM are among the most important data layers especially for vegetation management, view obstruction, solar radiation and cast shadows analysis, stormwater management as well as parks and recreation planning. At the pace urbanization is occurring, DSM datasets quickly become out of date limiting the accuracy DSM based products. Current methods to update statewide or countywide DSM at scale are costly because they require aircraft and LiDAR systems. These costs prohibit flight frequency producing DSM with low temporal resolution. To improve the temporal resolution of DSM datasets this study proposes a new method for rapid DSM updates. Automated change detection from high-resolution satellite imagery is used in order to deploy unmanned aerial systems (UAS) to update centralized DSM datasets. PlanetScope satellites provide high spatial (3m) and temporal (near daily) resolution multispectral (red, green, blue, near infrared) imagery (Marta, 2018) which can be used as input for per-pixel post-classification change detection to create a change detection map product, which identifies the temporal staleness of existing land cover data and indirectly indicates also the need for update in DSM datasets (Taubenböck et al., 2012, Jensen, 2015). Our paper describes the research focused on the process to create change detection maps that provide sufficient information about the change that it can be used for flight planning and UAS deployment to enable seamless DSM updates.

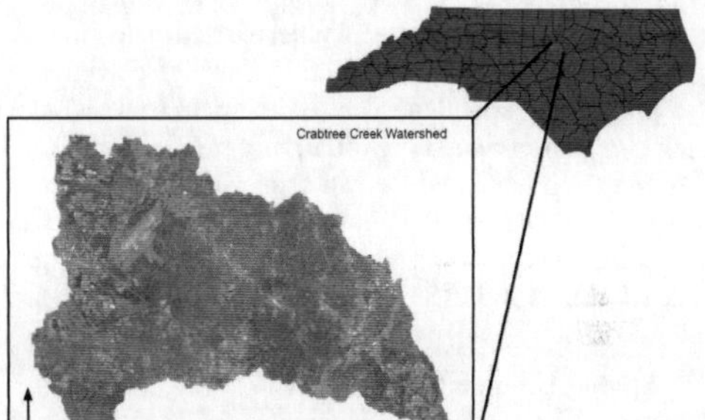

Figure 1: Crabtree Creek Watershed (376.25 km2) in the Research Triangle, North Carolina, USA.

2. STUDY AREA AND DATA

To develop and test our approach, we selected a study area in the Crabtree Creek watershed (376.25 km2) in the Research Triangle region of North Carolina in the United States (Figure 1). The area was chosen because of the rapid urbanization occurring in the two counties that make up the watershed. Between 2010 and 2018 Wake County's population grew 21.2% and Durham County's population grew 17.3% (U.S. Census Bureau QuickFacts: Durham County, North Carolina; North Carolina; Wake County, North Carolina; UNITED STATES, n.d.). As a consequence, the watershed experienced extensive new development including both commercial and residential buildings and updates to transportation infrastructure.

The data used in this study consisted of 3m resolution 4-band multispectral images (blue, green, red and near-infrared) from the PlanetScope constellation (Marta, 2018). PlanetScope temporal resolution is daily, however, this study does not require such a high temporal resolution because of the limited amount of surface change occurring in any given day. Instead, the images used for this study were captured on three cloud-free (1% cloud coverage) days across three months; February 14, 2019 (7 images), March 14, 2019 (9 images), and April 15, 2019 (6 images) and contain 100% area coverage. Mosaics of each image data set were created and clipped and masked to the study area bounds.

3. METHODS

Detecting change in remotely sensed images is a problem with multiple solutions depending on what type of change needs to be detected and what change information is important (Chen et al., 2012, Jensen, 2015). These change detection methods include analog, binary, thematic, and can be based on either pixel or object classifications (Jensen, 2015). Analog change detection produces useful visualizations of change on an image, but fail to provide quantitative results; Binary change detection produces good results but is only applicable if understanding the type of change is not important, while thematic change detection designates the from-to change based on predetermined classes (Jensen, 2015). Another consideration when deciding which type of change detection method to use is whether the change is best represented on a pixel by pixel basis or if object-based detection methods capture the desired change. Selecting a change detection method for a problem is also dependent on the data sources resolution, which can restrict what types of algorithms are available. In this study analog, binary, and thematic based change detection methods were evaluated to determine which method works best for identifying land cover change at the level of detail required for the development of UAS flight plans for DSM fusion.

3.1 Analog Change Detection

Analog change detection methods were used to provide visual clues into which image bands provided the most useful impervious surface change detection information for the study area. Near-infrared (NIR) and the calculated normalized difference vegetation index (NDVI) bands were both used to create analog change detection maps (Jensen, 2015, Lunetta et al., 2006). The NIR analog change map was created by placing the NIR band from February in the red memory bank and the NIR band from April in the blue and green memory banks of the displayed RGB image. The same method was used to generate the NDVI analog change detection map with the NIR band replaced with NDVI (Figure 2).

3.2 Binary Change Detection

Review of the analog change detection maps revealed that the NIR band was a good candidate for image differencing to extract binary change information. A statistical threshold was set for $\pm 3\sigma$ of the mean extracting areas of significant spectral change (Morisette and Khorram, n.d., Rosin, 2002, Jensen, 2015). The resulting change detection map contained three classes; Class -1 contained pixels that changed from noise, Class 0 contained no change, and class 1 contained land cover change (Table 1). The classes were then masked to exclusively contain class one generating a land cover change detection map (Figure 3).

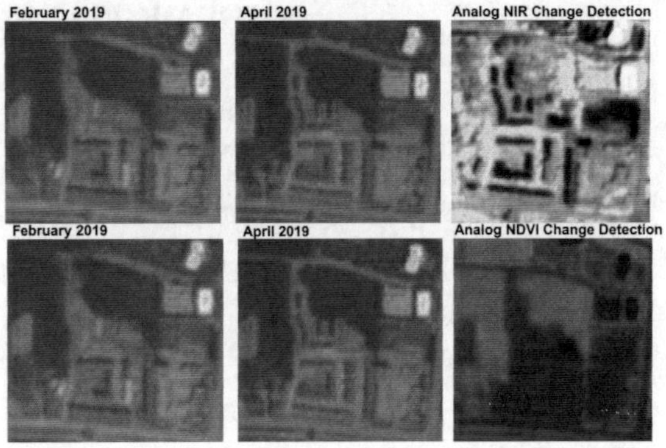

Figure 2: [Top Row] uses the near-infrared (NIR) band from February and April NIR (RGB(Feb NIR, Apr NIR, Apr NIR) to generate a change detection map showing new construction. [Bottom Row] uses NDVI from February and April in RGB memory banks RBG(Feb NDVI, Apr NDVI, Apr NDVI) to show disturbances in vegetation.

Class	Thresholds	Change Type
-1	[Min, μ - 3σ]	Change (Noise)
0	[μ - 3σ, μ +3σ]	No Change
1	[μ + 3σ, Max]	Land Cover Change

μ = -0.26, 3σ = -0.009

Table 1: Change thresholds set after band differencing the near-infrared band between February 2019 and April 2019 in the Crabtree Creek watershed, NC.

3.3 Thematic

Thematic change detection methods allow for detailed from-to data to be derived when evaluating multitemporal datasets (Jensen, 2015). This study used supervised classification of multiple composite images to perform a per-pixel post-classification comparison to extract thematic change (Rokni et al., 2016, Jensen, 2015). Due to a dearth of object-based classification methods currently implemented in Google Earth Engine, no object-based classification methods were attempted (Shelestov et al., 2017). Instead, a pixel-based method was implemented using a random forest.

3.3.1 Classification Supervised and unsupervised classification methods provide algorithms that enable thematic change detection in remote sensing data. The classification methods can be pixel or object-based both of which are applicable for change detection. Supervised learning classification methods require the development of training data and predefined classes and allow for the easy interpretation of the resulting classification. Unsupervised learning techniques require no training data and can identify previously unknown patterns in data. However, classes derived through unsupervised learning can be difficult to interpret and may not reflect the desired patterns of change. In this study, a random forest (Homer et al., 2015, Gislason et al., 2006) implemented in Google Earth Engine was used classify the images of the study area from February 2019 and April 2019 independently into six categories; water, roads, buildings, forest, grass, and barren. The random forests were set to generate 250 trees, each image mosaic contained 6 features (band); Red, green, blue, near-infrared, NDVI, and NDWI. Other features combinations using gray-level co-occurrence (GLCM) derived texture statistics such as second moment, contrast, and correlation bands were removed

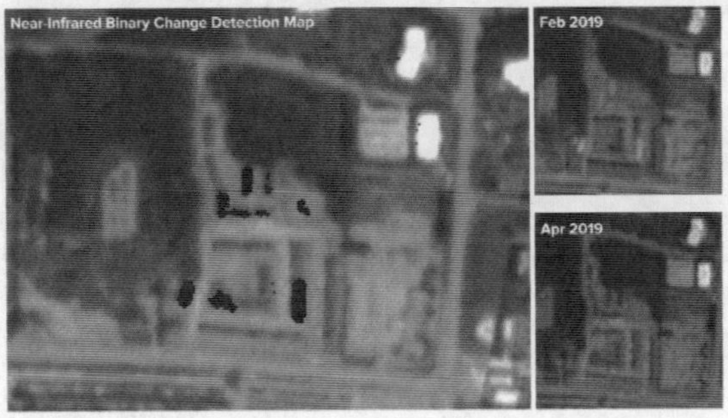

Figure 3: The purple patches in the left image represent the land-cover change (new construction at construction site) detected using binary change detection from image differencing and statistical thresholds set to select pixels greater than 3 std between February 2019 and April 2019 in Raleigh, NC using PlanetScope 3m multispectral data.

from the model after inspecting feature importance. However, it is important to note that random forest feature importance is approximated in Google Earth Engine by using the results of a single classification and regression tree (CART) because no current method is provided to view the trees generated by the random forest.

3.4 Future Work: Flight Planning for DSM Fusion

3.4.1 Flight Planning Once areas of change are detected, optimized UAS flight areas are generated using a GRASS GIS (GRASS Development Team, 2019) script that uses moving window resampling to generate 10 acre polygons extracted from the change detection maps, which are then imported into DroneDeploy (`www.dronedeploy.com`) UAS flight planning software (Reckling and Mitasova, 2018). The flight plan is then used to acquire overlapping imagery and generate updated orthoimagery and point cloud using structure from motion (Figure 4).

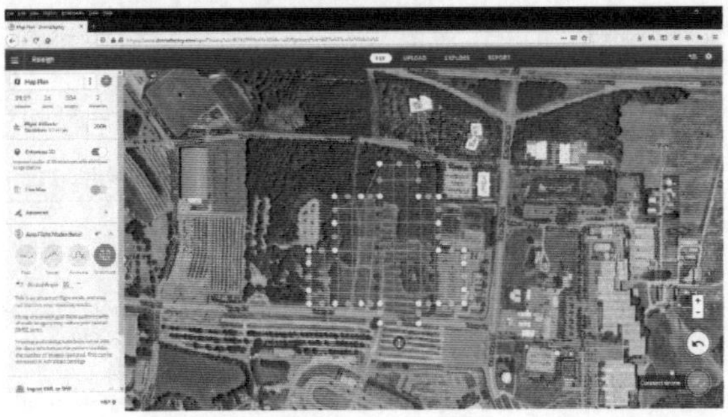

Figure 4: Flight area generated by using a GRASS GIS script, which utilizes a moving window resampling method to generate 10 acre polygons displayed in the DroneDeploy flight planning software.

3.4.2 Fusion With the captured data, OpenDroneMap (ODM) will be used to reconstruct the new DSM through WebODM user interface (`www.opendronemap.org`). This new, very high resolution DSM will then be seamlessly integrated with existing DSM or DEM products using smooth fusion method (Petrasova et al., 2017) to avoid any artifacts caused by misalignment along DEM edges. This workflow will be automated and accessible as a WebODM plugin (Figure 5).

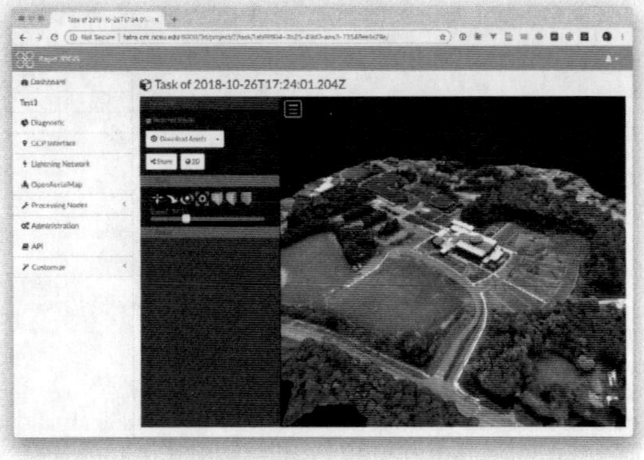

Figure 5: Generated orthoimagery and 3D model using structure from motion from UAS flight data, prepared for fusion in WebODM using the WebODM Fusion Plugin.

4. RESULTS

The classification map for the February image mosaic had a validated overall accuracy of 97% and a kappa of 96% and the classification of the April image mosaic had a validated overall accuracy of 98% and a kappa of 97%. The confusion matrices show that the grass and barren classes for both classifications contained the greatest amounts of misclassification (Figure 6).

Feb 2019	Water	Road	Building	Forest	Grass	Barren	CA
Water	42	0	0	0	0	0	1
Road	0	10	1	0	1	1	1
Building	0	0	33	0	0	0	0.97
Forest	0	0	0	199	0	0	0.96
Grass	0	0	0	3	19	1	0.95
Barren	0	0	0	0	0	18	0.88
PA	1	0.76	1	1	0.82	1	

Overall Accuracy	0.97
Kappa	0.96

April 2019	Water	Road	Building	Forest	Grass	Barren	CA
Water	41	0	0	0	0	0	1
Road	0	13	1	0	2	0	1
Building	0	0	48	0	0	0	0.97
Forest	0	0	0	191	0	0	1
Grass	0	0	0	0	16	2	0.84
Barren	0	0	0	0	1	13	0.86
PA	1	0.81	1	1	0.88	0.92	

Overall Accuracy	0.98
Kappa	0.97

Figure 6: Validated accuracy assessments of random forest classification for February 2019 (left) and April 2019 (right) image mosaics.

With the classified land cover map from February and April, a per-pixel post-classification comparison was performed by joining the two classified images together to generate a thematic (from-to) matrix (Figure 7). The change detection maps classes were reduced to five class; No Change, Natural to Impervious, Impervious to Natural, Anything to Water, and Natural. The resulting change detection maps show a large amount of change due to variations in the image and seasonal effects on vegetation (Figure 8).

To Class

Feb-Apr 2019 (From Class)	Water	Road	Building	Forest	Grass	Construction
Water	3768	375	286	1610	32	2
Road	630	58415	6305	842	19873	15833
Building	140	3167	7082	626	352	1408
Forest	96	9632	255	187033	106039	4330
Grass	19	6542	126	5147	46779	8983
Construction	3	1406	444	74	2184	5519

Figure 7: Thematic (from-to) changes from the classification outputs of random forests ran on images from February 2019 and April 2019.

Figure 8: Results from post classification pixel class change comparison between February and April 2019. Red indicates a change from a class to road, building or barren (impervious surface).

5. DISCUSSION

The current results from the post-classification change comparison between the classified February and April land cover maps contain too many changes that are related to variation in image quality and seasonal variation of vegetation. By comparing the number of pixels indicating a change from the post-classification change detection map (5% change) to the near-infrared change detection map (0.2% change) it shows that the post-classification change detection is overestimating the total amount of change. To improve the results of the post-classification change detection map a Hidden Markov Models (HMM) can be used to post-process the resulting classification in order to stabilize temporal pixel variation helping to identify noise from actual land cover change (Sulla-Menashe et al., 2019, Abercrombie and Friedl, 2016). Currently, no such method exists within Google Earth Engine, so the method will either need to be developed or data moved to another platform to perform the pixel stabilization. Other methods can also be explored to directly detect changes in DSM by using the minor variations in the PlanetScope constellation view angle to create a temporal DSM dataset (Ghuffar, 2018). Once the change detection process is finalized an automated process can be developed using both Planet and Google Earth Engine APIs to create a dataset that will prioritize UAS deployment plans to update DSM.

6. CONCLUSION

As urbanization and climate change continue to affect areas at an increasingly rapid rate the need for DSM datasets with high temporal resolution will become increasingly important. The method described in this paper provides a good starting point for the continued development of an optimized change detection methodology to enable flight planning for UAS to seamlessly update centralized DSM datasets. The current results provide a map of binary change detection which is limited in its ability to describe the thematic changes occurring across the landscape and a temporal per-pixel post-classification comparison of two land cover maps that overestimates the total amount of true land cover change. In future work that post-classification comparison can be improved by using a greater amount of temporal images to stabilized the pixels uses Hidden Markov Models to reduce the total amount of change detected from minor variations in the classified results.

ACKNOWLEDGMENTS

Thank you to Dr. Parker Abercrombie and Dr. Damien Sulla-Menashe for providing access to the Hidden Markov Model code they developed to create the MODIS C6 product (Sulla-Menashe et al., 2019), and to Dr. Josh Gray for general advice and the introduction to Dr. Abercrombie and Dr. Sulla-Menashe.

REFERENCES

2018 Revision of World Urbanization Prospects | Multimedia Library - United Nations Department of Economic and Social Affairs, n.d.

Abercrombie, S. P. and Friedl, M. A., 2016. Improving the Consistency of Multitemporal Land Cover Maps Using a Hidden Markov Model. IEEE Transactions on Geoscience and Remote Sensing 54(2), pp. 703–713.

Chen, G., Hay, G. J., Carvalho, L. M. T. and Wulder, M. A., 2012. Object-based change detection. International Journal of Remote Sensing 33(14), pp. 4434–4457.

Ghuffar, S., 2018. DEM Generation from Multi Satellite PlanetScope Imagery. Remote Sensing; Basel.

Gislason, P. O., Benediktsson, J. A. and Sveinsson, J. R., 2006. Random Forests for land cover classification. Pattern Recognition Letters 27(4), pp. 294–300.

GRASS Development Team, 2019. Geographic Resources Analysis Support System (GRASS GIS) Software, Version 7.6. Open Source Geospatial Foundation.

Homer, C., Dewitz, J., Yang, L., Jin, S., Danielson, P., Coulston, J., Herold, N., Wickham, J. and Megown, K., 2015. Completion of the 2011 National Land Cover Database for the Conterminous United States – Representing a Decade of Land Cover Change Information. PHOTOGRAMMETRIC ENGINEERING p. 11.

Jensen, J. R., 2015. Introductory Digital Image Processing: A Remote Sensing Perspective. 4th edn, Prentice Hall Press, Upper Saddle River, NJ, USA.

Lunetta, R. S., Knight, J. F., Ediriwickrema, J., Lyon, J. G. and Worthy, L. D., 2006. Land-cover change detection using multi-temporal MODIS NDVI data. Remote Sensing of Environment 105(2), pp. 142–154.

Marta, S., 2018. PLANET IMAGERY PRODUCT SPECIFICATIONS. p. 91.

Morisette, J. T. and Khorram, S., n.d. Accuracy Assessment Curves for Satellite-Based Change Detection. PHOTOGRAMMETRIC ENGINEERING p. 6.

Petrasova, A., Mitasova, H., Petras, V. and Jeziorska, J., 2017. Fusion of high-resolution DEMs for water flow modeling. Open Geospatial Data, Software and Standards 2(1), pp. 6.

Reckling, W. and Mitasova, H., 2018. Planning Optimal UAS Flight Areas for Rare Plant Monitoring in the Blue Ridge Mountains of North Carolina. In: AGU Fall Meeting Abstracts, Vol. 2018, pp. B33F–2735.

Rokni, K., Ahmad, A., Solaimani, K. and Hazini, S., 2016. A New Approach for Detection of Surface Water Changes Based on Principal Component Analysis of Multitemporal Normalized Difference Water Index. Journal of Coastal Research 32(2), pp. 443–451.

Rosin, P. L., 2002. Thresholding for Change Detection. Computer Vision and Image Understanding 86(2), pp. 79–95.

Shelestov, A., Lavreniuk, M., Kussul, N., Novikov, A. and Skakun, S., 2017. Exploring Google Earth Engine Platform for Big Data Processing: Classification of Multi-Temporal Satellite Imagery for Crop Mapping. Frontiers in Earth Science.

Sulla-Menashe, D., Gray, J. M., Abercrombie, S. P. and Friedl, M. A., 2019. Hierarchical mapping of annual global land cover 2001 to present: The MODIS Collection 6 Land Cover product. Remote Sensing of Environment 222, pp. 183–194.

Taubenböck, H., Esch, T., Felbier, A., Wiesner, M., Roth, A. and Dech, S., 2012. Monitoring urbanization in mega cities from space. Remote Sensing of Environment 117, pp. 162–176.

U.S. Census Bureau QuickFacts: Durham County, North Carolina; North Carolina; Wake County, North Carolina; UNITED STATES, n.d.

APPENDIX

Google Earth Engine Code `https://code.earthengine.google.com/038bdf4571b6cd76929053f98c6f19b8`

REFINEMENT OF CROPLAND DATA LAYER USING MACHINE LEARNING

Chen Zhang [1,2], Zhengwei Yang [3], Liping Di [1,2,*], Li Lin [1,2], Pengyu Hao [1]

[1] Center for Spatial Science and Systems, George Mason University, Fairfax, VA 22030, USA - (czhang11, ldi, llin2, phao)@gmu.edu
[2] Department of Geography and Geoinformation Science, George Mason University, Fairfax, VA 22030, USA
[3] Research and Development Division, U.S. Department of Agriculture National Agricultural Statistics Service, Washington, DC 20250, USA - Zhengwei.Yang@usda.gov

KEY WORDS: Cropland Data Layer, Machine Learning, Misclassification Correction, Crop Sequence Modeling, Raster Map Refinement

ABSTRACT:

As the most widely used crop-specific land use data, the Cropland Data Layer (CDL) product covers the entire Contiguous United States (CONUS) at 30-meter spatial resolution with very high accuracy up to 95% for major crop types (i.e., Corn, Soybean) in major crop area. However, the quality of early-year CDL products were not as good as the recent ones. There are many erroneous pixels in the early-year CDL product due to the cloud cover of the original Landsat images, which affect many follow-on researches and applications. To address this issue, we explore the feasibility of using machine learning technology to refine and correct misclassified pixels in the historical CDLs in this study. An end-to-end deep learning-based framework for restoration of misclassified pixels in CDL image is developed and tested. By feeding the CDL time series into the artificial neural network, a crop sequence model is trained and the misclassified pixels in an original CDL map can be restored. In the experiment with the 2005 CDL data of the State of Illinois, the misclassified pixels over Agricultural Statistics Districts (ASD) #1760 were corrected with a reasonable accuracy (>85%). The findings suggest that the proposed method provides a low-cost and reliable way to refine the historical CDL data, which can be potentially scaled up to the entire CONUS.

1. INTRODUCTION

Since its first release of a full state wide data product in 1997, the Cropland Data Layer (CDL) product of the U.S. Department of Agriculture (USDA) National Agricultural Statistics Service (NASS) has been widely used by growers, agricultural industry, governments, educators and students, and researchers world-wide for crop production, agricultural production planning and management, government policy formulation and decision making, teaching, and various research activities (Liknes et al., 2009; Thompson, Prokopy; Hao et al., 2015; Lark et al., 2015; Di et al., 2017). Currently, the CDL data covers the entire conterminous United States (CONUS) at 30-meter spatial resolution with a high accuracy up to 95% for classifying major crop types (i.e., Corn, Soybean, and Wheat). However, the quality of the early-year CDL products was not as good as recent years. In early years, there are many misclassified pixels in the CDL products because of cloud cover and lack of satellite images. Moreover, only a few states of CDL data were produced before 2008. For example, the year 2000 CDL covers only Illinois, Indiana, Mississippi, North Dakota, and a part of Arkansas and Iowa. Obviously, the earlier year CDLs' availability and low quality issues affect many follow-on Land Use and Land Cover (LULC) related researches and applications. Therefore, an effective method for refining and correcting the old CDL data is badly needed to improve the quality and accuracy of the historical CDL data.

It is well known that monocropping will result in degradation of soil, build-up of diseases and pests, and decline in productivity. Thus crop rotation becomes a common farming practice in U.S. Corn Belt. The crop rotation can significantly improve the soil condition, such as fertility and soil physical/chemical properties (Pikul et al., 2001; Karlen et al., 2006; Govaerts et al., 2007; Karlen et al., 2013; Van Eerd et al., 2014). Meanwhile, the crop sequence and cropping decision also have significant impact on crop yields and profitability (Temperly, Borges; Parajuli et al., 2013; Farmaha et al., 2016). Based on this common cropping practice, many crop mapping and yield estimation models and approaches were developed. Secchi et al. (2011) constructed an prediction model of future land use scenario in the state of Iowa based on the corn-soybean rotation and production costs. Schönhart et al. (2011) developed a crop sequence model to generate crop rotations based on agronomic criteria and observed data. Sahajpal et al. (2014) detected the pronounced shifts from grassland to cultivated area by modelling crop rotation in the U.S. Western Corn Belt. Hao et al. (2016) explored the crop classification based on the previous-year crop knowledge. Zhang et al. (2019a) produced a crop cover map of Nebraska State based on the common crop rotation patterns of corn, soybeans, winter wheat, and alfalfa. They further implemented a crop sequence-based machine learning framework for prediction of crop cover maps (Zhang et al., 2019b).

In this paper, we present a machine learning-based crop sequence model to refine the historical CDL data. The proposed model utilizes artificial neural network (ANN) to automatically learn crop sequence information from the CDL time series. The misclassified pixels in the crop cover map can be automatically identified and corrected using the trained model on the historical CDL.

The rest of the paper is organized as follows. Section 2 introduces the CDL data, the study area, and an end-to-end machine learning framework for the historical CDL data refinement. Section 3 demonstrates the experiment results and as-

*Corresponding author

sesses the refinement performance. Section 4 discusses the limitation of the current implementation and gives the conclusion.

2. METHODS

2.1 Cropland Data Layer

CDL is a raster formatted, geo-referenced, crop-specific land cover map produced by USDA NASS. It is an annual product covering the entire CONUS at 30-meter spatial resolution from 2008 to present and some states from 1997 to 2007. The production of CDL is mainly based on moderate resolution satellite imagery and extensive agricultural ground truth (Boryan et al., 2011). The misclassified pixels in the CDL refer to the pixels that are covered with "clouds" or "no data". These pixels are mainly existing in the CDL products before 2006 due to lack of high-quality satellite data and the algorithm limitation back then. Examples of the misclassified pixels in the early-year CDL are shown in Figure 1.

The CDL data products are freely downloaded from CropScape (https://nassgeodata.gmu.edu/CropScape/), which is developed and maintained in cooperation with Center for Spatial Information Science and Systems of George Mason University (Han et al., 2012; Zhang et al., 2019c). It provides an easy-to-use Web GIS application to visualize, analyse, and download CDL data. All data hosted on CropScape are disseminated via the OGC standards-compliant geospatial Web services, such as Web Map Service (WMS), Web Coverage Service (WCS), Web Feature Service (WFS), and Web Processing Service (WPS).

2.2 Study Area

The Agricultural Statistics District (ASD) #1760 of Illinois state is selected as the study area. The study area lies on the Central Corn Belt Plains Ecoregion, which is mainly covered by corn, soybeans, grassland, and forest as shown in Figure 2. It can be seen that the 2005 CDL contains a considerable number of pixels are labelled as "clouds or no data" over the study area. The purpose of this study is to restore those misclassified pixels in the study area of 2005 CDL using the machine-learned crop sequence model.

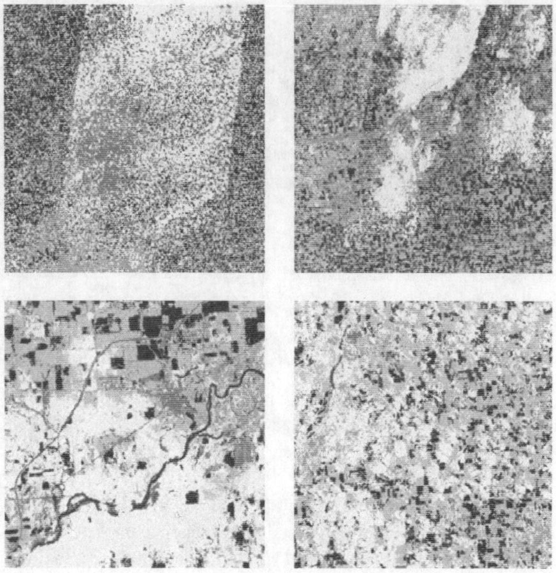

Figure 1. Examples of misclassified pixels in the early-year CDL data.

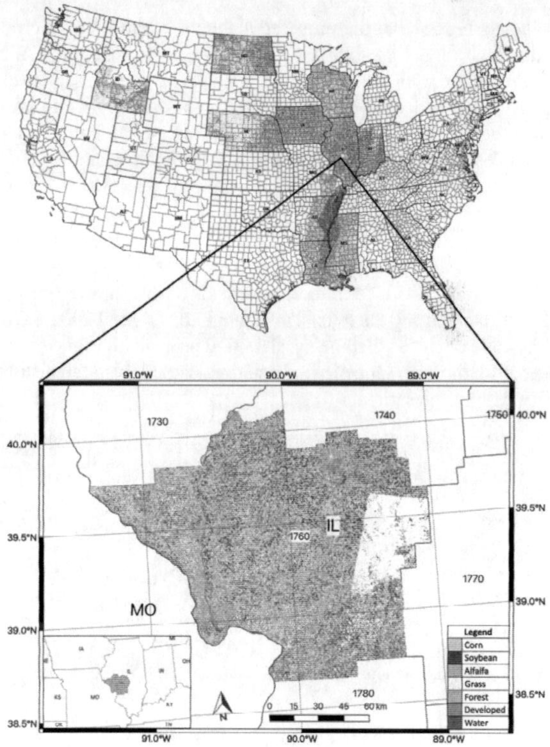

Figure 2. Study area with 2005 CDL as the experiment data (data available from CropScape).

2.3 Machine Learning Framework

To automatically correct the misclassified pixels in CDL, an end-to-end machine learning framework is proposed in this paper. The proposed framework is composed of four major components: data preparation, model training, classification, and evaluation.

2.3.1 Data Preparation: In data preparation, the CDLs from 2006 to 2018 are stacked sequentially to form CDL time series. All pixels of the CDL time series are arranged into a 2-D array of samples. Each row of the data set array represents a pixel consisting of a sequence of crop type values of different years. Training and validation data sets are randomly sampled from the "good pixels" in the study area and labelled with 2005 CDL. The experiment data set includes all pixels corresponding to those misclassified pixels in the study area without labels.

2.3.2 Model Training: The crop sequence model is trained by feeding the training set into the artificial neural network, which contains one input layer, multiple hidden layers, and one output layer. The input layer contains a group of neurons corresponding to the same pixel of the CDL time series. Each input pixel represents a specific value of its crop type. There are mul-

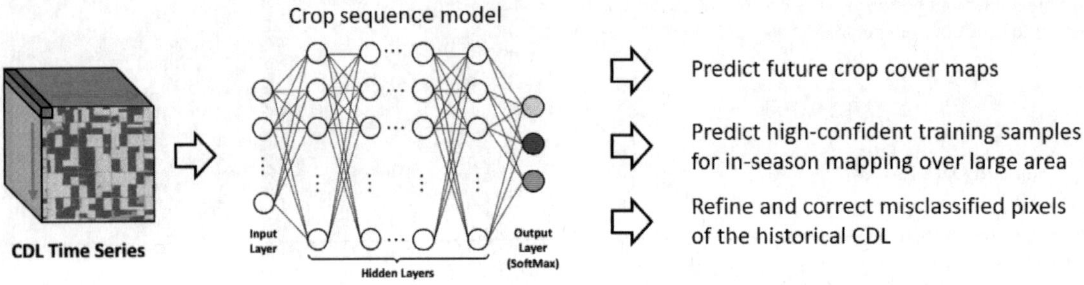

Figure 3. Applications of machine learning-based crop sequence model.

tiple hidden layers between the input layer and the output layer. The output layer uses SoftMax to estimate the probability of each crop type.

2.3.3 Classification and Validation: By feeding the experiment data set to the well-trained crop sequence model, the misclassified pixels in the original CDL can be refined. To validate the refinement performance, we applied the same crop sequence model to the validation set. Then we measured the model by calculating the agreement of the classified label and the original label of the validation set.

The applications of the proposed machine learning-based crop sequence model are illustrated in Figure 3. In this study, the crop sequence model is used to restore the historical crop cover map. This model, on the other hand, can be also applied to predict the future crop cover maps with the high-confident training samples for early-season and in-season crop mapping.

3. RESULTS

The refined 2005 CDL data of ASD #1760 is illustrated in Figure 4. Comparing the refined result with the original 2005 CDL data, we observed that the misclassified pixels had been corrected with the crop sequence information learned from the historical CDL time series.

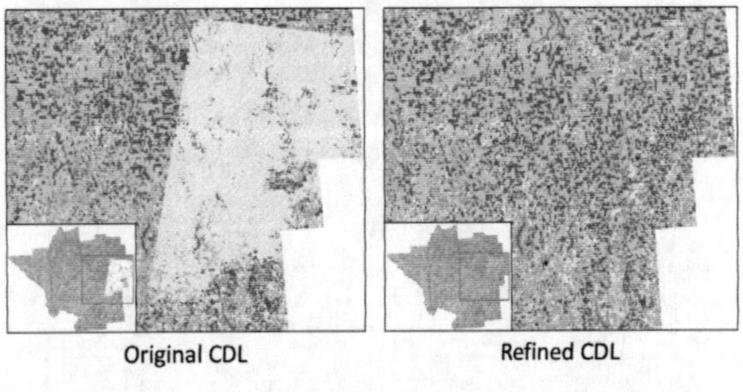

Figure 4. Refined 2005 CDL of ASD #1760.

The overall accuracy of the refined pixels is unable to be accessed directly due to lack of ground reference data. Instead, we utilized the validation data set, derived from the "good pixels" in the study area of 2005 CDL to indirectly measure the performance of the model. The overall accuracy of validation based on the validation sample set is over 85%. Therefore, the actual overall accuracy of the refined pixels may vary. To further validate the performance of refinement, the ground reference data are required.

4. CONCLUSION

This study investigated the feasibility of using machine learning technology to refine CDL data. An end-to-end ANN-based framework was proposed and tested to correct the misclassified pixels in the historical CDL data. The preliminary experiment result indicates that the misclassified pixels over the ASD #1760 could be corrected with reasonable accuracy (>85%). The findings suggest that the proposed machine learning approach is effective and low-cost for correcting the misclassified pixels, and has great potential for refining the historical CDL over large geographic area. More experiments and validation will be conducted in the future.

ACKNOWLEDGEMENTS

This work is supported by the U.S. Department of Agriculture National Agricultural Statistics Service.

REFERENCES

Boryan, C., Yang, Z., Mueller, R., Craig, M., 2011. Monitoring US Agriculture: The US Department of Agriculture, National Agricultural Statistics Service, Cropland Data Layer Program. *Geocarto International*, 26(5), 341–358. doi:10.1080/10106049.2011.562309.

Di, L., Yu, E. G., Kang, L., Shrestha, R., Bai, Y., 2017. RF-CLASS: A Remote-Sensing-Based Flood Crop Loss Assessment Cyber-Service System for Supporting Crop Statistics and Insurance Decision-Making. *Journal of Integrative Agriculture*, 16(2), 408-423. doi:10.1016/S2095-3119(16)61499-5.

Farmaha, B. S., Eskridge, K. M., Cassman, K. G., Specht, J. E., Yang, H., Grassini, P., 2016. Rotation Impact on On-Farm Yield and Input-Use Efficiency in High-Yield Irrigated Maize–Soybean Systems. *Agronomy Journal*, 108(6), 2313-2321. doi:10.2134/agronj2016.01.0046.

Govaerts, B., Mezzalama, M., Unno, Y., Sayre, K. D., Luna-Guido, M., Vanherck, K., Dendooven, L., Deckers, J., 2007. Influence of Tillage, Residue Management, and Crop Rotation on Soil Microbial Biomass and Catabolic Diversity. *Applied Soil Ecology*, 37(1), 18-30. doi:10.1016/j.apsoil.2007.03.006.

Han, W., Yang, Z., Di, L., Mueller, R., 2012. CropScape: A Web Service Based Application for Exploring and Disseminating US Conterminous Geospatial Cropland Data Products for Decision Support. *Computers and Electronics in Agriculture*, 84, 111–123. doi:10.1016/j.compag.2012.03.005.

Hao, P., Wang, L., Zhan, Y., Wang, C., Niu, Z., Wu, M., 2016. Crop Classification Using Crop Knowledge of the Previous-Year: Case Study in Southwest Kansas, USA. *European Journal of Remote Sensing*, 49(1), 1061-1077. doi:10.5721/EuJRS20164954.

Hao, P., Zhan, Y., Wang, L., Niu, Z., Shakir, M., 2015. Feature Selection of Time Series MODIS Data for Early Crop Classification Using Random Forest: A Case Study in Kansas, USA. *Remote Sensing*, 7(5), 5347-5369. doi:10.3390/rs70505347.

Karlen, D. L., Cambardella, C. A., Kovar, J. L., Colvin, T. S., 2013. Soil Quality Response to Long-Term Tillage and Crop Rotation Practices. *Soil and Tillage Research*, 133, 54-64. doi:10.1016/j.still.2013.05.013.

Karlen, D. L., Hurley, E. G., Andrews, S. S., Cambardella, C. A., Meek, D. W., Duffy, M. D., Mallarino, A. P., 2006. Crop Rotation Effects on Soil Quality at Three Northern Corn/Soybean Belt Locations. *Agronomy Journal*, 98(3), 484-495. doi:10.2134/agronj2005.0098.

Lark, T. J., Salmon, J. M., Gibbs, H. K., 2015. Cropland Expansion Outpaces Agricultural and Biofuel Policies in the United States. *Environmental Research Letters*, 10(4), 044003. doi:10.1088/1748-9326/10/4/044003.

Liknes, G. C., Nelson, M. D., Gormanson, D. D., Hansen, M., 2009. The Utility of the Cropland Data Layer for Forest Inventory and Analysis. *Proceedings of the Eighth Annual Forest Inventory and Analysis Symposium*, 259–264. doi:10.2737/WO-GTR-79.

Parajuli, P. B., Jayakody, P., Sassenrath, G. F., Ouyang, Y., Pote, J. W., 2013. Assessing the Impacts of Crop-Rotation and Tillage on Crop Yields and Sediment Yield Using a Modeling Approach. *Agricultural Water Management*, 119, 32-42. doi:10.1016/j.agwat.2012.12.010.

Pikul, J. L., Carpenter-Boggs, L., Vigil, M., Schumacher, T. E., Lindstrom, M. J., Riedell, W. E., 2001. Crop Yield and Soil Condition under Ridge and Chisel-Plow Tillage in the Northern Corn Belt, USA. *Soil and Tillage Research*, 60(1), 21-33. doi:10.1016/S0167-1987(01)00174-X.

Sahajpal, R., Zhang, X., Izaurralde, R. C., Gelfand, I., Hurtt, G. C., 2014. Identifying Representative Crop Rotation Patterns and Grassland Loss in the US Western Corn Belt. *Computers and Electronics in Agriculture*, 108, 173-182. doi:10.1016/j.compag.2014.08.005.

Schönhart, M., Schmid, E., Schneider, U. A., 2011. CropRota – A Crop Rotation Model to Support Integrated Land Use Assessments. *European Journal of Agronomy*, 34(4), 263-277. doi:10.1016/j.eja.2011.02.004.

Secchi, S., Kurkalova, L., Gassman, P. W., Hart, C., 2011. Land Use Change in a Biofuels Hotspot: The Case of Iowa, USA. *Biomass and Bioenergy*, 35(6), 2391-2400. doi:10.1016/j.biombioe.2010.08.047.

Temperly, R. J., Borges, R., 2006. Tillage and Crop Rotation Impact on Soybean Grain Yield and Composition. *Agronomy Journal*, 98(4), 999-1004. doi:10.2134/agronj2005.0215.

Thompson, A. W., Prokopy, L. S., 2009. Tracking Urban Sprawl: Using Spatial Data to Inform Farmland Preservation Policy. *Land Use Policy*, 26(2), 194-202. doi:10.1016/j.landusepol.2008.02.005.

Van Eerd, L. L., Congreves, K. A., Hayes, A., Verhallen, A., Hooker, D. C., 2014. Long-Term Tillage and Crop Rotation Effects on Soil Quality, Organic Carbon, and Total Nitrogen. *Canadian Journal of Soil Science*, 94(3), 303-315. doi:10.4141/cjss2013-093.

Zhang, C., Di, L., Lin, L., Guo, L., 2019a. Extracting Trusted Pixels from Historical Cropland Data Layer Using Crop Rotation Patterns: A Case Study in Nebraska, USA. *2019 8th International Conference on Agro-Geoinformatics (Agro-Geoinformatics)*, 1–6. doi:10.1109/Agro-Geoinformatics.2019.8820236.

Zhang, C., Di, L., Lin, L., Guo, L., 2019b. Machine-Learned Prediction of Annual Crop Planting in the U.S. Corn Belt Based on Historical Crop Planting Maps. *Computers and Electronics in Agriculture*, 166, 104989. doi:10.1016/j.compag.2019.104989.

Zhang, C., Di, L., Yang, Z., Lin, L., Yu, E. G., Yu, Z., Rahman, M. S., Zhao, H., 2019c. Cloud Environment for Disseminating NASS Cropland Data Layer. *2019 8th International Conference on Agro-Geoinformatics (Agro-Geoinformatics)*, 1–5. doi:10.1109/Agro-Geoinformatics.2019.8820465.

AUTHOR INDEX